Lecture Notes in Control and Information Sciences

Edited by M. Thoma and A. Wyner

Vol. 81: Stochastic Optimization
Proceedings of the International Conference,
Kiew, 1984
Edited by I. Arkin, A. Shiraev, R. Wets
X, 754 pages, 1986.

Vol. 82: Analysis and Algorithms
of Optimization Problems
Edited by K. Malanowski, K. Mizukami
VIII, 240 pages, 1986.

Vol. 83: Analysis and Optimization
of Systems
Proceedings of the Seventh International
Conference of Analysis and Optimization
of Systems
Antiba, Juno 26-27, 1086
Edited by A. Bensoussan, J. L. Lions
XVI, 901 pages, 1986.

Vol. 84: System Modelling
and Optimization
Proceedings of the 12th IFIP Conference
Budapest, Hungary, September 2–6, 1985
Edited by A. Prékopa, J. Szelezsán, B. Strazicky
XII, 1046 pages, 1986.

Vol. 85: Stochastic Processes
in Underwater Acoustics
Edited by Charles R. Baker
V, 205 pages, 1986.

Vol. 86: Time Series and
Linear Systems
Edited by Sergio Bittanti
XVII, 243 pages, 1986.

Vol. 87: Recent Advances in
System Modelling and
Optimization
Proceedings of the IFIP-WG 7/1
Working Conference
Santiago, Chile, August 27-31, 1984
Edited by L. Contesse, R. Correa, A. Weintraub
IV, 199 pages, 1987.

Vol. 88: Bruce A. Francis
A Course in H_∞ Control Theory
XI, 156 pages, 1987.

Vol. 88: Bruce A. Francis
A Course in H_∞ Control Theory
X, 150 pages, 1987.
Corrected - 1st printing 1987

Vol. 89: G. K. H. Pang/A. G. J. McFarlane
An Expert System Approach to
Computer-Aided Design
of Multivariable Systems
XII, 223 pages, 1987.

Vol. 90: Singular Perturbations
and Asymptotic Analysis
in Control Systems
Edited by P. Kokotovic,
A. Bensoussan, G. Blankenship
VI, 419 pages, 1987.

Vol. 91 Stochastic Modelling
and Filtering
Proceedings of the IFIP-WG 7/1
Working Conference
Rome, Italy, Decembre 10-14, 1984
Edited by A. Germani
IV, 209 pages, 1987.

Vol. 92: L. T. Grujić, A. A. Martynyuk,
M. Ribbens-Pavella
Large-Scale Systems Stability Under
Structural and Singular Perturbations
XV, 366 pages, 1987.

Vol. 93: K. Malanowski
Stability of Solutions to Convex
Problems of Optimization
IX, 137 pages, 1987.

Vol. 94: H. Krishna
Computational Complexity
of Bilinear Forms
Algebraic Coding Theory and
Applications to Digital
Communication Systems
XVIII, 166 pages, 1987.

Vol. 95: Optimal Control
Proceedings of the Conference on
Optimal Control and Variational Calculus
Oberwolfach, West-Germany, June 15-21, 1986
Edited by R. Bulirsch, A. Miele, J. Stoer
and K. H. Well
XII, 321 pages, 1987.

Vol. 96: H. J. Engelbert/W. Schmidt
Stochastic Differential Systems
Proceedings of the IFIP-WG 7/1
Working Conference
Eisenach, GDR, April 6-13, 1986
XII, 381 pages, 1987.

For information about Vols. 1–80 please contact your bookseller or Springer-Verlag.

Lecture Notes in Control and Information Sciences

Edited by M. Thoma and A. Wyner

147

J. P. Zolésio (Editor)

Stabilization of Flexible Structures

Third Working Conference
Montpellier, France, January 1989

Springer-Verlag Berlin Heidelberg GmbH

Editor
J. P. Zolésio
CNRS & INLN
Faculté des Sciences
University of Nice, Parc Valrose
06034 Nice Cedex,
France

ISBN 978-3-540-53161-6 ISBN 978-3-540-46731-1 (eBook)
DOI 10.1007/978-3-540-46731-1

Originally published by Springer-Verlag Berlin Heidelberg New York in 1990.
Offsetprinting: Mercedes-Druck, Berlin

61/3020-543210 Printed on acid-free paper

FOREWORD

This volume contains the papers presented during the third working conference "Stabilization of Flexibles Structures " held in Montpellier,January 1989.The three conferences,Nice June 1987, Montpellier January 1988,Montpellier January 1989, were sponsored by the following French institutions working on control and stabilization:

Centre National de la Recherche Scientifique (CNRS) ,

Centre de Mathématiques Appliquées (CMA),which a department of l'Ecole Nationale Supérieure des Mines de Paris,located in Sophia-Antipolis

Aérospatiale , Cannes - la - Bocca

The collaboration of these three institutions was initiated in 1986.As a result there exist now several joint works and a regular seminar "Stabilization of Flexible Structures " .

This volume is divided in three parts:

Examples of Flexibles Structures

Mathematics concerning Stability,Wave Equation,Non Cylindrical Domain

Shape Variation in Hyperbolic Problems

J would like to express my thanks to Professor M.THOMA who has accepted again to publish these proceedings in the Lecture Notes in Control and Information Sciences.

J.P. ZOLESIO

CONTENTS

RECENT WORK ON THE SCOLE MODEL

Walter Littman
University of Minnesota
Minneapolis, MN 55455, USA

1. Introduction.

In a number of papers (see for example [BT]) Balakrishnan and Taylor introduced the "SCOLE" (Spacecraft Control Laboratory Experiment) model for a vibrating flexible mast, which at one end is attached to a spaceship, and at the other end to an antenna reflector. Mathematically, the system consists, essentially, of three uncoupled partial differential equations, two of which are the Euler Bernoulli beam equation while the third is the one dimensional wave equation. At the "left" end "clamped" boundary conditions are imposed. At the "right" end control forces and torques are imposed, yielding complicated non homogeneous boundary conditions which are nonlinear and in which the unknown functions - representing beam deflections and the torsion angle about the beam axis - are coupled.

Two problems present themselves: one is the "open loop" exact controllability of the system: can an initial disturbance - in an appropriated function space - be exactly controlled to rest in a finite time by applying the forces and torques at the right end in an appropriate fashion? The second question is one of closed loop stabilization: Can the (inhomogeneous) control forces and torques at the "right" end be chosen as functions of the velocities and angular velocities at that end in such a way that the energy of the system approaches zero asymptotically as $t \to \infty$. In that case can this decay be made to exponential? In this note we shall discuss some recent work dealing with the first question.

2. The open loop problem.

In [LM1] the "reduced SCOLE" system is considered, consisting of a single Euler Bernoulli beam equation, arising from the plane motion of a beam.

Consider the mixed problem:

$$\begin{array}{cccc} w_{tt} + w_{xxxx} = 0 & 0 \le x \le 1, \quad t \ge 0 & \\ w(0,t) = w_x(t,0) = 0 & & \\ w_{tt} - \beta_1 w_{xxx} = f_1(t) & w_{xtt} + \beta_2 w_{xx} = f_2(t) & \text{at } x = 1 \\ w(x,0) = w_0(x) & w_t(x,0) = w_1(x) & 0 \le x \le 1. \end{array}$$

(Here $\beta_1 > 0, \beta_2 > 0$).

The control problem: Given initial conditions $w_0(x)$ and $w_1(x)$ (possibly satisfying some compatibility conditions at $x = 0$), can we find functions $f_1(t)$ and $f_2(t)$ such that the resulting solution of the mixed problem vanishes for $t \ge T$?

An answer was given in [LM1]: (Here the H's refer to Sobolev spaces)

Given initial data in $H^6 \times H^4$ on $0 \le x \le 1$, with compatibility conditions

$$\begin{array}{cc} w_0(0) = w_0'(0) = 0, & w_1(0) = w_1'(0) = 0 \\ w_0^{(4)}(0) = w_0^{(5)}(0) = 0; & \end{array}$$

then for each positive duration T, there exist two controllers $f_1(t)$ and $f_2(t)$ continuous in [0,T] and C^∞ on (0, T] such that the corresponding solution, $w(x,t)$ to the mixed problem vanishes from $t \ge T$. Furthermore the functions $f_1(t)$ and $f_2(t)$ are given by explicit formulas.

Note: In the proof it actually suffices for the initial data to be in $H^{5\frac{1}{2}} \times H^{3\frac{1}{2}}$.

3. Improvements.

There are several directions in which the method of [LM1] can be extended and improved.

First of all, although the original three dimensional SCOLE system seems much more complicated, the methods of [LM1] encompass essentially all mathematical difficulties, and the exact controllability of the three dimensional model can be achieved by a minor modification of the method. The only difference is that since one of the equations in the full system is a wave equation, the time T is not arbitrarily small, but is governed by the time it takes to control the one dimensional wave equation.

Secondly, to what extent is the high degree of smoothness of the initial data really necessary? It follows from a result of Triggiani that an initial disturbance assumed only to have *finite energy* can not be controlled by *locally integrable* f_1 and f_2 (see the discussion in [LM2]). The method described above yields f_1 and f_2 which are C^∞ for *positive* t but which may have singularities at $t = 0$, in the case where the initial data is assumed to be merely L^2, or have finite energy. *It can be shown however that if we merely require* f_1 *and* f_2 *to be in* $L^2[0,T]$, *rather than be continuous at zero, it suffices to take the initial data merely in* $H^{4\frac{1}{2}} \times H^{2\frac{1}{2}}$. This has the advantage that the compatibility conditions $w_0^{(4)}(0) = 0$ and $w_0^{(5)}(0) = 0$ can now be dispensed with.

Finally it is of interest to consider the case where the material properties of the beam vary from point to point. This problem has been recently solved by Steven Taylor, a Ph.D. student at Minnesota, who in the course of providing the solution has obtained a number of related results of independent interest. We describe Taylor's work in the next section.

4. The work of Steven Taylor.

The main ingredient in the work is the establishment of a certain degree of regularity in t of solutions of a class of equations with variable coefficients on the semi infinite interval $0 \le x < \infty$. Consider the equation

$$Lw = \frac{\partial^2 w}{\partial t^2} + \sum_{0 \le i \le 4} b_i(x) \frac{\partial^i w}{\partial x^i} = 0 \qquad 0 \le x < \infty, \quad t \ge 0.$$

We assume $b_4 > 0, b_4 \in C^4, b_3 \in C^3, b_2, b, b_0$ each is C^2, and that these coefficients are constant outside of a bounded set. If w is a solution to the initial value problem consisting of the above equations and boundary condition

$$w(0,t) = 0 \qquad w_x(0,t) = 0$$

and initial conditions (satisfying appropriate compatibility conditions at $x = 0$)

$$w(x,0) = w_0(x) \in H^2([0,\infty)) \text{ with compact support}$$

$$w_t(x,0) = w_1(x) \in L^2([0,\infty)) \quad \text{with compact support,}$$

then the solution $w(x,t)$ belongs to the Gevrey class $\gamma^{(2)}$ in the variable t. Loosely speaking, a C^∞ function $f(t)$ is in the Gevrey class $\gamma^{(\delta)} (\delta \ge 1)$ if for each $\theta > 0$ one can find a constant $C_\theta > 0$ such that $|f^{(n)}(t)| < C_\theta \theta^n n^{n\delta}$ for all $n = 0, 1, 2, \cdots$ In case of functions $w(x,t)$ it is understood that the stated estimates hold uniformly in compact subsets of $x \geq 0$, $t > 0$.

Once the Gevrey regularity result is established, the controllability result follows as in [LM1]. We briefly outline the procedure. Suppose the basic interval is $0 \leq x \leq 1$.

We write $L = T + A$, where $T = \frac{\partial^2}{\partial t^2}$. We extend the initial data as smoothly as possible to have compact support in the larger interval $0 \leq x \leq 1 + \epsilon$ and solve the resulting mixed problem for the half line $x \geq 0$. By the Gevrey regularity result the resulting solution $W(x,t)$ will be of a class $\gamma^{(2)}$ for $t > 0$. Letting T_1 be an arbitrary positive number one can find a function $h(t)$ of class $\gamma^{(2)}$ such that $h(t) = 1$ for $t < T_1/2$ and $h(t) = 0$ for $t > T_1$. Then $h(t)W(x,t)$ is of class $\gamma^{(2)}$ in t and satisfies $(T + A)h(t)W(x,t) \equiv F(x,t)$, where $F(x,t)$ is of class $\gamma^{(2)}$ in t, vanishes for $t \geq T_1$ and $0 \leq t \leq T_1/2$. Then one constructs a solution Z of $(T + A)Z = F$ vanishing for $t \geq T_1, 0 \leq t \leq T_1/2$. The solution is given by the formula

$$Z(x,t) = \sum_{0 \leq j < \infty} (-1)^j T^j A^{-(j+1)} F(x,t),$$

where A^{-1} is an appropriately defined inverse of operator A. It can be shown that the series for W converges because $F(x,t)$ is of class $\gamma^{(2)}$ in t. One then sets $w(x,t) = W(x,t)h(t) - Z(x,t)$. Then $w(x,t)$ satisfies the equation $Lw = o$, the correct initial conditions in $0 \leq x \leq 1$, and the correct boundary conditions at $x = 0$, and furthermore vanishes for $t \geq T_1$. The boundary controllers are then read off from the traces of the correct boundary operators at $t = 1$.

This method has recently been extended by Taylor to equations of the form

$$\frac{\partial^2 w}{\partial t^2} + \sum_{0 \leq i \leq 4} b_i(x) \frac{\partial^i w}{\partial x_i} + \sum_{0 \leq i \leq 2} a_i(x) \frac{\partial^i w}{\partial t \partial x_i} = 0$$

where $b_4(x) > 0$ and $a_2(x) \leq 0$.

BIBLIOGRAPHY

[BT] Balakrishnan, A. V., and Taylor, L. W., A Mathematical Problem and a Spacecraft Control Laboratory Experiment (SCOLE) Used to Evaluate Control Laws for Flexible Spacecraft, Proceedings of NASA SCOLE workshop, Dec. 1984.

[LM1] Littman, W., Markus, L. Exact Boundary Controllability of a Hybrid System of Elasticity, Archive for Rational Mechanics and Analysis, 103 #3 1988, 193-236.

[LM2] Littman, W., Markus, L., Remarks on Exact Controllability and Stabilization of a Hybrid System in Elasticity Through Boundary Damping, Proceedings of Symposium in Santiago Spain (1987), Springer Lecture Notes on Control and Information Sciences #114, 1989.

Acknowledgement: This research was in part sponsored by NSF Grant 8722402.

MATHEMATICAL STUDY OF LARGE SPACE STRUCTURES

D.CIORANESCU – J.SAINT JEAN PAULIN

The structures we study are made of identical cells periodically distributed. The material is concentrated along layers (honeycomb structures) or along bars (reinforced structures) or in some directions along layers and in others along bars.

For instance a tower is a honeycomb structure and a crane is a reinforced structure. In these cases we have periodicity in only one direction. We can also consider large space structures with periodicity in two or in three directions. The period can be cubic, triangular, hexagonal. As a rule, they are perforated structures with very big holes and very little material.

There is no symmetry assumption on the distribution of the material in the cell. Thus non diagonal oblique bars or layers can be considered. The thickness $\varepsilon\delta$ of the material is small compared with the period ε, which is itself small compared with the global magnitude of the structure.

More complex structures are for instance the gridworks in which the global thickness is $e = \varepsilon\eta$ where η is a third small parameter.

The aim of our mathematical study is to give the global thermal or elastic behaviour of these structures with Dirichlet or Neumann ([1],[2],[4]) or Fourier boundary conditions on the boundary of the holes or eigenvalue problems([6]).Though we know the characteristic constants of the material, the numerous cells and great thickness of the bars or layers make a direct calculation of the solution $u_{\varepsilon\delta}$ of the problem far too long and expensive.

We get rid of the first difficulty by using homogenization methods (such as multiple scale method or variational method) in the directions in which we have periodicity. This means that we approximate the real oscillating material by a homogeneous one whose constant characteristic coefficients are calculated on a representative period.

However the very small thickness of the material in the period still makes the computation very costly. To overcome this second difficulty we give a perturbation method which enables to give explicitly a good approximation of the real material . In this step the thickness of the material is the essential parameter.

Our method applies to more complicated structures (such as those containing oblique bars or layers) as it consists in superposing results obtained independently in each direction in which the material is distributed.

For gridworks, we have one more step in our asymptotic study because they are thin in one direction. We give the dependence in this third small parameter by using plate techniques ([5],[7]). For tall structures such as towers and cranes we also apply rod techniques ([3],[8]).

We can even give the explicit value of the global coefficients and estimate the error between $u_{\varepsilon\delta}$ and the limit result $u^{\star}$.

We list here some results (for detailed proofs see the references).

Fig.1

Fig.2

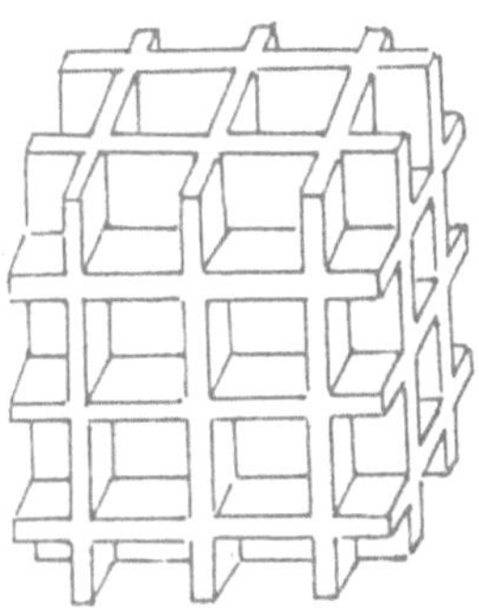

Fig.3

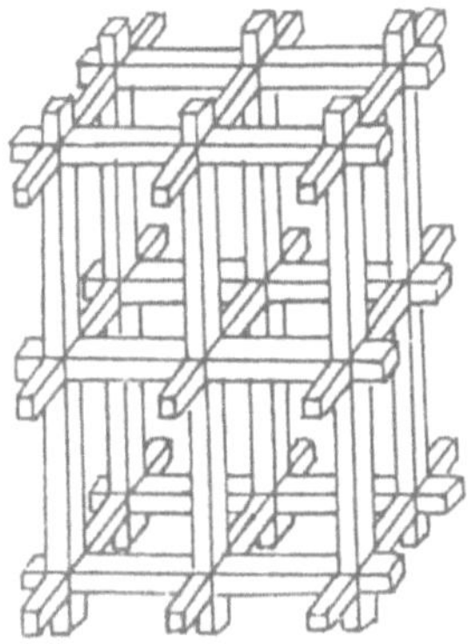

Fig.4

Fig.5

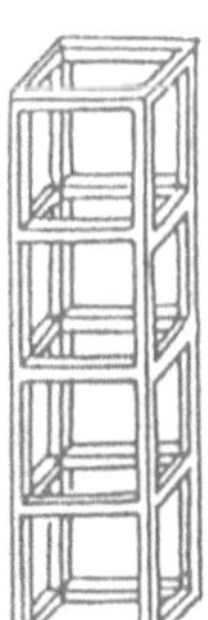

Fig.6

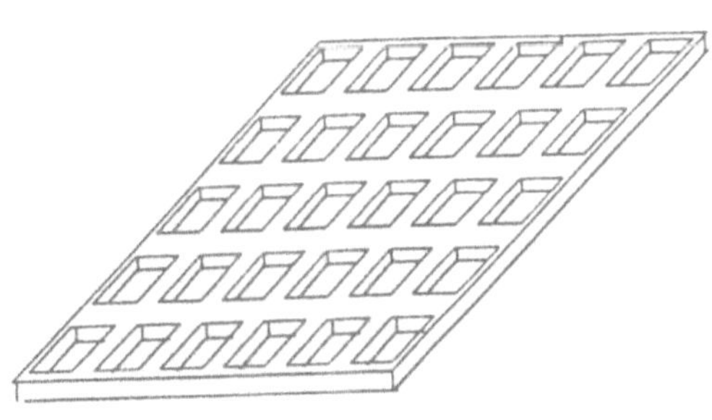

Fig.7

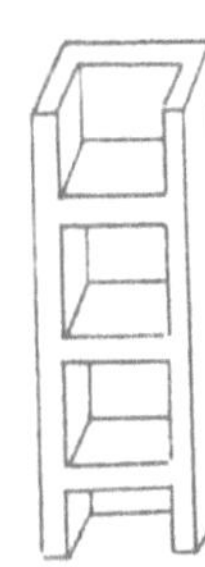

Fig.8

1. THERMAL PROBLEM

The structure we study occupies a domain Ω .We denote by $\Omega_{\varepsilon\delta}$ the part of Ω corresponding to the material and by $T_{\varepsilon\delta} = \Omega - \Omega_{\varepsilon\delta}$ that corresponding to the holes. Consider the problem:

$$\begin{cases} -\dfrac{\partial}{\partial x_i}(a_{ij}\dfrac{\partial u_{\varepsilon\delta}}{\partial x_j}) = 0 & \text{in } \Omega_{\varepsilon\delta} \quad \text{(part occupied by the material)} \\ u_{\varepsilon\delta} = 0 & \text{on } \partial\Omega \quad \text{(the exterior boundary of the domain)} \\ a_{ij}\dfrac{\partial u_{\varepsilon\delta}}{\partial x_j} n_j = 0 & \text{on } \partial T_{\varepsilon\delta} \quad \text{(the boundary of the holes)} \end{cases}$$

We make the following assumptions:

(*i*) f in $L^2(\Omega)$

(*ii*) the coefficients a_{ij} are constants $(i, j = 1, ..., n)$

(*iii*) There exists a positive number A such that

$$a_{ij}(y)\zeta_i\zeta_j \geq A\zeta_i\zeta_i \quad \text{for any} \quad \zeta \in \mathbb{R}^n$$

We make successively $\varepsilon \to 0$ and $\delta \to 0$

1.1. A bidimensional case without oblique bars

We consider the structure in fig. 1. The limit problem is ([1]):

$$\begin{cases} -\dfrac{1}{2}q_{ij}\dfrac{\partial^2 u^\star}{\partial x_i \partial x_j} = f & \text{in } \Omega \quad \text{(the whole domain occupied by the structure)} \\ u^\star = 0 \quad \text{on } \partial\Omega \end{cases}$$

with

$$q_{ij} = 2a_{ij} - \Sigma_{k=1}^{2}\frac{a_{ik}a_{kj}}{a_{kk}}.$$

This solution is unique. The matrix q_{ij} is diagonal and positive definite. If $(A_{ij}) = (a_{ij})^{-1}$ denotes the inverse matrix of the matrix (a_{ij}), one also has:

$$\begin{cases} q_{ii} = \dfrac{1}{A_{ii}} & \text{(no summation in i)} \\ q_{ij} = 0 \quad \text{if} \quad i \neq j. \end{cases}$$

In the particular case where $a_{ij} = \delta_{ij}$, the limit problem is:

$$-\frac{1}{2}\Delta u^\star = f \quad \text{in} \quad \Omega.$$

1.2. A bidimensional case with oblique bars

When the geometry is that of fig. 2, the limit problem is ([1]):

$$\begin{cases} -\dfrac{1}{(2+\sqrt{2})} q_{ij} \dfrac{\partial^2 u^\star}{\partial x_i \partial x_j} = f & \text{in } \Omega \\ u^\star = 0 \quad \text{on } \partial\Omega \end{cases}$$

with

$$q_{ij} = a_{ij} - \frac{1}{(2+\sqrt{2})}\left[\frac{a_{i1}a_{1j}}{a_{11}} + \frac{a_{i2}a_{2j}}{a_{22}} + \sqrt{2}\,\frac{(a_{i1}-a_{i2})(a_{j1}-a_{j2})}{a_{11}-a_{12}-a_{21}+a_{22}}\right].$$

The same method applies to structures with non diagonal bars and it leads again to explicit coefficients.

1.3. A three dimensional case for honeycomb structures

We consider the honeycomb structure in fig. 3. The limit problem is ([1]):

$$\begin{cases} -\dfrac{1}{3} q_{ij} \dfrac{\partial^2 u^\star}{\partial x_i \partial x_j} = f & \text{in } \Omega \\ u^\star = 0 \quad \text{on } \partial\Omega \end{cases}$$

with

$$q_{ij} = 3a_{ij} - \Sigma_{k=1}^{3} \frac{a_{ik}a_{kj}}{a_{kk}}.$$

The matrix (q_{ij}) is coercive and in general non diagonal. In the case where $a_{ij} = \delta_{ij}$, the limit problem is:

$$\begin{cases} -\dfrac{2}{3}\Delta u^\star = f & \text{in } \Omega \\ u^\star = 0 \quad \text{on } \partial\Omega \end{cases}$$

1.4. A three dimensional case for reinforced structures

We now study a reinforced structure (see fig. 4). The limit problem is ([1]):

$$\begin{cases} -\frac{1}{3} q_{ij} \frac{\partial^2 u^\star}{\partial x_i \partial x_j} = f & \text{in } \Omega \\ u^\star = 0 & \text{on } \partial\Omega \end{cases}$$

with

$$\begin{cases} q_{ii} = \frac{1}{A_{ii}} & \text{(no summation in i)} \\ q_{ij} = 0 \quad \text{if} \quad i \neq j \end{cases}$$

where $(A_{ij}) = (a_{ij})^{-1}$ is the inverse matrix of the matrix (a_{ij}). The matrix (q_{ij}) is coercive and diagonal. In the case where $a_{ij} = \delta_{ij}$, the limit problem is:

$$\begin{cases} -\frac{1}{3} \Delta u^\star = f & \text{in } \Omega \\ u^\star = 0 & \text{on } \partial\Omega \end{cases}$$

Note that in the two dimensional case (fig 1) there is no distinction between reinforced and honeycomb structures.

1.5.Error estimate

In all these cases, we show that $u^\star$ is a good approximation of $u_{\varepsilon\delta}$. More precisely, we prove that if f is sufficiently smooth, then :

$$(\star) \qquad \frac{1}{(meas\Omega_{\varepsilon\delta})^{\frac{1}{2}}} \parallel u_{\varepsilon\delta} - u^\star \parallel_{H^1(\Omega_{\varepsilon\delta})} \quad \leq \quad C \quad (\delta^{\frac{1}{2}} + \varepsilon^{\frac{1}{2}}).$$

where C is a constant independant of ε and δ.

1.6. Tall structures

In the bidimensional case (see fig. 5), the limit problem obtained after an appropriate rescaling is ([3]) :

$$\begin{cases} -\frac{1}{2} q \frac{\partial^2 u^\star}{\partial {x_2}^2} = f(0, x_2) & \text{in } (0, L) \\ q \frac{\partial u^\star}{\partial x_2}(L) = 0 \\ u^\star(0) = 0 \end{cases}$$

with

$$q = a_{22} - \frac{a_{21}a_{12}}{a_{11}}.$$

In the three dimensional case (see fig. 6), after an appropriate rescaling the limit problem is:

$$\begin{cases} -\frac{1}{3}q\frac{\partial^2 u^\star}{\partial {x_3}^2} = f(0,0,x_3) \quad \text{in } (0,L) \\ q\frac{\partial u^\star}{\partial x_3}(L) = 0 \\ u^\star(0) = 0 \end{cases}$$

with

$$q = \frac{1}{A_{33}}$$

where $(A_{ij}) = (a_{ij})^{-1}$ is the inverse matrix of the matrix (a_{ij}). We also have an error estimate analogous to that given by $(\star)$.

1.7.Gridworks

We study now a gridwork (see fig.7).The temperature $u^{e\varepsilon\delta}$ is solution of the system :

$$\begin{cases} -\frac{\partial}{\partial x_i}(a_{ij}\frac{\partial u^{e\varepsilon\delta}}{\partial x_j}) = 0 & \text{in } \Omega_{e\varepsilon\delta}\text{(part occupied by the material)} \\ a_{3j}\frac{\partial u^{e\varepsilon\delta}}{\partial x_j}n_3 = -g^{\pm} & \text{on the top and bottom surfaces} \\ a_{ij}\frac{\partial u^{e\varepsilon\delta}}{\partial x_j}n_j = 0 & \text{on the boundary of the holes} \\ u^{e\varepsilon\delta} = 0 & \text{on the exterior lateral boundary of the gridwork} \end{cases}$$

We prove that if $(e,\varepsilon,\delta) \to (0,0,0)$ then:

$$u^{e\varepsilon\delta} \sim e^{-1}u^\star$$

(with convergence in a H^1 space), where $u^\star$ is independent of x_3 and satisfies :

$$\begin{cases} -\dfrac{1}{A_{11}}\dfrac{\partial^2 u^\star}{\partial {x_1}^2} - \dfrac{1}{A_{22}}\dfrac{\partial^2 u^\star}{\partial {x_2}^2} = 2(g^+ + g^-) & \text{in } \omega = (0, l_1) \times (0, l_2) (\text{cross section of } \Omega) \\ u^\star = 0 \quad \text{on} \quad \partial\omega \end{cases}$$

and (A_{ij}) is the inverse matrix of (a_{ij}).

2. ELASTICITY PROBLEMS.

Assume that the structure contained in the domain Ω is clamped on its exterior boundary and subjected to applied body forces f. The displacement $u^{\varepsilon\delta}$ is given by solving the system :

$$\begin{cases} -\dfrac{\partial}{\partial x_i}(a_{ijkh}\dfrac{\partial u_k^{\varepsilon\delta}}{\partial x_h}) = f_i & \text{in } \Omega_{\varepsilon\delta} \quad (\text{part occupied by the material}) \\ u^{\varepsilon\delta} = 0 & \text{on } \partial\Omega \\ (a_{ijkh}\dfrac{\partial u_k^{\varepsilon\delta}}{\partial x_h})n_j = 0 & \text{on the boundary of the holes} \end{cases}$$

The elasticity constants a_{ijkh} satisfy the usual hypotheses of symmetry and coercitivity. For sake of simplicity we suppose that they are Lamé constants:

$$a_{ijkh} = \lambda\delta_{ij}\delta_{kh} + \mu(\delta_{ik}\delta_{jh} + \delta_{ih}\delta_{jk}).$$

2.1. Honeycomb structures

For a honeycomb structure (see fig.3) the limit problem is ([4]) :

$$\begin{cases} -\dfrac{1}{3} q_{ijkh} \dfrac{\partial^2 u_k^\star}{\partial x_j \partial x_h} = f_i \quad \text{in} \quad \Omega \\ u^\star = 0 \quad \text{on} \quad \partial\Omega. \end{cases}$$

The coefficients (q_{ijkh}) satisfy the symmetries of elasticity and are defined by :

$$\begin{cases} q_{1111} = q_{2222} = q_{3333} = \dfrac{8\mu(\lambda+\mu)}{(\lambda+2\mu)} \\ q_{1122} = q_{2233} = q_{3311} = \dfrac{2\lambda\mu}{(\lambda+2\mu)} \\ q_{1212} = q_{2323} = q_{3131} = \mu \end{cases}$$

Moreover

$$\frac{1}{(meas\Omega_{\varepsilon\delta})^{\frac{1}{2}}} \quad \| u^{\varepsilon\delta} - u^* \|_{H^1(\Omega_{\varepsilon\delta})} \quad \leq \quad C \quad (\delta^{\frac{1}{2}} + \varepsilon^{\frac{1}{2}}).$$

where C is a constant independant of ε and δ.

2.2. Reinforced structures

We study now a reinforced structure (see fig 4). The limit coefficients are ([4]) :

$$\begin{cases} q_{1111} = q_{2222} = q_{3333} = \dfrac{\mu(3\lambda + 2\mu)}{(\lambda + \mu)} = E \\ q_{ijkh} = 0 \quad \text{in all the other cases.} \end{cases}$$

The matrix (q_{ijkh}) is not coercive any more, so we do not have a limit system corresponding to it. Generally it is known that spatial reinforced structures are not stable.

2.3. Gridworks

We consider an anisotropic material satisfying the equations of linearized elasticity in a gridwork (see fig.7):

$$\begin{cases} -\dfrac{\partial}{\partial x_j}(a_{ijkh}\dfrac{\partial u_k^{e\varepsilon\delta}}{\partial x_h}) = F_i^e \qquad \text{in } \Omega_{e\varepsilon\delta} \quad \text{(part occupied by the material)} \\ a_{i3kh}\dfrac{\partial u_k^{e\varepsilon\delta}}{\partial x_h} n_3 = G_i^{e\pm} \qquad \text{on the top and bottom surfaces} \\ a_{i\alpha kh}\dfrac{\partial u_k^{e\varepsilon\delta}}{\partial x_h} n_\alpha = 0 \qquad \text{on the boundary of the holes} \\ u_{e\varepsilon\delta} = 0 \qquad \text{on the exterior lateral boundary of the gridwork} \end{cases}$$

where $u^{e\varepsilon\delta} = (u_1^{e\varepsilon\delta}, u_2^{e\varepsilon\delta}, u_3^{e\varepsilon\delta})$ is the displacement, $F^e = (F_1^e, F_2^e, F_3^e)$ is the volumic density of the applied body forces and $G^{e\pm} = (G_1^{e\pm}, G_2^{e\pm}, G_3^{e\pm})$ is the density of surfaces forces. The greek indices take values in $\{1,2\}$ and the latin indices in $\{1,2,3\}$.There is no applied force on the boundary of the holes and the plate is supposed to be clamped on its exterior lateral boundary.

We make an apppropriate rescaling on the displacements and on the data and let successively $e \to 0, \varepsilon \to 0$ and $\delta \to 0$. The limit problem for an isotropic material is :

- for the deflection $u_3^\star$

$$\begin{cases} \frac{E}{6}[\frac{\partial^4 u_3^\star}{\partial z_1^4} + \frac{\partial^4 u_3^\star}{\partial z_2^4}] + \frac{4\mu}{3}\frac{\partial^4 u_3^\star}{\partial z_1^2 \partial z_2^2} = 2(\int_{-\frac{1}{2}}^{\frac{1}{2}} f_3^\star dz_3) + (g_3^{\star +} + g_3^{\star -}) + \\ \quad + 2\frac{\partial}{\partial z_\alpha}[(\int_{-\frac{1}{2}}^{\frac{1}{2}} z_3 f_\alpha^\star dz_3) + \frac{1}{2}(g_\alpha^{\star +} - g_\alpha^{\star -})] \qquad \text{in} \omega = (0,l_1) \times (0,l_2) (\text{cross section of } \Omega) \\ u_3^\star = 0 \quad \text{on} \quad \partial\omega \\ \frac{\partial u_3^\star}{\partial n} = 0 \quad \text{on} \quad \partial\omega \end{cases}$$

- for the lateral displacements $u_\alpha^\star$

$$u_\alpha^\star = -z_3 \frac{\partial u_3^\star}{\partial z_\alpha} + w_\alpha^\star$$

with

$$\begin{cases} -E\frac{\partial^2 w_\alpha^\star}{\partial x_\alpha^2} = 2(\int_{-\frac{1}{2}}^{\frac{1}{2}} f_\alpha^\star dz_3) + (g_\alpha^{\star +} + g_\alpha^{\star -}) \quad \text{in} \quad \omega \\ w_\alpha^\star n_\alpha = 0 \quad \text{on} \quad \partial\omega \quad (\text{no summation on} \quad \alpha). \end{cases}$$

2.4.Towers or cranes

We consider the system of linearized elasticity for a tall structure occupying the domain $\Omega_{e\varepsilon\delta}$ (see fig.8 for a tower and fig.6 for a crane) :

$$\begin{cases} -\frac{\partial}{\partial x_j}(a_{ijkh}\frac{\partial u_k^{e\varepsilon\delta}}{\partial x_h}) = F_i^e \quad \text{in } \Omega_{e\varepsilon\delta} \quad (\text{part occupied by the material}) \\ a_{i3kh}\frac{\partial u_k^{e\varepsilon\delta}}{\partial x_h} n_3 = 0 \quad \text{on the top of the structure} \\ u_{e\varepsilon\delta} = 0 \quad \text{on the bottom of the structure} \\ a_{i\alpha kh}\frac{\partial u_k^{e\varepsilon\delta}}{\partial x_h} n_\alpha = G^{e\varepsilon\delta} \quad \text{on the rest of the boundary} \end{cases}$$

We let successively $e \to 0, \varepsilon \to 0, \delta \to 0$. After rescaling, the limit displacement has the form ([8]) :

$$\begin{cases} u_3^\star = -z_1 \dfrac{\partial V_1^\star(z_3)}{\partial z_3} - z_2 \dfrac{\partial V_2^\star(z_3)}{\partial z_3} + V_3^\star(z_3) \\ u_\alpha^\star = V_\alpha^\star(z_3) \end{cases}$$

with

$$\begin{cases} q \dfrac{\partial^4 V_\alpha^\star}{\partial z_3^4} = -\mathcal{F}_\alpha + \dfrac{\partial \mathcal{G}_\alpha}{\partial z_3} \quad \text{in} \quad (0, L) \\ V_\alpha^\star(0) = \dfrac{\partial V_\alpha^\star}{\partial z_3}(0) = 0 \\ \dfrac{\partial^2 V_\alpha^\star}{\partial z_3^2}(L) = \dfrac{\partial^3 V_\alpha^\star}{\partial z_3^3}(L) = 0 \end{cases}$$

and

$$\begin{cases} q_3 \dfrac{\partial^2 V_3^\star}{\partial z_3^2} = -\mathcal{F}_3 \quad \text{in} \quad (0, L) \\ V_3^\star(0) = \dfrac{\partial V_3^\star}{\partial z_3}(L) = 0 \end{cases}$$

where $\mathcal{F}_\alpha$ and $\mathcal{G}_\alpha$ are limits of integrals of the given forces and of their first order moments. Of course the value of q and q_3 are different for towers and for cranes.

- For towers (see fig.8) , we have :

$$\begin{cases} q = \dfrac{E}{3} \\ q_3 = 2E \end{cases}$$

- For cranes (see fig.6) , we have :

$$\begin{cases} q = \dfrac{E}{2} \\ q_3 = E \end{cases}$$

2.5. Conclusion

We point out that for gridworks and tall structures, we start with the real three dimensional system of linearized elasticity. In the first step of the proof, we make $e \to 0$ and we use plate or rod techniques to get second order and fourth order systems satisfied by the displacements. The following step, $\varepsilon \to 0$, is the classical homogenization process in perforated domains. The last step, $\delta \to 0$, gives the possibility to compute explicitly the overall coefficients.

REFERENCES

[1] D.Cioranescu - J.Saint Jean Paulin, Reinforced and honeycomb structures, J.Math.Pures et Appl. 65 (1986), 403-422

[2] D.Cioranescu - J.Saint Jean Paulin, Problèmes de Neumann et de Dirichlet dans des structures réticulées de faible épaisseur, Comptes-Rendus de l'Académie des Sciences, I, 303 (1986),7-10

[3] D.Cioranescu - J.Saint Jean Paulin, Tall structures - Problems of towers and cranes, in Proc. of International Conf. Appl. of Multiple Scaling in Mechanics, edit . P.G.Ciarlet and E.Sanchez-Palencia, R.M.A. 4, Masson, Paris (1987), 77-92

[4] D.Cioranescu - J.Saint Jean Paulin, Elastic behaviour of very thin cellular structures, in Material Instabilities in Continuum Mechanics, edit.J.M.Ball, Clarendon Press, Oxford (1988), 64-75

[5] D.Cioranescu - J.Saint Jean Paulin, Global behaviour of very thin cellular structures - Applications to networks, in Trends in Applications of Mathematics to Mechanics, Proc.of 7th Symposium , edit. J.Besseling and W.Eckhauss, Springer Verlag (1988), 26-34

[6] D.Cioranescu - J.Saint Jean Paulin, Conditions de Fourier et problèmes de valeurs propres pour des structures réticulées, Publication du Laboratoire d'Analyse Numérique n° 88054, Paris (1988)

[7] D.Cioranescu - J.Saint Jean Paulin, Asymptotic Analysis of elastic wireworks (to appear)

[8] D.Cioranescu - J.Saint Jean Paulin, Towers and cranes in linearized elasticity: an asymptotic study (to appear)

Doina Cioranescu
Laboratoire Analyse Numérique- CNRS
4 Place Jussieu
75252 Paris Cedex 05
France

Jeannine Saint Jean Paulin
Université de Metz-Mathématiques
Ile de Saulcy
57045 Metz Cedex 01
France

Symbolic Formulation of Dynamic Equations for Interconnected Flexible Bodies: the GEMMES Software

C. GARNIER

Aérospatiale
Division Systèmes Stratégiques et Spatiaux
100 Bd du Midi, BP 99
06322 Cannes-la-Bocca – FRANCE

Abstract

This article presents a new tool: symbolic computation, used for the generation of dynamic equations for poly-articulated flexible structure. We begin with a presentation of the formulation used and of the modelling of flexible bodies. Then the capabilities of the symbolic computation software: GEMMES are presented, and we describe some examples of space structures analyzed using GEMMES. Advantages of such method on pure numerical ones are emphasized.

1 INTRODUCTION

The present evolution of space technology calls for accurate dynamic modelling of large flexible structures or structures with a complex geometry. In many cases, we can overlook neither the flexibility of the structures' various components and their coupling with the rigid motions, nor the non-linear terms of the mechanical equations. Since we can no longer be satisfied with a simple 2D model, we need a complete, generally complex dynamic model of these structures.

The computation of such models is so complicated that it calls for the use of a computer to be efficient. The present solution is to use softwares such as CONSTRUCT, MIDAS or ADAMS, which compute the dynamic equations in a numerical way (using floating point expressions). This method is fraught with two main drawbacks:

- it can be used only to make numerical simulations of the structure's motion and cannot be used to obtain the system's dynamic equations,

- as the dynamic model is time-dependent, this method requires a new computation of the model at each new time step.

The emergence of symbolic computation software provides an alternative to these numerical programs. With help of the MAPLE software, we wrote the GEMMES program which computes the dynamic equations of a structure in a symbolic form (using the name of the data instead of their numerical values). The GEMMES-generated equations are just like hand-written ones but, due to the use of the computer, these equations are faultless and very quickly derived. Such symbolic model can be used in two different manners:

- the equations can be linearized around a nominal state in order to obtain a model useful to design control laws,
- a dedicated numerical simulator can be derived from the equations. As it is directly dependent upon the system under analysis, this simulator is more efficient than the ones provided by the general numerical softwares.

We first present GEMMES capabilities, and then continues with some examples of space structures analyzed thanks to GEMMES.

2 GEMMES

2.1 An overview

Some previous experiences in dynamic modelling of flexible spacecraft[1] has shown the difficulty to obtain hand-written, faultless, symbolic equations of motions and the corresponding numerical simulator. The idea of using symbolic computation for dynamic modelling and simulator generation emerged from such experiences, and is now concretized in GEMMES.

The main specifications for such software were the following:

- 3D capabilities
- rigid/flexible bodies

[1] L. Passeron and C. Garnier "*Attitude Control for a Data Relay Satellite: a Decentralized Approach*", IAF 85-229 36th IAF Congress, Stockholm October 1985

- classical mechanical hinges (revolute, universal, planar, translationnal ...)
- any kind of force/torque

A choice has been made at the beginning of the project between using a general computer algebra system as MAPLE, MACSYMA, REDUCE or writing a mechanics' dedicated one. We have choosen to use a commercial computer algebra system in order to avoid to write an evaluator/simplificator which is a complicated task and also to be able to use all predefined functions and data structures. The choice was made on MAPLE which runs on "small" computers.

GEMMES is broken in four parts which are:

- a functionnal pre-processor used to describe the topology and the main features of the studied system,
- an equation computation program,
- a library for display, simplification, evaluation, linearization of the equations,
- a program for FORTRAN and numerical simulators generation.

We will now describe these functionnalities.

2.2 Computations of the equations

2.2.1 Equations of a free body

Dynamic behavior of one free body is described by 3 matricial equations for translations, rotations and deformations. This formulation plus use of quasi-coordinates allows to describe the system in a very compact form. Another advantage of this matricial representation is the possibility to use a symbolic value for the number of modes of deformations.

Let $\mathcal{P}_i$, $\mathcal{Q}_i$ be respectively the position and velocity parameters of body i. If we link a frame $\mathcal{R}_i$ with origin O_i to body i, the velocity variables of body i are:

- $\dot{D}_i$ which is the first derivative w.r.t. time of vector $\overrightarrow{OO_i}$,
- Ω_i which is the instantaneous rotation vector of frame $\mathcal{R}_i$ w.r.t. frame $\mathcal{R}_0$

- $\dot{Q}_i$ which is the first derivative w.r.t. time of the vector of modal coordinates Q_i

Thus the relation $\mathcal{Q} = \dot{\mathcal{P}}$ is not always valid, for a free body, its rotation is described by three angles and its rotationnal velocity by the instantaneous rotation vector.

With these notations, the dynamic equations of body i are

$$M_i(\mathcal{P}_i)\,\dot{\mathcal{Q}}_i = S_i(\mathcal{P}_i, \mathcal{Q}_i) + F_i \tag{1}$$

$M_i(\mathcal{P}_i)$ is the mass matrix, $S_i(\mathcal{P}_i, \mathcal{Q}_i)$ the vector of Coriolis and centrifugal forces called "second member", F_i is the vector of applied forces and torques.

2.2.2 Equations of an articulated structure with multipliers

The basis is well-known because it's the Lagrange equations with multipliers for joints accomodation.We define $\mathcal{P}$ and $\mathcal{Q}$ which are respectively the position and velocity parameters obtained by concatenation of $\mathcal{P}_i$ and $\mathcal{Q}_i$.

The final equation 2 is easily derived, because it corresponds to the equations of n independent free bodies.

$$\begin{cases} M(\mathcal{P})\,\dot{\mathcal{Q}} = S(\mathcal{P}, \mathcal{Q}) + F + \Delta_V(\mathcal{P})^T\Gamma \\ \Delta_V(\mathcal{P})\,\mathcal{Q} = 0 \end{cases} \tag{2}$$

F is the vector of external forces and torques applied to the bodies. $\Delta_V(\mathcal{P})$, the constraint matrix, expresses the relations between the velocity parameters imposed by the joints while Γ is the vector of Lagrange multipliers.

2.2.3 Multipliers elimination

To get the final equations, we want to eliminate the Lagrange multipliers from equation 2. Then we must find a kernel of matrix $\Delta_V(\mathcal{P})$ which is a matrix B such that:

$$\Delta_V(\mathcal{P})\,B = 0 \tag{3}$$

Then we write the final equations with a new velocity parameter q such that

$$\mathcal{Q} = B\,q \tag{4}$$

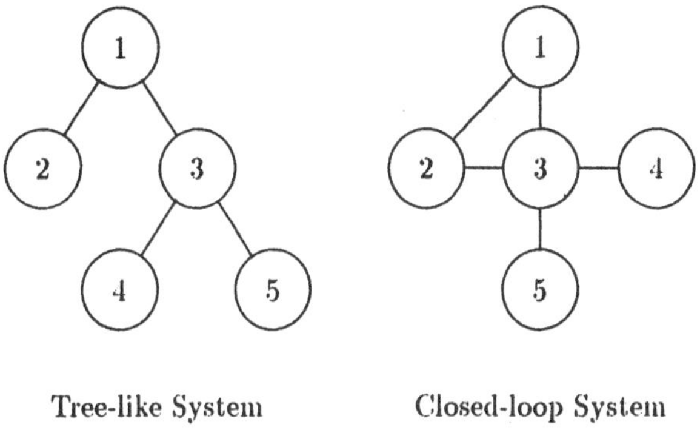

Figure 1: Example of tree-like and closed-loop systems

and a new position parameter p. So the final equation will be:

$$\underbrace{B^T\ M(p)\ B}_{\text{Final mass matrix}}\ \dot{q} = \underbrace{B^T\ S(p,q)\ -\ B^T\ M(p)\dot{B}\ q}_{\text{Final second member}} + B^T\ F \tag{5}$$

The complexity of computing kernel B depends upon whether the structure is in topological tree or has closed loops. In a tree-like structure there is only one path to go from a body to another. On the contrary, in a closed-loop structure, there can be various paths from a body to another. Figure 1 shows two examples. In the closed-loop system of this figure there are for instance two different paths to go from body 1 to body 3.

In the case of tree-like systems, the kernel can be easily computed from a mere analysis of the joints between the bodies by retaining joints' degrees of freedom. Writing position and velocity of each 'son' body with respect to its 'father' and to the connecting hinge degrees of freedom gives directly the kernel of the constraint matrix.

The case of closed-loop systems is far more complex and is not actually completely solved. The main problem is computing the kernel of the constraint matrix. No algorithm is already implemented to solve this problem. In the case of a closed-loop system, GEMMES acts in the following way:

- first, it automatically detects the presence of a closed loop in the system,

- then, it cuts as many joints as necessary to get a tree-like system. The tree obtained is such that the lengths of branches are minimal.
- then, it computes the equations of the tree-like system. There remain Lagrange multipliers corresponding to the cut joints. Therefore, GEMMES also compute the remaining constraint matrix.

The Lagrange multipliers must be at this moment hand-eliminated if a solution of the problem is known or easy to compute. But, most of the time, the simulator will include a set of algebraic equations connected with the differential ones.

2.3 Modelling of flexible bodies

The flexibility is represented by means of modal coordinates. For each flexible body, displacement and rotation due to deformation are described by:

$$\eta(M,t) = \sum_{n=1}^{N} \phi_n(M) q_n(t)$$

$$\beta(M,t) = \sum_{n=1}^{N} \alpha_n(M) q_n(t)$$

ϕ_n and α_n (the "modes") are functions depending only upon the space variable. They can be computed by solving Partial Differential Equation or by Finite Element Analysis of each body. We define the vector of modal coordinates $Q_i = \begin{pmatrix} q_{i_1} \\ \vdots \\ q_{N_i} \end{pmatrix}$ where N_i is the "number of modes" retained.

The mass matrix and vector of Coriolis and centrifugal effects for one flexible free body are

$$\mathcal{M}_i = \begin{pmatrix} m_i I_3 & -m_i \tilde{y}_i & B_i \\ m_i \tilde{y}_i & In_i & C_i + (Q_i^T \odot I_3)\, H_i \\ B_i^T & C_i^T + H_i^T (Q_i \odot I_3) & A_i \end{pmatrix}$$

$$
S_i = \begin{pmatrix}
-\left(m_i\,\widetilde{\Omega_i}\,\widetilde{\Omega_i}\,y_i \;+\; 2\,\widetilde{\Omega_i}\,B_i\,\dot{Q}_i\right) \\
\cdots\cdots\cdots\cdots\cdots\cdots\cdots\cdots\cdots\cdots\cdots\cdots \\
-\widetilde{\Omega_i}\,In_i\,\Omega_i \;-\; (\dot{Q}_i^T \otimes I_3)\,II_i\,\dot{Q}_i \;-\; \widetilde{\Omega_i}\,(C_i \;+\; (Q_i{}^T \otimes I_3)\,II_i)\,\dot{Q}_i \\
-\left[(\Lambda_i{}^T \;+\; (Q_i{}^T \otimes I_3)\,\Pi_i)\,(\dot{Q}_i \otimes I_3) \;+\; (\dot{Q}_i^T \otimes I_3)\,\Pi_i\,(Q_i \otimes I_3)\right]\Omega_i \\
\cdots\cdots\cdots\cdots\cdots\cdots\cdots\cdots\cdots\cdots\cdots\cdots \\
(I_{n_i} \otimes \Omega_i{}^T)\left[II_i\,\dot{Q}_i \;+\; (\frac{\Lambda_i}{2} \;+\; \Pi_i\,(Q_i \otimes I_3))\,\Omega_i\right] \;-\; II_i{}^T\,(\dot{Q}_i \otimes I_3)\,\Omega_i
\end{pmatrix}
$$

The basic formulas for μ_i, k_i, B_i, C_i, II_i^j, Π_i^j and Λ_i are

$$\mu_i = \int_S \phi_i(M)^T \phi_i(M)dm \;,\; k_i = \mu_i(2\Pi f_i)^2$$

$$B_i = \int_S \phi_i(M)dm,\; C_i = \int_S \widetilde{r_0(M)}\phi_i(M)dm$$

$$II_i^j = \int_S \widetilde{\phi_j(M)}\phi_i(M)dm,\; \Pi_i^j = -\int_S \widetilde{\phi_i(M)}\widetilde{\phi_j(M)}dm$$

$$\Lambda_i = -\int_S \widetilde{r_0(M)}\widetilde{\phi_i(M)} + \widetilde{\phi_i(M)}\widetilde{r_0(M)}dm$$

2.4 The library

The GEMMES library is a set of miscellaneous MAPLE functions which are helpful in symplifying or computing the matricial expressions. Due to the use of matricial expressions, it has been necessary to write new functions to cope with this particular representation. Most of these functions have their equivalent in MAPLE for scalar expressions. These functions include:

- display of equations
- expansion (distribution of products over sums),
- factorisation (with non-commutative multiplication of matrices),

- extraction of sub-expressions,
- simplification of cross products,
- linearization of expressions.

That library allows to compute simplified equations[2] with respect to the "rough" ones which are first generated.

2.5 Numerical interface

As one of the goal of dynamic modelling is to derive simulators from the equations, a FORTRAN interface has been written. This interface allows the translation of any expression (scalar or matricial) in the FORTRAN instructions that compute its numerical value.

Using this basic facility, a function which automatically generates a simulator from the mass matrix, the second member and the force vector of a mechanical system has been written. This function includes some optimization features as taking into account the fact that the mass matrix is constant or not, or searching of some common or constant sub-expressions. This functionalities which may substantially shorten the computation time is not so easy to implement in a purely numerical software.

3 Applications

Ge give in this section some examples of poly-articulated structures which have been or are studied using GEMMES.

3.1 AMADEUS experiment

AMADEUS is an experiment which has flyed on the Soviet MIR Space Station in November 1988. The goal of this experiment was to test new kinds of deployment mechanism and also to demonstrate validity of simulations for complex 3D deployment analysis.

[2]in the sens of symbolic computations, i.e. to find an equivalent but more compact representation of the same object

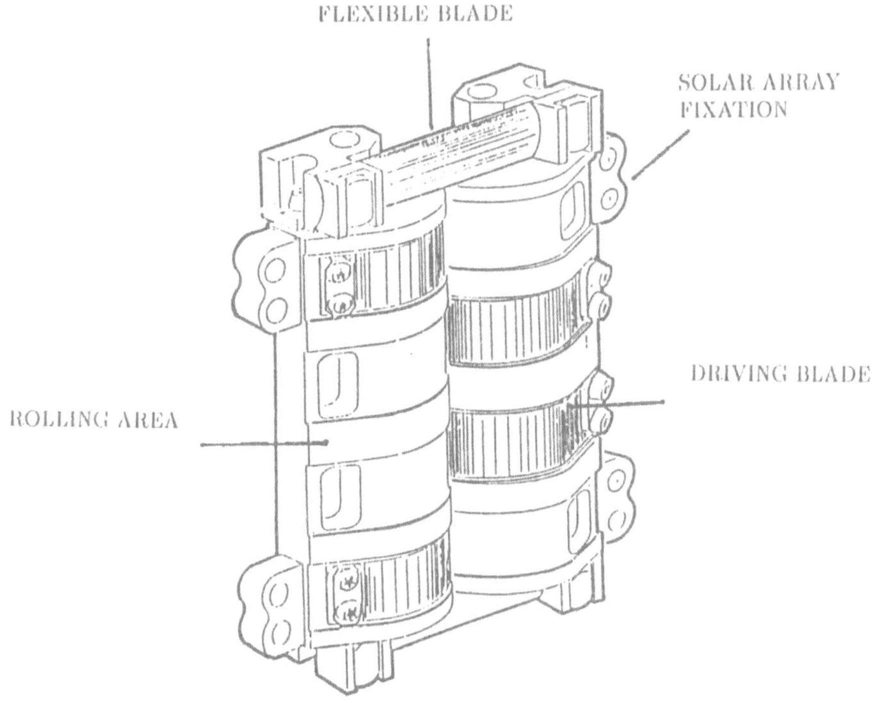

Figure 2: Deployment mechanism

These new rotation mechanisms don't use revolute joints, but 2 cylinders which roll without gliding (figure 2). This allow to be frictionless, as the contact points between the two cylinders have a relative velocity which equals 0. The motorization is obtained by a thick flexible metallic blade which is constrained in stowed configuration. During experiments, four beams were connected using such mechanisms in 2D (see figure 3) and 3D configurations. It was possible to add some dummy masses in order to modify inertia characteristics.

The corresponding simulator were built using GEMMES on a MicroVAX II computer. The main difficulty was the modelling of the rotation mechanism which is not so classical in mechanics. For example, the contact point is not the same at each instant or the instantaneous rotation point is moving. The first attempt to modelize such mechanism uses an auxiliary body without mass and inertia which connect the centers of the two cylinders. Then a constraint between rotation angles (first/auxiliary and second/auxiliary bodies) was added.

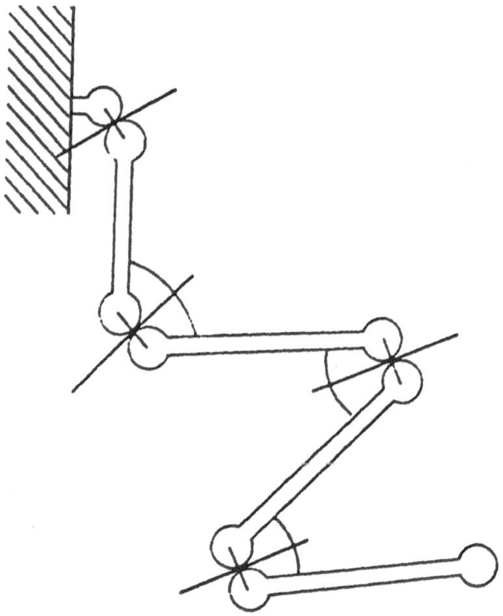

Figure 3: AMADEUS Experiment

Finally a new joint were created and added in GEMMES joint library which avoids use of the auxiliary body. This allows consequent savings in computer memory and CPU-time.

Very good correlations between simulation and experiment on the MIR station have been established on criteria such as position versus time or occurences of shocks. Simulations with the dedicated generated simulator appear to be very fast (10 times more at minimum) compared to ADAMS software ones.

3.2 Micro-vibrations computations

Another application of GEMMES has been the computation of micro-vibration environment (frequency distribution and levels in the range [10-100] Hz) on a telecom satellite. Such knowledge is very important for future optical communications between spacecraft which asks for very fine pointing (better than .3 μradian). The design of the control law must include the perturbations in order to give the needed rejection.

Spacecraft has been modelled as the collection of six bodies: one main body (flexible), two solar arrays (flexible), two antennas (flexible) and one optical package (rigid). Bodies were connected together using 2 revolute and 3 universal joints. Actuators were momemtum wheel, solar array and antennas drive mechanism, thrusters. The final model was represented by a set of near 200 differential equations.

The advantages of using a symbolic formulation for model building are the following:

- possibility to conduct parametric studies (modification of solar array orientation or modification of decenter of rotating bodies) without the need of building a new model,
- possibility to obtain synthetic results (transfer function, PSD ...) instead of only simulations results, as the equations are explicitely computed and linearized,
- possibility to modify one sub-structure (antenna modes, mass of the optical package ...) without building a new model

4 Conclusion

GEMMES appears very powerful in order to achieve two tedious, error-beset yet indispensable tasks:

- equations computations of flexible poly-articulated structure,
- building of numerical simulators.

It's now possible to construct symbolic dynamic model of flexible structures with complex geometry: 3D models with large number of flexible modes ... These models can be used for simulations, but as they are derived in a symbolic way instead of a numeric one, they can also be used for control purpose in a linearized or non-linear form as algebraic equations are eliminated. The main advantages are essentially CPU-time saving during simulation or parametric studies and exact linearization of equations w.r.t. numerical differentiation.

Adaptive Optics - Shape Control of an Adaptive Mirror

C. TRUCHI
AEROSPATIALE, Division Systèmes Stratégiques et Spatiaux. 100, Boulevard du Midi, BP 99
06322 CANNES-LA-BOCCA CEDEX, FRANCE.

Abstract

In the first part of the study, we explain the principle of adaptive optics for ground-based telescopes. Then, we introduce the Zernike polynomials which correspond to systematic optical aberrations such as defocus, coma, astigmatism... In optics, these functions are used to determine the aberrated wavefront.

By using the finite element method, we give a model of the adaptive mirror corresponding to the COME-ON project (developed by ONERA, CGE, ESO and Observatoire de Meudon). On a reduced model, we develop a quasi-static control law to determine the optimal voltage inputs, to compensate the atmospheric phase shift.

1 Adaptive optics for ground-based telescopes

1.1 Principle of adaptive optics

In practice, the imaging quality of ground-based telescopes is degraded by the transmission of the light, from the astronomical object through the turbulent atmosphere. The reason for this degradation is a random spatial and temporal wavefront perturbation induced by the turbulences in the different layers of the atmosphere. It is possible to correct the phase shift of the wavefront with a technique called adaptive optics (for high spatial and temporal frequencies). The basic principle of adaptive optics is to use a phase shifting optical element, which can be controlled in space and time, in order to compensate the atmospheric phase shift.

An adaptive optical system (see figure 1) contains four basic elements: an optical train and image detector, a wavefront sensor (Shack-Hartmann), a servo-control system and a phase-shifting optical element. The distortion of the received wavefront is usually compensated by reflecting the light beam on a deformable mirror. The surface of this mirror is adjusted in real time

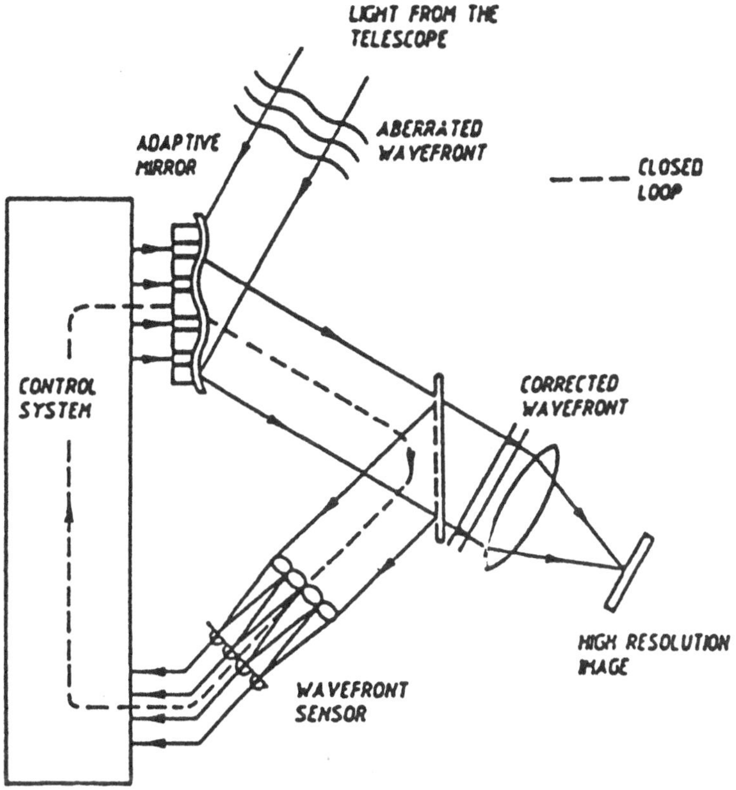

Figure 1: Principle of the application of adaptive optics in astronomy

(with piezoelectric actuators), to compensate the path length aberrations. The information required to deform the mirror is obtained by analyzing the light beam with a wavefront sensor. A map of wavefront errors (δ_{rms}) is then derived at each instant of time. Using this error map, the control system determines the signals required to drive the phase shifting optical element and to null the phase aberrations by closing the adaptive loop. The complexity and design of an adaptive system depends of the aperture size D of the telescope, the direction of the optical path specified by the zenith angle γ, the wavelength λ, and the atmospheric conditions (Fried's parameter r_0, called atmospheric correlation or coherence diameter).

Active optics is used to compensate:

- residual aberrations.
- aberrations due to gravitational, thermal and wind effects on the telescope.

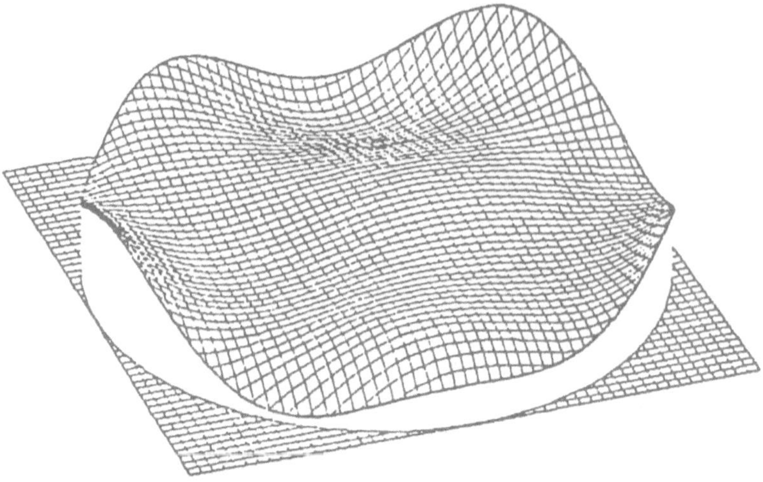

Figure 2

- mechanical aberrations.

These aberrations have low spatial and temporal frequencies (10^{-4} Hz - a few Hz). The figure 2 is an example of an aberrated wavefront to compensate in active optics.

Adaptive optics is used to compensate the atmospheric turbulence: correspond to high spatial and temporal frequencies (a few Hz - 100 Hz and more). The figure 3 is an example of an aberrated wavefront to compensate in adaptive optics.

The active and adaptive optics will correspond to two separate systems.

Adaptive optics now offers the possibility to obtain the diffraction limited resolution at near infrared wavelengths with the largest existing optical telescope. Currently several laboratories in Europe are working on components for such systems.

In 1986 a collaboration between four laboratories was started for the realization of the so-called COME-ON project (CGE, Observatoire de Meudon, ESO, ONERA) (see [1]). The goal is to build an instrument based on available technologies in order to gather experience with these new techniques for astronomical applications.

The proposed device contains two active components, a tip-tilt mirror and a 19 actuators deformable mirror which are driven in real time by the commands computed from visible wavefront data measured with a Shack-

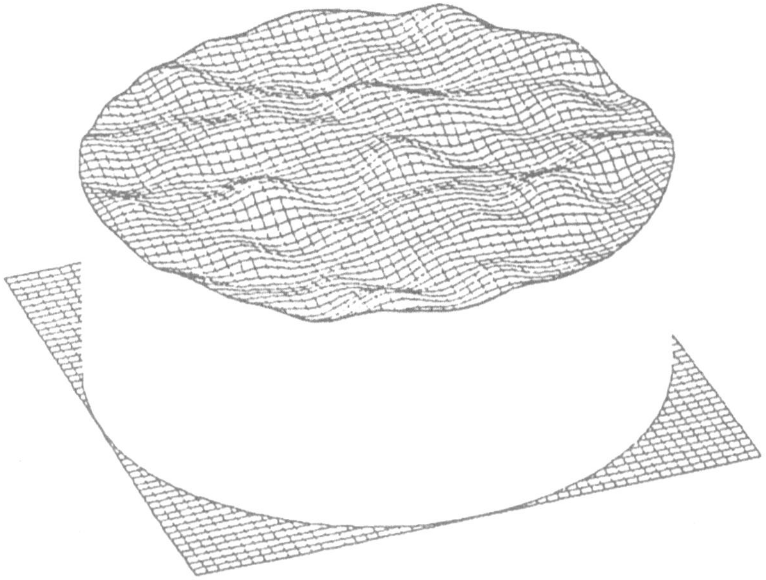

Figure 3

Hartmann type wavefront sensor.

The COME-ON project is a first step with the aim to produce diffraction limited images at the focus of a 3.6 m telescope for wavelength longer than 3 μm. The expected results are:

- To find the gain and the limitations of an adaptive optical system in astronomy.
- To gather new astronomical results with high angular resolution for the 2 to 5 μm wavelength range.
- To join together European engineers and scientists specialists in this area to define adaptive optical systems required by future VLT projects (the ESO Very Large Telescope, see [2]).

1.2 The Zernike polynomials

In optics, the Zernike plynomials are used to determine the aberrated wavefront (see [3]). Let $y(t,r,\theta)$ denote the phase shift of the wavefront (given by the wavefront sensor + integration), then :

$$y(t,r,\theta) = \sum_{i=1}^{N} a_i(t)\ z_i(r,\theta) \tag{1}$$

$z_i(r,\theta)$ represent the Zernike polynomials, which correspond to systematic optical aberrations such as defocus, coma, astigmatism..., N = number of modes.

If we consider that the atmospheric perturbations have a temporal frequency of 20 Hz, the frequency associated to the measures is : $10 \times 20 = 200$ Hz. So, the components $a_i(t)$ for $i = 1 \ldots N$ are given every 5 ms.

The Zernike polynomials are defined as follows:

Let:

$$R_{m,p}(r) = \sum_{i=0}^{p} \frac{(-1)^{p-i}\,(m+p+i)!}{(p-i)!\,(m+i)!\,i!}\, r^{m+2i}$$

Now, for each $m,\ p \in \mathbf{N},\ m \neq 0$:

$$zc_{m,p} = \sqrt{2(m+2p+1)}\; R_{m,p}(r)\; cos(m\theta)$$

$$zs_{m,p} = \sqrt{2(m+2p+1)}\; R_{m,p}(r)\; sin(m\theta)$$

and:

$$z_{0,p} = \sqrt{2p+1}\; R_{0,p}(r) \qquad m = 0$$

In the example we shall study (COME-ON project), the aberrated wavefront is developed as the expression (1) with $N = 22$. We take no notice of the first mode (bias): this mode does not really correspond to an aberration. The tilt's aberrations will be corrected by a special mirror (tip-tilt mirror). So, we only have 19 modes (from number 4 to number 22):

$z_1 = 1$ bias (constant)
$z_2 = 2\,r\,cos(\theta)$ tilt (lateral position)
$z_3 = 2\,r\,sin(\theta)$ tilt
$z_4 = \sqrt{3}\,(2r^2 - 1)$ defocus
$z_5 = \sqrt{6}\,r^2\,sin(2\theta)$ astigmatism (3rd order)
$z_6 = \sqrt{6}\,r^2\,cos(2\theta)$ astigmatism (3rd order)
$z_7 = 2\sqrt{2}\,(3r^3 - 2r)\,sin(\theta)$ coma (3rd order)
$z_8 = 2\sqrt{2}\,(3r^3 - 2r)\,cos(\theta)$ coma (3rd order)
$z_9 = 2\sqrt{2}\,r^3\,sin(3\theta)$ trefoil
$z_{10} = 2\sqrt{2}\,r^3\,cos(3\theta)$ trefoil
$z_{11} = \sqrt{5}\,(6r^4 - 6r^2 + 1)$ spherical (3rd order)
$z_{12} = \sqrt{10}\,(4r^4 - 3r^2)\,cos(2\theta)$ astigmatism (5th order)
$z_{13} = \sqrt{10}\,(4r^4 - 3r^2)\,sin(2\theta)$ astigmatism (5th order)
$z_{14} = \sqrt{10}\,r^4\,cos(4\theta)$ tetrafoil
$z_{15} = \sqrt{10}\,r^4\,sin(4\theta)$ tetrafoil
$z_{16} = 2\sqrt{3}\,(10r^5 - 12r^3 + 3r)\,cos(\theta)$ coma (5th order)
$z_{17} = 2\sqrt{3}\,(10r^5 - 12r^3 + 3r)\,sin(\theta)$ coma (5th order)
$z_{18} = 2\sqrt{3}\,(5r^5 - 4r^3)\,cos(3\theta)$
$z_{19} = 2\sqrt{3}\,(5r^5 - 4r^3)\,sin(3\theta)$
$z_{20} = 2\sqrt{3}\,r^5\,cos(5\theta)$

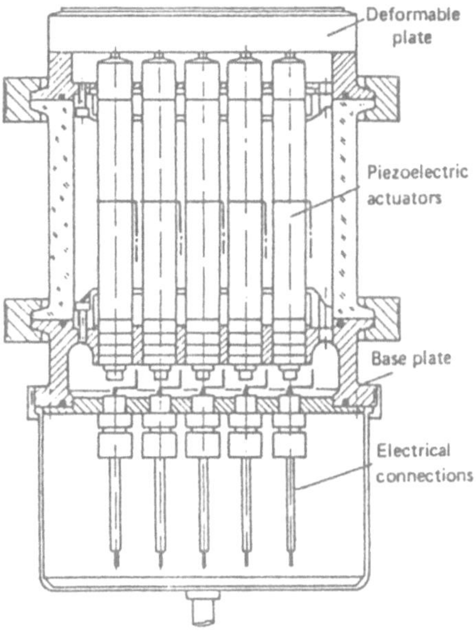

Figure 4: Deformable mirrors developed at CGE (France) (cross section)

$$z_{21} = 2\sqrt{3}\, r^5 \, sin(5\theta)$$
$$z_{22} = \sqrt{7}\,(20r^6 - 30r^4 + 12r^2 + 1) \qquad \text{spherical (5th order)}$$

2 Mathematical and numerical model of an adaptive mirror

2.1 The adaptive mirror characteristics

In this paper, we shall study the model of the adaptive mirror contains in the COME-ON project, by using a finite element method. This mirror (developed at Laboratoires de Marcoussis CGE, see figure 4) has the following characteristics:

- Number of piezoelectric actuators: 19 PZT
- Real diameter of the mirror: 100 mm
- Useful diameter of the mirror: 70 mm
- Distance between two actuators: 17.5 mm (triangular grid on the useful area: it corresponds to the optimal locations of support points in the case of active optics)

- Stroke of the actuators: $\pm 7.5\ \mu m$ (the value of the influence radius is about 30 mm)
- Control voltage: ± 1500 Volts
- Front plate: Silicon, 1 mm thick, coated with silver
- Optical flatness: $\dfrac{\lambda}{4}$ at $0.6\ \mu m$
- Bandwidth of the atmospheric perturbations: $20 - 30$ Hz
- The goal of the system is to achieve an error RMS (Root-Mean-Square) on the corrected wavefront less than $\dfrac{\lambda}{14}$ with $\lambda = 3.5\ \mu m$

2.2 The mathematical model

Let $w(t, r)$ denote the mirror displacement at position (vector) r, at time t. Then, the displacement $w(t, r)$ due to applied voltage u_i at actuator location r_i satisfies the partial differential equation:

$$\rho \frac{\partial^2 w}{\partial t^2} + D\Delta^2 w = K_1 \sum_{i=1}^{N_A} [\, K_u u_i - w(t, r)\,]\, h(r - r_i) \qquad (2)$$

for $0 < |r| < R$. R is the radius of the mirror. h is supported by a neighborhood of r_i:

$$h(x) = \begin{cases} 1/\pi a^2 & \text{if } \| x \| < a \\ 0 & \text{if } \| x \| \geq a \end{cases}$$

D : flexural rigidity of the plate
ρ : density per unit area
N_A : number of actuators

K_1 and K_u are respectively the elastic and the piezoelectric constant of the actuators, and a is the diameter of the surface occupied by each actuator.

The mirror is supported by the actuators and we shall take the initial data to be zero. In [4] we prove the existence and unicity of the solution of the problem (2) and we obtain the following result:

Theorem: Let $f \in L^1(0, T, L^2(\Omega))$; the unique element w of the problem (2) is such that Δw, $\dot{w}$, $(\sum_{i=1}^{N_A} h_i)^{1/2} w$ belong to $C^0([0, T], L^2(\Omega))$.

with $\Omega = \{x \in \mathbf{R}^2 \ / \ |x| < R\}$ and $Q = [0, T] \times \Omega$ the evolution cylinder.

So, we have now all the data for the finite element analysis of the adaptive mirror.

6.909278E+07	8.312207E+03	1.322929E+03	1.000000E+00	6.909278E+07
2.990966E+08	1.729441E+04	2.752491E+03	9.999999E-01	2.990963E+08
2.990966E+08	1.729441E+04	2.752491E+03	1.000000E+00	2.990966E+08
7.932590E+08	2.816486E+04	4.482574E+03	9.999999E-01	7.932588E+08
7.932593E+08	2.816486E+04	4.482574E+03	9.999999E-01	7.932590E+08
1.036409E+09	3.219331E+04	5.123723E+03	9.999999E-01	1.036409E+09
1.666140E+09	4.081837E+04	6.496445E+03	9.999999E-01	1.666139E+09
1.670273E+09	4.086897E+04	6.504496E+03	9.999999E-01	1.670273E+09
2.425169E+09	4.924600E+04	7.837742E+03	9.999999E-01	2.425169E+09
2.425169E+09	4.924600E+04	7.837742E+03	9.999999E-01	2.425169E+09
3.049452E+09	5.522185E+04	8.788828E+03	9.999999E-01	3.049452E+09
3.049454E+09	5.522186E+04	8.788832E+03	9.999999E-01	3.049454E+09
4.598886E+09	6.781506E+04	1.079310E+04	9.999999E-01	4.598882E+09
4.598915E+09	6.781531E+04	1.079314E+04	9.999999E-01	4.598911E+09
5.070643E+09	7.120844E+04	1.133317E+04	9.999999E-01	5.070639E+09
5.070655E+09	7.120856E+04	1.133319E+04	9.999999E-01	5.070651E+09
5.133648E+09	7.164950E+04	1.140337E+04	9.999999E-01	5.133644E+09
7.736267E+09	8.795606E+04	1.399864E+04	9.999999E-01	7.736263E+09
7.740293E+09	8.797894E+04	1.400228E+04	9.999999E-01	7.740289E+09
7.848931E+09	8.859419E+04	1.410020E+04	9.999999E-01	7.848927E+09
ω^2	ω	N	m	k

Figure 5

2.3 The results of the finite element analysis

The results of the finite element analysis of the mirror are given for the 20 first eigenmodes (see figure 5). For every eigenmode, we obtain the pulsation ω, the eigenfrequency N, the generalized mass m and the generalized stiffness k. We also have, for every eigenmode, the associated eigenvector, which correspond to the vertical displacement of the mirror. We notice that the deformations associated to the eigenvectors can be easily developed on the Zernike modes: it will be useful to obtain the compatibility between the control model developed on the eigenmodes of the mirror and the optical data developed on the Zernike modes.

Thus, we have the following model (in modal coordinates):

$$[M]\,\ddot{Q} \;+\; [K]\,Q \;=\; [E]\,F \tag{3}$$

with:

$[M]$: The generalized mass matrix, diagonal, 20×20. In fact: $[M]$=Identity ($m_{i,i} = 1$, see figure 5)

$[K]$: The generalized stiffness matrix, diagonal, 20 × 20. The coefficients correspond to the last column of figure 5.

$[E]$: The load matrix, 20 × 19. The terms are obtained with the eigenvectors: but, only the values which correspond to actuation points.

F : The actuation forces (19 × 1)

Q : The modal coefficients (20 × 1)

With the model (3) we obtain, now, the first order system (size 40):

$$\begin{cases} \dot{X} = AX + BF \\ Y = CX \\ X(0) = 0 \end{cases} \tag{4}$$

with:

$$A = \begin{pmatrix} 0 & Id_{20\times 20} \\ -K & 0 \end{pmatrix} \qquad B = \begin{pmatrix} 0 \\ E \end{pmatrix}$$

$$X = \begin{pmatrix} Q \\ \dot{Q} \end{pmatrix} \qquad C = \begin{pmatrix} Id_{20\times 20} & 0 \end{pmatrix}$$

2.4 A reduced model

The aim of this section is to reduce the previous system (4) and the control law will be developed on this reduced system. We have to define the interaction matrix between the measures (decomposed on the Zernike modes) and the eigenmodes obtained with the finite element method. This interaction matrix will give us the most contributing eigenmodes, so it will be possible to reduce the size of the system (4).

We recall the measures obtained with the Shack-Hartmann wavefront sensor:

$$y_i(r,\theta) = \sum_{j=1}^{19} a_{ij}\, z_j(r,\theta)$$

for $i = 1 \ldots M$ (M= number of measures, we have $M = 10$, frequency of the measures: 200 Hz)

Let $C(M \times 20)$ denote the interaction matrix, so we have to minimize:

$$RMS_i = \sum_{k=1}^{N} [y_i^k - \sum_{j=1}^{20} c_{ij}\, E_j^k]^2 \quad , \quad i = 1 \ldots 10 \tag{5}$$

with:

N= number of nodes of the finite element mesh

$y_i^k = y_i(r_k, \theta_k)$ = value of the measure i at the point (r_k, θ_k)

$E_j^k = E_j(r_k, \theta_k)$ = value of the eigenmode j at the point (r_k, θ_k)

By derivating the expression (5):

$$\begin{cases} \dfrac{\partial RMS_i}{\partial c_{il}} = 2\sum_{k=1}^{N} [y_i^k - \sum_{j=1}^{20} c_{ij}\, E_j^k]\, E_l^k = 0 \\ \forall\, i = 1 \ldots M \\ \forall\, l = 1 \ldots 20 \end{cases}$$

So, we have to solve the following linear system:

$$CA = B \qquad (6)$$

with:

$$A(20 \times 20) \;:\; A_{ij} = \sum_{k=1}^{N} E_i^k \; E_j^k$$

$$B(M \times 20) \;:\; B_{ij} = \sum_{k=1}^{N} y_i^k \; E_j^k$$

With the decompositions of the 10 measures ($M = 10$) on the eigenmodes, obtained with the matrix C, we will be able to study the modes E_j, $j = 1 \ldots 20$ which will contribute in those expansions.

For every eigenmode E_j, $j = 1 \ldots 20$, we compute a mean value of the coefficients c_{ij}, for $i = 1 \ldots 10$:

$$\begin{cases} m_j = \dfrac{1}{10} \sum_{i=1}^{10} |c_{ij}| \\ j = 1 \ldots 20 \end{cases}$$

We shall only consider the modes which correspond to the highest mean values: among the 20 modes, we keep the numbers 1, 2, 3, 6, 9, 10 and 17, for the others the mean values are almost equal to zero.

In the next section we shall study a control law on the model (4) reduced to the seven selecting modes: new size of the first order system=14.

3 A quasi-static control law

3.1 Theoretical approach

Having developed a model for the adaptive mirror displacement, we go on now to determine the 19 actuator input voltages (equivalent to actuator input forces) to produce a given displacement profile so as to correct the aberrated wavefront. More specifically, we shall compute the root-mean-square error between the specified mirror response (given by the wavefront sensor) and the model driven mirror response for a fixed actuator configuration (triangular grid on the useful area of the mirror). The optical specification, to have a good image quality is that this root-mean-square error has to be less than $\lambda/14$ with $\lambda = 3.5\ \mu$m.

From the measures y_i, $i = 1 \ldots M$ of the wavefront sensor, we compute the desired profile of the adaptive mirror $Xref_i$, $i = 1 \ldots M$ (with $M = 10$):

$$\begin{cases} Xref_1 = -y_1 & \\ Xref_{i+1} = -y_{i+1} + Xref_i & i = 1 \ldots 9 \end{cases}$$

because the sensor analyse the wavefront after its reflexion on the adaptive mirror and to cancel the phase aberrations the mirror has to follow the opposite deformation (multiply by 1/2 because of the reflexion).

The discrete root-mean-square error is:

$$RMS(t) = \left[\frac{1}{N}\sum_{i=1}^{N}\left(\sum_{j=1}^{7}(Y_j(t) - Xref_j(t))\ E_j(r_i,\theta_i)\right)^2\right]^{1/2} \tag{7}$$

with:

N: number of nodes of the finite element mesh.

Y: the modal coefficients, solutions of the system (4) reduced to the seven selecting modes.

$Xref$: the desired displacement, given in modal coefficients.

$E_j(r_i,\theta_i)$: value of the eigenmode j at the point (r_i,θ_i).

Let $RMSM$, denote the mean value in time of $RMS(t)$, then we require that:

$$RMSM = \left(\frac{1}{IT}\sum_{i=1}^{IT} RMS(t_i)^2\right)^{1/2} < \frac{\lambda}{14} = \frac{3.5}{14}\ \mu\text{m} = 0.25\ \mu\text{m} \tag{8}$$

with IT = the number of time iterations.

We study a quasi-static control law with a time filtering of the reference (to smooth the desired deformation of the mirror, to smooth the applied forces on the actuators):

$$F = K_s\ (Xref)_f \tag{9}$$

with:

K_s: an interaction matrix, 19×7

$(Xref)_f$: the filtered reference

Remark: The relation between the input voltages U and the input forces F is:

$$F = K_1 K_u U$$

$$K_u = 5.10^{-9}\ \text{m/volt} \qquad \text{and} \qquad K_1 = 10^6\ \text{N/m}$$

We have now to compute the matrix K_s. We only consider the static part of the system (4):

$$Kx = EF$$

then:

$$F = E^T(EE^T)^{-1}Kx$$

and:

$$K_s = E^T(EE^T)^{-1}K \tag{10}$$

With the system (4), the equalities (9), (10) and by introducing the equations associated to the analogic filter of the reference, we have all the informations for the numerical results.

3.2 Numerical results

We note the system (4) reduced to the seven selecting modes in the following way:

$$\begin{cases} \dot{X}_1 = AX_1 + BF \\ Y = CX_1 \end{cases} \tag{11}$$

Then, we introduce the system (size 14) associated to a second order filter (see [5]) of the reference $Xref$, such that:

$$\begin{cases} \dot{X}_2 = A_2X_2 + B_2(Xref) \\ (Xref)_f = CX_2 \end{cases} \tag{12}$$

with:

$$A_2 = \begin{pmatrix} 0 & Id_{7\times 7} \\ -\omega_0^2\, Id_{7\times 7} & -2\xi_0\omega_0\, Id_{7\times 7} \end{pmatrix} \qquad B_2 = \begin{pmatrix} 0 \\ \omega_0^2\, Id_{7\times 7} \end{pmatrix}$$

ω_0 = pulsation, ξ_0 = damping ratio (generally $\xi_0 = \frac{\sqrt{2}}{2}$)

$$X_2 = \begin{pmatrix} (Xref)_f \\ (\dot{Xref})_f \end{pmatrix} \qquad C = (\, Id_{7\times 7} \quad 0\,)$$

With the systems (11) and (12), the relation (9), we obtain the first order system (size 28):

$$\begin{cases} \dot{X} = LX + G(Xref) \\ Y = DX \end{cases} \tag{13}$$

with:

$$L = \begin{pmatrix} A & BK_sC \\ 0 & A_2 \end{pmatrix} \qquad G = \begin{pmatrix} 0 \\ B_2 \end{pmatrix}$$

$$X = \begin{pmatrix} X_1 \\ X_2 \end{pmatrix} \qquad\qquad D = (C \quad 0)$$

The time discretization of the system (13) is achieved by a Runge-Kutta algorithm.

The minimum value of $RMSM$ is 0.179 μm for $N_0 = 500$ Hz (frequency associated to the filter: $\omega_0 = 2\pi N_0$).

If $N_0 < 500$ Hz and N_0 decreases, then $RMSM$ increases, because the frequency N_0 is not high enough to follow the reference.

If $N_0 > 500$ Hz and N_0 increases, then $RMSM$ increases. In that case, the mechanical modes (≥ 1323 Hz, see figure 5) are too much excited by the actuators.

The best numerical results have been obtained with $N_0 = 250$ Hz, in that case:

$$RMSM = 0.233\ \mu\text{m} < \frac{\lambda}{14} \qquad \text{(see the expression (8))}$$

the optical specification is satisfied and the mechanical modes are not too much excited by the actuators.

The applied forces on the actuators are varying between -2 Newtons and 2 Newtons. These values are compatible with the actuators possibilities.

Figure 6 represent the variations of $RMS(t)$ (see the expression (7)). The peaks correspond to an important variation of the reference.

Figure 7 to 13 represent respectively the coefficients of the seven modes. Three curves are superimposed:

1/ The reference (the steps)

2/ The filtered reference

3/ The solution Y of the system (13) (with some oscillations, only for the two first modes)

Remark: In fact, in this study the bandwidth of the atmospheric perturbations is higher than $20-30$ Hz (about 150 Hz, see the reference $Xref$). With a bandwidth of $20-30$ Hz, the value of $RMSM$ will decrease: the peaks resulting from an important variation of the reference will disappear.

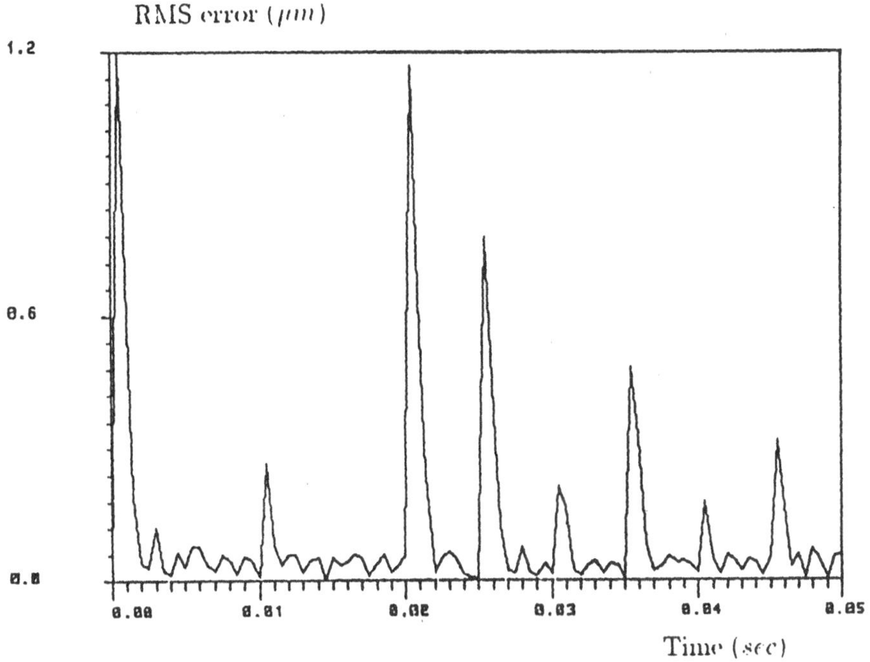

Figure 6

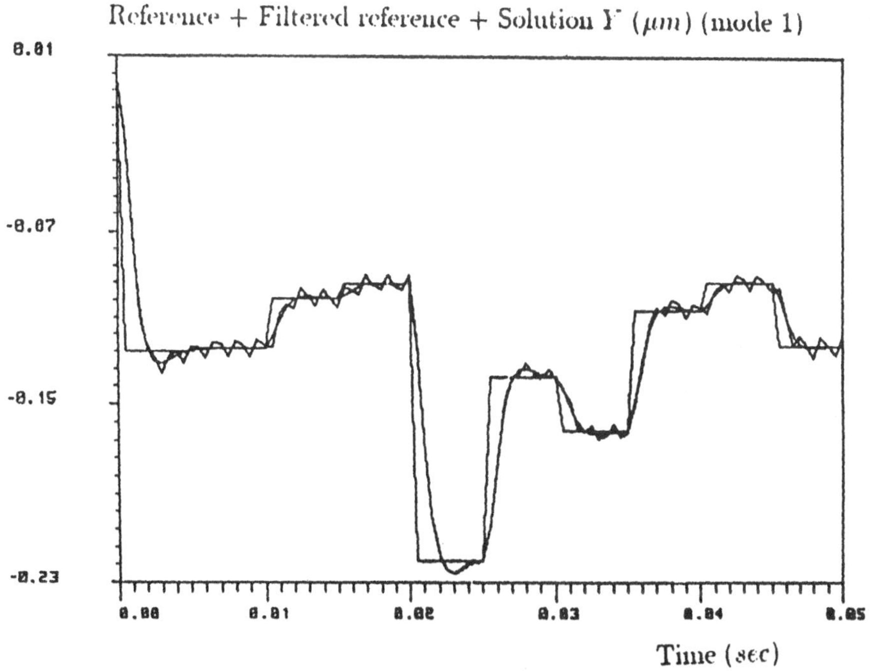

Figure 7

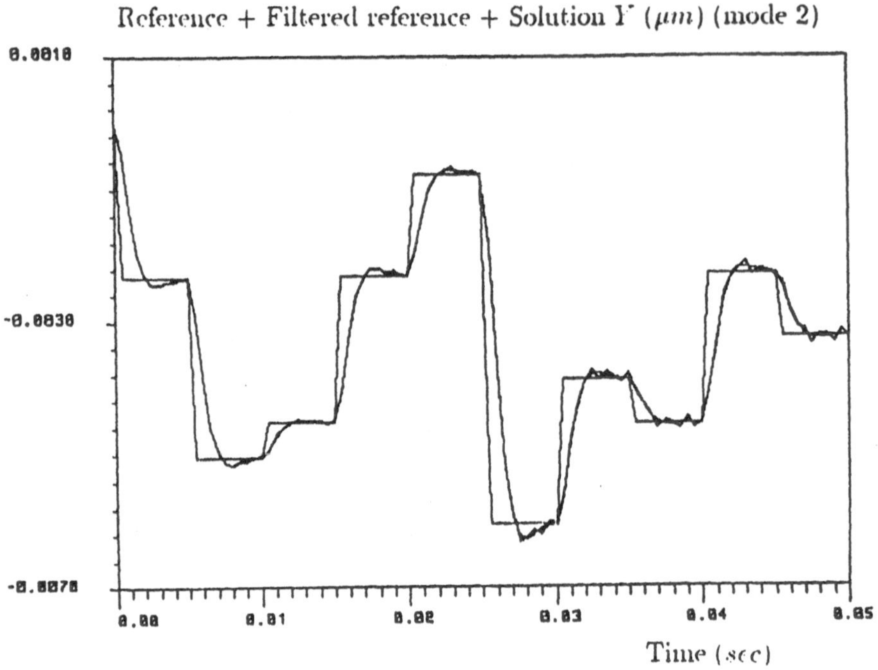

Figure 8

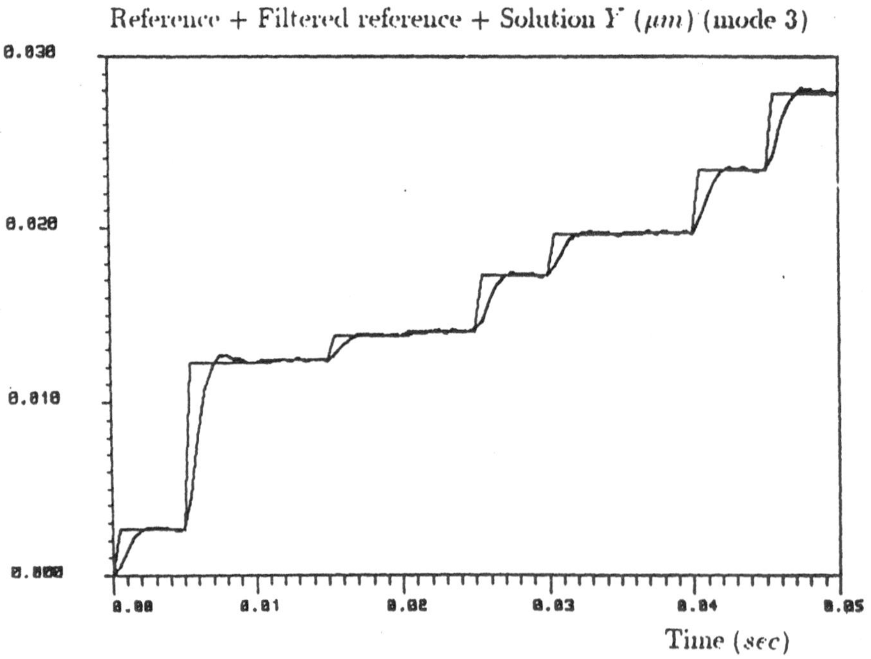

Figure 9

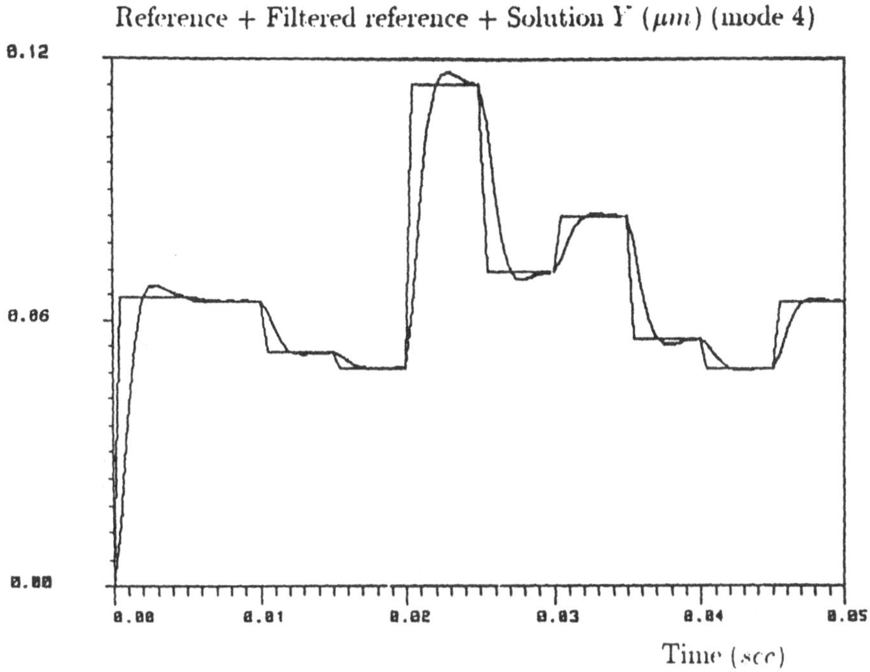

Figure 10

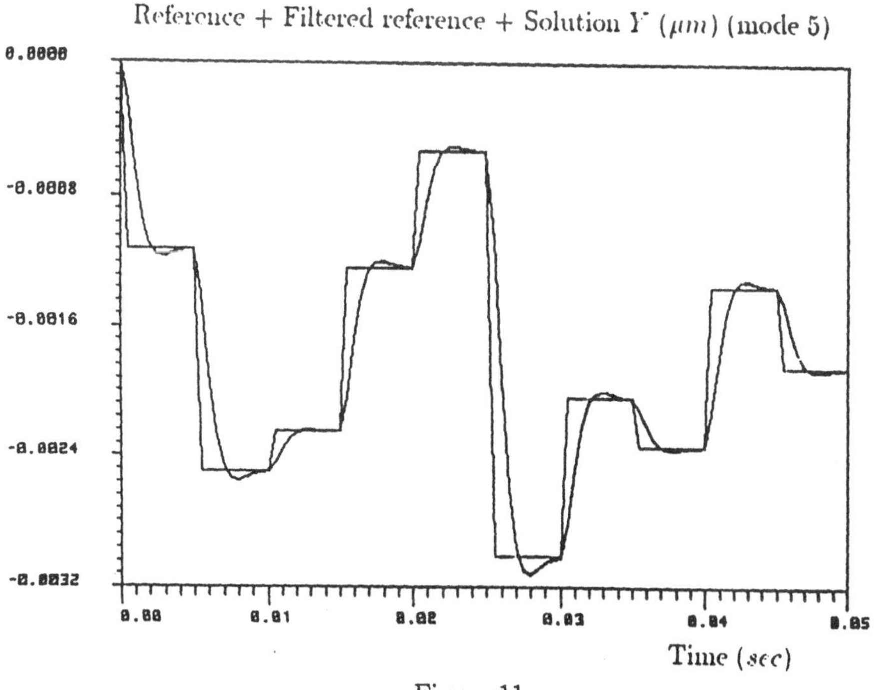

Figure 11

Reference + Filtered reference + Solution Y (μm) (mode 6)

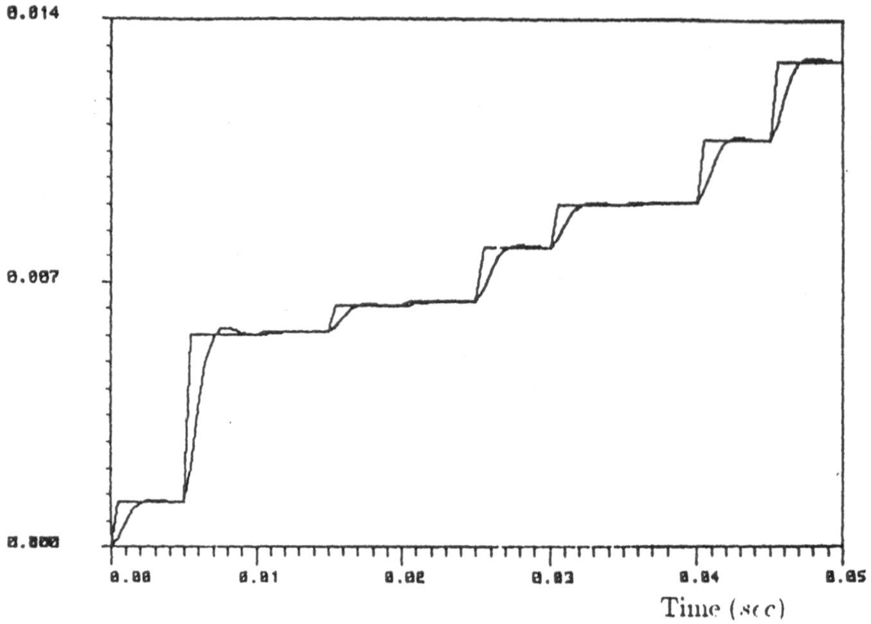

Figure 12

Reference + Filtered reference + Solution Y (μm) (mode 7)

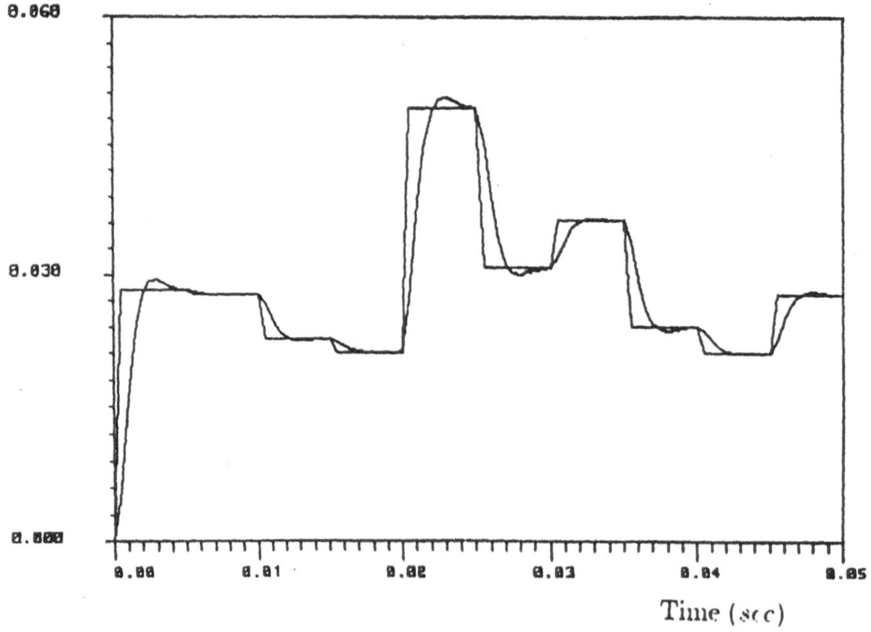

Figure 13

4 Conclusion

We have developed the shape control of an adaptive mirror with a model based on the finite element method. With a quasi-static control law, a second order filter on the reference, we obtain a root-mean-square error on the wavefront less then $\lambda/14$ with $\lambda = 3.5$ μm, which corresponds to a good image quality.

References

[1] P. KERN[1,4], P. LENA[1], G. ROUSSET[2], J.C. FONTANELLA[2], F. MERKLE[3], J.P. GAFFARD[4], *"Prototype of an adaptive optical system for infrared astronomy"*

1 : Université Paris 7 et Observatoire de Paris, Laboratoire associé au CNRS n° 325, Place J.Janssen, 92190 Meudon, France.

2 : Office National d'Etudes et de Recherches Aérospatiales (ONERA), 29, av. de la Division Leclerc, 92322 Chatillon sous Bagneux, France.

3 : European Southern Observatory (ESO), Karl-Schwarzschild-Str. 2, 8046 Garching bei München, Federal Republic of Germany.

4 : Laboratoire de Marcoussis (CGE), Route de Nozay, 91460 Marcoussis, France.

[2] F. MERKLE, *"Adaptive optics for ESO's Very Large Telescope (VLT) project"*, (May 1986).

[3] Victor L. GENBERG, *"Optical surface evaluation"*, Eastman Kodak Company, Rochester, NY 14650.

[4] L. PASSERON, C. TRUCHI, J.P. ZOLESIO, *"Dynamic modeling, control theory and stabilization for flexible structures: industrial applications at Aérospatiale Cannes"*, Proceedings of the IFAC Workshop on stabilization of flexible structures, Montpellier (December 1987)

[5] M. LABARRERE, J.P. KRIEF, B. GIMONET, *"Le filtrage et ses applications"*, Cepadues Editions.

ENERGY DECAY ESTIMATES FOR A BEAM WITH NONLINEAR BOUNDARY FEEDBACK

F. Conrad

Université de Nancy 2 et U.A. CNRS 750, BP 239,
54506 Vandoeuvre les Nancy, France

J. Leblond and J.P. Marmorat

Centre de Mathématiques Appliquées, ENSMP,
Sophia-Antipolis, 06565 Valbonne, France

Abstract

We obtain decay estimates for an Euler-Bernoulli beam which is clamped at one end and controlled at the other end by a point force that is a nonlinear function of the observed transversal velocity. Numerical simulations show that the estimates are fairly accurate.

INTRODUCTION

We consider an homogeneous Euler-Bernoulli beam clamped at the left end and controlled at the right end by a point force $f_0(t)$ and a point bending moment $f_1(t)$.
Let $u(x,,t)$ denote the transverse deflection of the beam. After a rescaling, the equations for the open-loop system are

$$\begin{cases} u_{tt} + u_{xxxx} = 0 & 0 < x < 1 \,;\; t > 0 \\ u(0,t) = 0 \,;\; u_x(0,t) = 0 & t > 0 \\ -u_{xx}(1,t) = f_1(t) & t > 0 \\ u_{xxx}(1,t) = f_0(t) & t > 0 \\ + \text{ initial data } u(.,0)\,,\; u_t(.,0) & 0 < x < 1 \end{cases} \tag{1}$$

The elastic energy associated with any solution of Equation (1) is

$$E(t) = \frac{1}{2}\int_0^1 u_t^2(x,t)\,dx + \frac{1}{2}\int_0^1 u_{xx}^2(x,t)\,dx$$

and, formally

$$\frac{dE}{dt}(t) = -f_1(t)\,u_{xt}(1,t) - f_0(t)\,u_t(1,t).$$

In order to damp the free vibrations of the beam, it is natural to introduce transverse and/or angular velocity feedback, in a way such that $\frac{dE}{dt}(t) \leq 0\,,\ \forall t$. It is then expected that $E(t)$ decreases to zero.

Actually, in the case of linear feedback, it has been proved that uniform exponential stability is achieved with $u_{xxx}(1,t) = \alpha\, u_t(1,t)\,,\ -u_{xx}(1,t) = \beta\, u_{xt}(1,t)$ with $\alpha > 0$, $\beta \geq 0$ [2]. The proof is based on the construction of a suitable Lyapunov functional and uses energy multipliers. It works also for a system of connected beams. A similar result has been established for $\alpha = 0$, $\beta > 0$ by means of a frequency domain method [3].

It is of interest, from a practical point of view, to investigate the case of nonlinear feedback. Specifically, in this paper, we consider control by a point force only, which is a nonlinear function of the observation taken as the transverse velocity at the right end:

$$u_{xxx}(1,t) = g(u_t(1,t))$$

where $g : \mathrm{R} \to \mathrm{R}$ is some monotone function with $g(0) = 0$.
The model case corresponds to $g(\xi) \sim |\xi|^{p-1}\,\xi$ as $\xi \to 0$, with $0 < p \leq 1$.

Therefore the equations of the closed-loop system we study here are

$$\begin{cases} u_{tt} + u_{xxxx} = 0 \\ u(0,t) = 0;\ u_x(0,t) = 0 \\ -u_{xx}(1,t) = 0 \\ u_{xxx}(1,t) = g(u_t(1,t)) \\ +\ \text{initial data } u(.,0)\,,\ u_t(.,0) \end{cases} \tag{2}$$

In a first part of the paper, we set up decay estimates for the energy of any solution of (2), given the behaviour of g.
In a second part we show, by numerical experiment, that the theoretical estimates are fairly accurate, at least for the model case mentioned above.

We begin with some preliminary results which are usefull in establishing the decay estimates.

PRELIMINARY RESULTS

We set $H = L^2(0,1)$, with its usual norm denoted by $|\ |_H$,
$V = \{v \in H^2(0,1)/v(0) = v_x(0) = 0\}$ with norm $\|\ \|_V$ defined by
$\|v\|_V^2 = \int_0^1 v_{xx}^2(x)\,dx$.

We consider first the uncontrolled system ($f_0 = f_1 = 0$ in (1)):

$$\begin{cases} u_{tt} + u_{xxxx} = 0 \\ u(0,t) = 0\,;\ u_x(0,t) = 0 \\ -u_{xx}(1,t) = u_{xxx}(1,t) = 0 \\ u(.,0) = u_0 \in V\,,\ u_t(.,0) = u_1 \in H \end{cases} \tag{3}$$

This problem is well posed in $C(0,T;V \times H)\ \forall T > 0$.

Proposition 1 . $\forall T_0 > 0,\ \exists C_0 > 0$ *such that*

$$|u_1|_H^2 + \|u_0\|_V^2 \le 2\,C_0 \int_0^{T_0} u_t^2(1,t)\,dt \tag{4}$$

for any solution u *of* (3).

Proof. Let ω_k^2, $k = 1,2,\cdots$ be the eigenvalues of $\frac{d^4}{dx^4}$ with the boundary conditions of (3), and ϕ_k, $k = 1,2,\cdots$ be the Hilbert basis of associated eigenvectors such that $|\phi_k|_H = 1$, $\forall k$. Then any solution u of (3) is given by an expansion

$$u(x,t) = \sum_{k=1}^{\infty} [a_k \cos\omega_k t + b_k \sin\omega_k t]\,\phi_k(x)$$

where a_k and b_k are well-defined by the initial conditions. Then Inequality (4) is equivalent to

$$\sum_{k=1}^{\infty} \omega_k^2\,(a_k^2 + b_k^2) \le 2\,C_0 \int_0^{T_0} \left[\sum_{k=1}^{\infty} \omega_k\,(-a_k \sin\omega_k t + b_k \cos\omega_k t)\,\phi_k(1)\right]^2 dt$$

or, with
$c_k = b_k + i\,a_k$ if $k \ge 1$ ($c_0 = 0$),
$c_k = \overline{c_{-k}}$, $\omega_k = -\omega_{-k}$, $\phi_k = \phi_{-k}$ if $k \le -1$, to

$$\sum_{k\in Z} \omega_k^2\,|c_k|^2 \le C_0 \int_0^{T_0} \left[\sum_{k\in Z} \omega_k\,c_k\,e^{i\omega_k t}\,\phi_k(1)\right]^2 dt$$

and, with $d_k = \omega_k\,c_k$, to

$$\sum_{k\in Z} |d_k|^2 \le 4\,C_0 \int_0^{T_0} \left[\sum_{k\in Z} d_k\,e^{i\omega_k t}\right]^2 dt \tag{5}$$

since $\phi_k(1) = \pm 2$ [4].
By the well known asymptotic behaviour of the ω_k's,
$\liminf_{|k|\to\infty} \omega_{k+1} - \omega_k = +\infty$.
We conclude that (5) is true for any $T_0 > 0$ by an extension of an inequality originally due to Ingham [8], [1]. Note that C_0 depends on T_0.

Remark 1 . Inequality (4) is just an observability condition.

Next, we consider the open-loop system $(1)_a$ that is, (1) with $f_1 = 0$, $f_0 \neq 0$. We recall a result which has been established in [9]

Proposition 2 . *Let* $T > 0$. *For any* $u(.,0) \in V$, $u_t(.,0) \in H$ *and* $f_0 \in L^2(0,T)$, $(1)_a$ *admits a unique solution* $(u, u_t) \in C(0,T; V \times H)$, $u_t(1,t) \in L^2(0,T)$, *and the mapping*

$$\begin{array}{rcl} V \times H \times L^2(0,T) & \to & C(0,T; V \times H) \times L^2(0,T) \\ \{u(.,0), u_t(.,0), f_0\} & \mapsto & \{u, u_t, u_t(1,.)\} \end{array}$$

is continuous.

Finally, we consider the closed-loop system (2). The following well-posedness result is just a particular case of a more general one obtained in [6].

Proposition 3 . *Assume* g *is continuous, monotone, and* $g(0) = 0$. *Then* (2) *defines a nonlinear semi-group of contractions on* $V \times H$ *with generator* $\mathcal{A}$, *and* $\text{dom}(\mathcal{A})$ *is dense in* $V \times H$.

As a consequence of Proposition 3 and general results on semi-groups [11] we have the following:
let $u(.,0)$, $u_t(.,0) \in \text{dom}(\mathcal{A})$ and consider the elastic energy as usually

$$E(t) = \frac{1}{2}|u_t(.,t)|_H^2 + \frac{1}{2}\|u(.,t)\|_V^2.$$

Since in that case $(u, u_t) \in \text{dom}(\mathcal{A})$ and $t \mapsto (u(.,t), u_t(.,t)) \in V \times H$ is Lipschitz continuous, the following formula makes sense:

$$\begin{array}{rcl} \dfrac{dE}{dt}(t) & = & -\left[u_t(x,t)\, u_{xxx}(x,t)\right]_0^1 + \left[u_{xt}(x,t)\, u_{xx}(x,t)\right]_0^1 \\ & = & -g(u_t(1,t))\, u_t(1,t) \end{array}$$

and by integration

$$E(0) - E(t) = \int_0^t g(u_t(1,\tau))\, u_t(1,\tau)\, d\tau \tag{6}$$

$\forall t \geq 0$, $\forall\, (u(.,0), u_t(.,0)) \in \text{dom}(\mathcal{A})$.

DECAY ESTIMATES

Here we need more precise hypotheses on the behaviour of g. We assume g continuous, monotone as before and

$$g(\xi)\,\xi \geq \alpha\,|\xi|^r + \beta\,|\xi|^2 \tag{7}$$

$$g(\xi)\,\xi + \gamma\,|\xi|^s \geq \delta\,g(\xi)^2 \tag{8}$$

with $\alpha, \beta, \gamma, \delta > 0$; $0 < s \leq 2\,(r-1) \leq r \leq 2$.

For instance, we can consider the model case: $g(\xi) \sim |\xi|^{p-1}\,\xi$ as $\xi \to 0$, with $p \in (0,1]$, and g asymptotically linear as $\xi \to \infty$. Actually, any function of the form $g(\xi) = a\,|\xi|^{p-1}\,\xi + b\,\xi$ with a, $b > 0$ satisfies assumptions (7) and (8) with $s = 2\,p$, $r = 1 + p$.

Theorem 4 . *Assume* (7) *and* (8). *Then, for any solution of* (2) *with initial data in* $V \times H$
(i) if $s = r$: $E(t) \leq M_0\,e^{-\mu t}\,E(0)$ *where* $M_0 > 0$, $\mu > 0$ *are constants;*
(ii) if $s < r$: $E(t) \leq M_1\,\left(\frac{1}{t+T_1}\right)^{\frac{s}{r-s}}\,E(0)$ *where* $M_1 > 0$ *and* $T_1 > 0$ *depend continuously on* $E(0)$.

Remark 2 . If $s = r$, then (7) and (8) imply that $s = r = 2$ and g is "almost" linear: $b\,|\xi| \leq |g(\xi)| \leq B\,|\xi| \;\; \forall\,\xi$.

Remark 3 . In the model case, $g(\xi) \sim |\xi|^{p-1}\,\xi$ as $\xi \to 0$, we have $s = 2\,p$, $r = 1+p$, thus $E(t) \leq M_1\,\left(\frac{1}{t+T_1}\right)^{\frac{2p}{1-p}}\,E(0)$ if $p \in (0,1)$ and exponential decay if $p = 1$.

For sublinear g, the energy decays more and more rapidly as g approaches the limit linear case.

Proof of Theorem 4. We follow a technique used in [7] for abstract second order systems with distributed feedback on the velocity.

Step 1. Let u be a solution of the closed-loop system (2). By the energy formula (6)

$$\int_0^T g(u_t(1,\tau))\,u_t(1,\tau)\,d\tau \leq E(0) \quad \forall T \geq 0. \tag{9}$$

Let $T_0 > 0$ be fixed and choose $T = k\,T_0$, $k \in \mathrm{N}$. We deduce from (9) that there exists $p \in \{0, \cdots, k-1\}$ such that

$$\int_{pT_0}^{(p+1)T_0} g(u_t(1,\tau))\,u_t(1,\tau)\,d\tau \leq \frac{E(0)}{k}. \tag{10}$$

Step 2. Let v be the solution of the uncontrolled system (3) with initial conditions $v(.,pT_0) = u(.,pT_0)$, $v_t(.,pT_0) = u_t(.,pT_0)$. Then $w = u - v$ satisfies

$$\begin{cases} w_{tt} + w_{xxxx} = 0 \\ w(0,t) = w_x(0,t) = 0 \\ w_{xx}(1,t) = 0 \\ w_{xxx}(1,t) = u_{xxx}(1,t) = g(u_t(1,t)) \\ w(.,pT_0) = 0\,; \ w_t(.,pT_0) = 0 \end{cases} \tag{11}$$

which is an open-loop controlled system of type (1)a. By (10), (7) and (8) $g(u_t(1,t)) \in L^2(pT_0,(p+1)T_0)$. We apply Proposition 2 to (11)

$$\int_{pT_0}^{(p+1)T_0} w_t^2(1,\tau)\,d\tau \le C \int_{pT_0}^{(p+1)T_0} g(u_t(1,\tau))^2\,d\tau$$

$$\le \frac{C}{\delta} \int_{pT_0}^{(p+1)T_0} [g(u_t(1,\tau))\,u_t(1,\tau) + \gamma\,|u_t(1,\tau)|^s]\,d\tau \quad \text{by (8)}$$

$$\le \frac{C}{\delta}\frac{E(0)}{k} + \frac{C\gamma}{\delta}\left[\int_{pT_0}^{(p+1)T_0} u_t^r(1,\tau)\,d\tau\right]^{\frac{s}{r}} T_0^{\frac{r-s}{s}}$$

by (10) and Hölder's inequality.
Finally, using (7) and again (10) for the second term, we get

$$\int_{pT_0}^{(p+1)T_0} w_t^2(1,\tau)\,d\tau \le C\,\frac{E(0)}{k} + C\left[\frac{E(0)}{k}\right]^{\frac{s}{r}} \tag{12}$$

where C stands for various constants depending on $T_0, \alpha, \beta, \gamma, \delta$.

Step 3. Since $w = u - v$, we have $v_t(1,t) = u_t(1,t) - w_t(1,t)$ so that

$$\int_{pT_0}^{(p+1)T_0} v_t^2(1,\tau)\,d\tau \le 2\int_{pT_0}^{(p+1)T_0} u_t^2(1,\tau)\,d\tau + 2\int_{pT_0}^{(p+1)T_0} w_t^2(1,\tau)\,d\tau.$$

Using (7) and (10) for the first term of the r.h.s. and (12) for the second term we get

$$\int_{pT_0}^{(p+1)T_0} v_t^2(1,\tau)\,d\tau \le C\,\frac{E(0)}{k} + C\left[\frac{E(0)}{k}\right]^{\frac{s}{r}} \tag{13}$$

with similar constants.

Step 4. We recall that v is the solution of an uncontrolled system. Applying Proposition 1 to v, and (13) we get

$$\begin{aligned} E(pT_0) &= \frac{1}{2}|u_t(.,pT_0)|_H^2 + \frac{1}{2}\|u(.,pT_0)\|_V^2 \\ &= \frac{1}{2}|v_t(.,pT_0)|_H^2 + \frac{1}{2}\|v(.,pT_0)\|_V^2 \\ &\leq C\frac{E(0)}{k} + C\left[\frac{E(0)}{k}\right]^{\frac{s}{r}} \end{aligned}$$

with similar constants.
Since $E(.)$ is nonincreasing, this implies ($p \leq k$)

$$E(kT_0) \leq C\frac{E(0)}{k} + C\left[\frac{E(0)}{k}\right]^{\frac{s}{r}} \quad \forall k \in \mathrm{N} \tag{14}$$

with constants depending only on $\alpha, \beta, \gamma, \delta$ and T_0. (14) is the basic inequality which leads to the desired estimates.

Step 5.
(a) if $s = r (= 2)$, (14) is equivalent to

$$E(kT_0) \leq C\frac{E(0)}{k}$$

and estimate (i) follows in a standard way
(b) if $s < r$, (14) implies that there exists a constant C depending now also on $E(0)$ such that

$$E(kT_0) \leq C\left[\frac{E(0)}{k}\right]^{\frac{s}{r}}. \tag{15}$$

From (15) one obtains estimate (ii) after some algebra (see [5] for details).

NUMERICAL SIMULATION

It is of interest to know wether the estimates obtained in Theorem 4, and in particular the exponents, are significant of the true decay of the energy, or not.

Therefore numerical calculation has been performed to solve (2) with various initial conditions, and nonlinearities corresponding to the model case.

Approximation of the beam equation

Starting from a variational form of System (2)

$$\begin{cases} \int_0^1 u_{tt}(x,t)v(x)dx + g(u_t(1,t))v(1) + \int_0^1 u_{xx}(x,t)v_{xx}(x)dx = 0 \\ (u,\dot{u}) \in L^\infty(0,T;V\times H)\,;\ \forall v \in V \\ +\ \text{initial data in } V\times H \end{cases}$$

the beam equation is approximated w.r.t. space by a finite element method. Hermite cubics have been used; on each mesh, $u(.,t)$ is approximated by a combination of the four polynomials
$(\xi-1)^2(2\xi+1)$, $\xi^2(3-2\xi)$, $(\xi-1)^2\xi$, $\xi^2(\xi-1)$
for $\xi \in (0,1)$ (up to an affine transform of $(0,1)$ unto the adequate interval).

This leads to a system of $2N$ second order in time differential equations of the form

$$\begin{cases} M\,U_{tt} + C(U_t) + K\,U = 0 \\ U(0) \text{ and } U_t(0) \text{ are given} \end{cases} \tag{16}$$

where the $2N$ vector function U is an approximation of the unknown functions $u(.,t)$ and $u_x(.,t)$ at the nodes $x_j = j\,h$, $j = 1,\cdots N$, $N\,h = 1$;
M is the mass matrix obtained by assembling elementary mass matrices and K is the stiffness matrix obtained in a similar way [10];
${}^TC(U_t) = (0,\cdots 0, g(U_{2N-1,t}), 0)$.
The energy at time t is then $E(t) = \frac{1}{2}{}^TU_tMU_t + \frac{1}{2}{}^TUKU$.

The evolution problem (16) is solved by a Newmark method [12] which gives the approximations U^n and $\dot{U}^n$ of U and U_t at time $n\,\Delta t$ as the solution of the following iterative scheme

$$\begin{cases} (M+\theta\,\Delta t^2\,K)\,U^{n+1} + \theta\,\Delta t^2\,C(\dot{U}^{n+1}) \\ \quad = M\,(U^n + \Delta t\,\dot{U}^n) + (\theta - \frac{1}{2})\,\Delta t^2\,(K\,U^n + C(\dot{U}^n)) \\ M\,\dot{U}^{n+1} + \delta\,\Delta t\,C(\dot{U}^{n+1}) + \delta\,\Delta t\,K\,U^{n+1} \\ \quad = M\,\dot{U}^n + (\delta - 1)\,\Delta t\,(K\,U^n + C(\dot{U}^n)) \end{cases} \tag{17}$$

with $U^0 = U(0)$, $\dot{U}^0 = U_t(0)$. We choose $\theta = \frac{1}{4}$, $\delta = \frac{1}{2}$ which guaranties unconditional stability in case $C \equiv 0$. The method is second order accurate in time.

Note that (17) is a nonlinear implicit system in U^{n+1}, $\dot{U}^{n+1}$. In order to get the solution, the term $g(\dot{U}_{2N-1}^{n+1})$ involved in the expression of $C(\dot{U}^{n+1})$ is written as $\tilde{g}(\dot{U}_{2N-1}^{n+1})\,\dot{U}_{2N-1}^{n+1}$ ($\tilde{g}(z) = \frac{g(z)}{z}$); a delay is then introduced in the nonlinear terms by the splitting $g(\dot{U}_{2N-1}^{n+1}) \rightsquigarrow \tilde{g}(\dot{U}_{2N-1,i}^{n+1})\,\dot{U}_{2N-1,i+1}^{n+1}$ (and $U^{n+1} \rightsquigarrow U_{i+1}^{n+1}$, $\dot{U}^{n+1} \rightsquigarrow \dot{U}_{i+1}^{n+1}$ everywhere else) so that for each i the pair $(U_{i+1}^{n+1}, \dot{U}_{i+1}^{n+1})$ is solution of a linear system and converges well in almost all cases tested.

Results and conclusion

Numerical tests on the linear model case proved that $N = 5$ is sufficient to get a good accuracy in the space variable, but time step has to be taken small enough (up to 10^{-3}).

Though the theoretical estimates need an assumption on the global behaviour of g, it is observed numerically that the asymptotic behaviour of the energy decay rate does not depend on the coefficient b involved in the sublinear feedback function $g(\xi) = |\xi|^{p-1}\,\xi + b\,|\xi|$; therefore we choose $b = 0$ and consider the model problem with $g(\xi) = |\xi|^{p-1}\,\xi$, $p \in (0,1]$.

We take as initial conditions either (a): $u(x,0) = \frac{x^2}{2}$ or (b): $u(x,0) = \frac{x^3}{\sqrt{12}}$, and $u_t(x,0) = 0$ ($E(0) = \frac{1}{2}$) in order to test the decay (the exponent if it exists) and the influence of initial data, at least on the constants involved in the estimation.

Figures 1 to 8 represent the energy $E(t)$ for p from 0.2 to 0.9 and for the two above initial conditions (a) and (b). These figures are drawn in a log-log scale. According to the theory, the energy decreases more and more rapidly as p goes up to 1. It has a linear asymptotic behaviour in this log-log representation.
Figure 9 shows the mean slopes of these asymptotes for each value of p. We note a good adequacy between these experimental exponents and the theoretical one set up in Theorem 4 (see also Remark 3). Moreover, the fact that the initial condition does not affect the exponent is numerically verified.
Finally, figure 10 shows the behaviour of $E(t)$ in the limit case of a linear feedback ($p = 1$), in a lin-log representation.

In conclusion, these experimental results are in conformity with the theoretical estimates of Theorem 4. It will be interesting to make an equivalent analysis in the case of a superlinear feedback ($p > 1$).

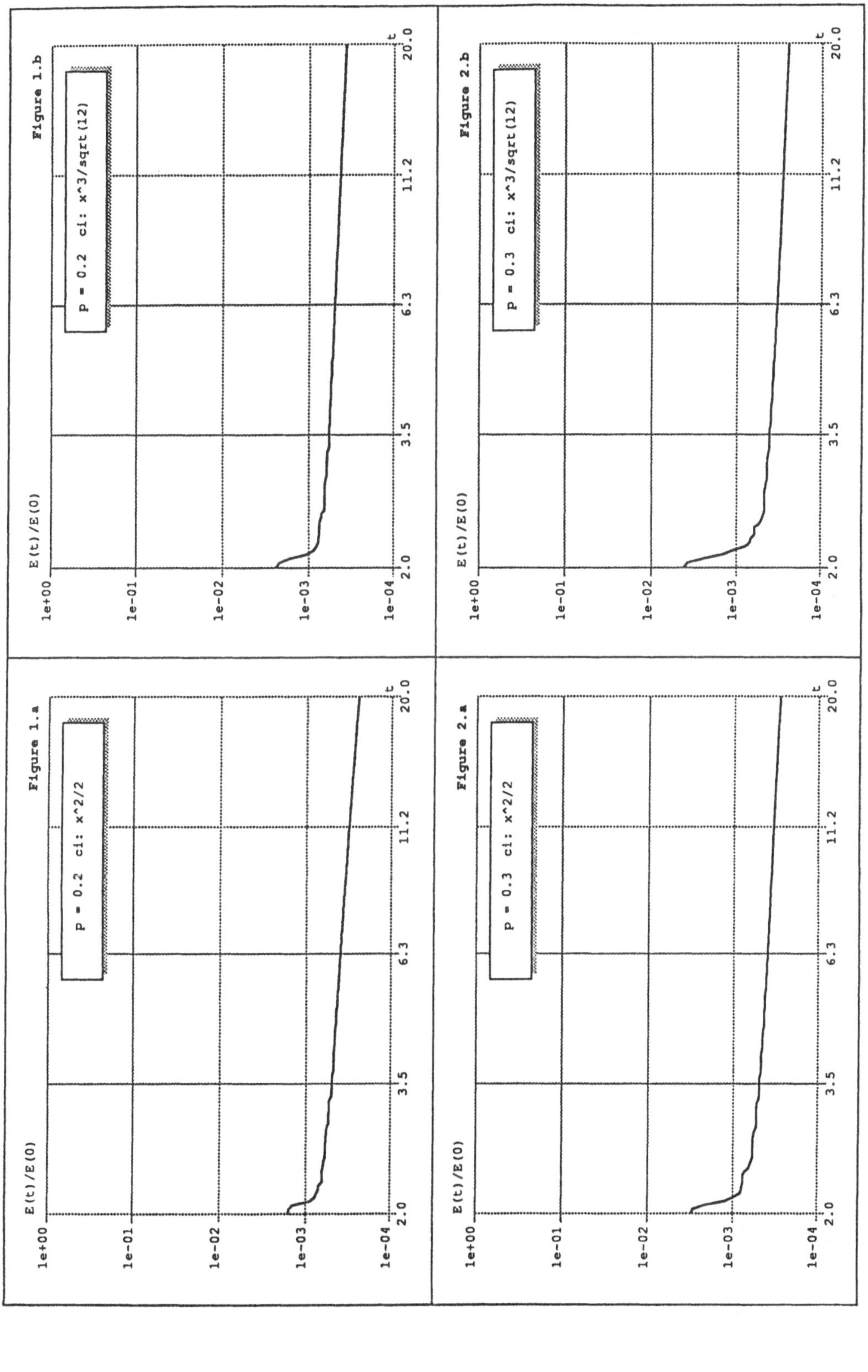

Figure 1.a
p = 0.2 ci: x^2/2
Figure 1.b
p = 0.2 ci: x^3/sqrt(12)
Figure 2.a
p = 0.3 ci: x^2/2
Figure 2.b
p = 0.3 ci: x^3/sqrt(12)
E(t)/E(0)
1e+00
1e-01
1e-02
1e-03
1e-04
2.0
3.5
6.3
11.2
20.0
t

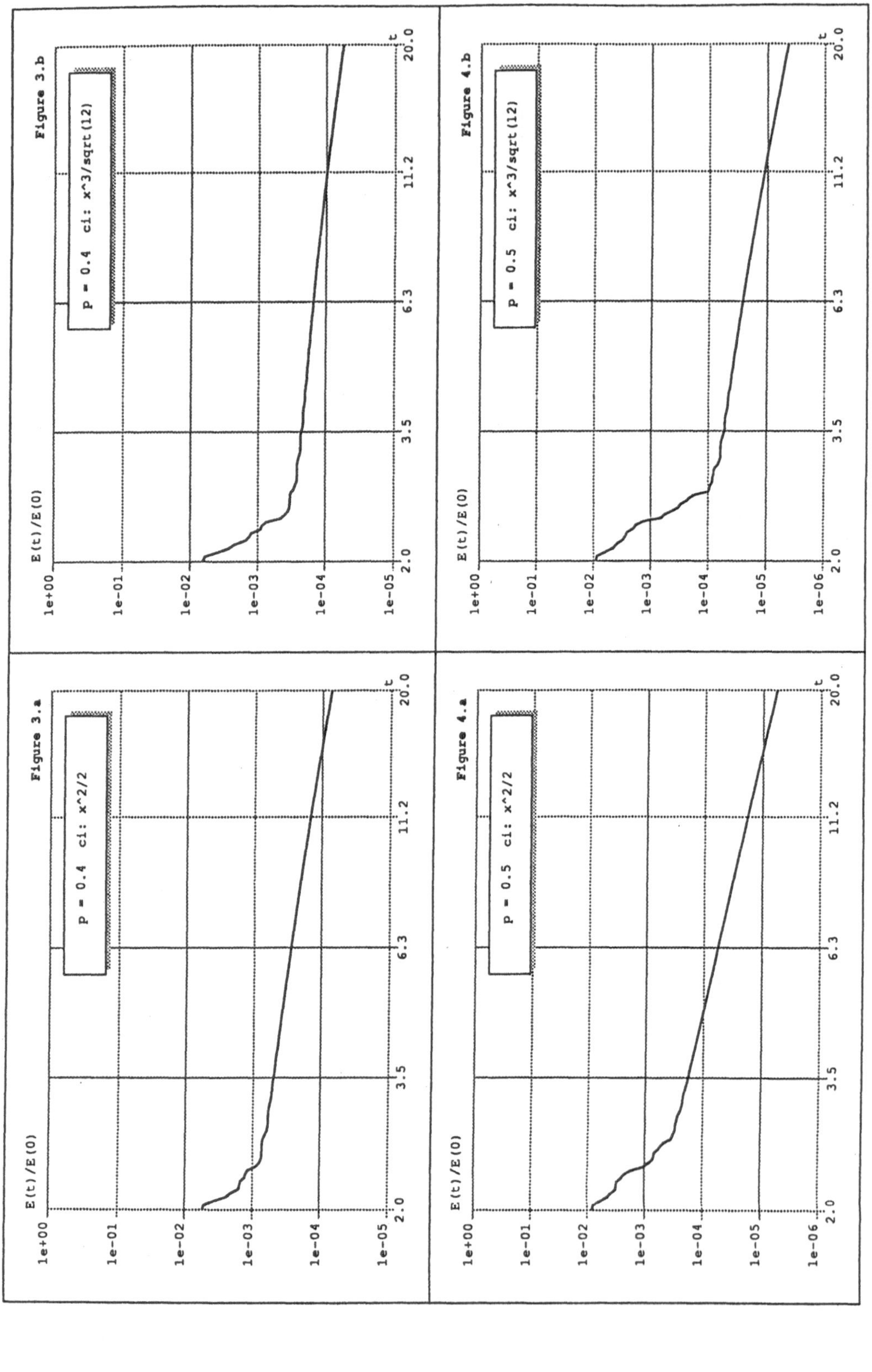
Figure 3.a
p = 0.4 ci: x^2/2
E(t)/E(0)
1e+00
1e-01
1e-02
1e-03
1e-04
1e-05
2.0
3.5
6.3
11.2
20.0
t
Figure 3.b
p = 0.4 ci: x^3/sqrt(12)
E(t)/E(0)
1e+00
1e-01
1e-02
1e-03
1e-04
1e-05
2.0
3.5
6.3
11.2
20.0
t
Figure 4.a
p = 0.5 ci: x^2/2
E(t)/E(0)
1e+00
1e-01
1e-02
1e-03
1e-04
1e-05
1e-06
2.0
3.5
6.3
11.2
20.0
t
Figure 4.b
p = 0.5 ci: x^3/sqrt(12)
E(t)/E(0)
1e+00
1e-01
1e-02
1e-03
1e-04
1e-05
1e-06
2.0
3.5
6.3
11.2
20.0
t

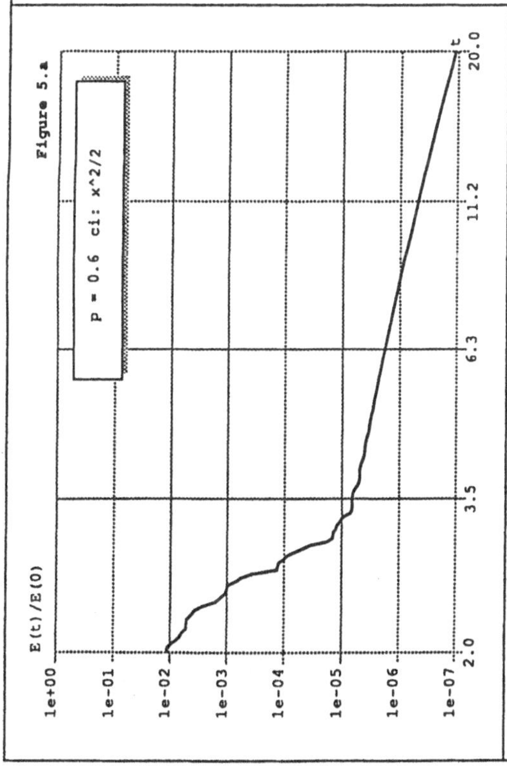

Figure 5.a

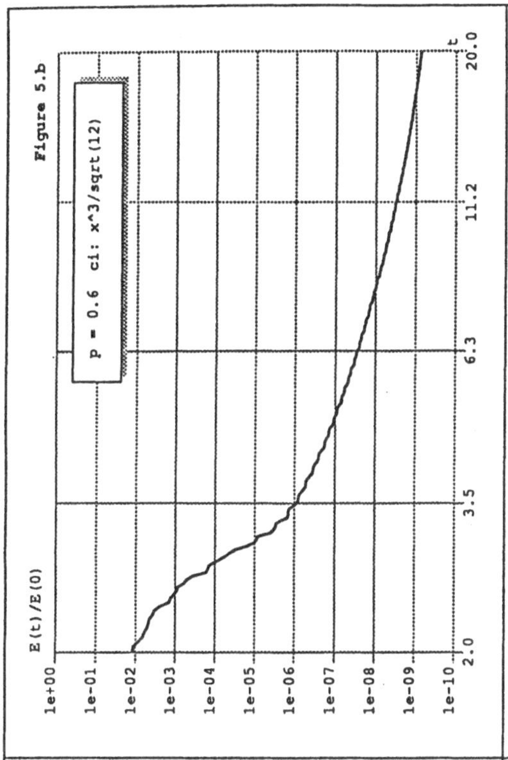

Figure 5.b

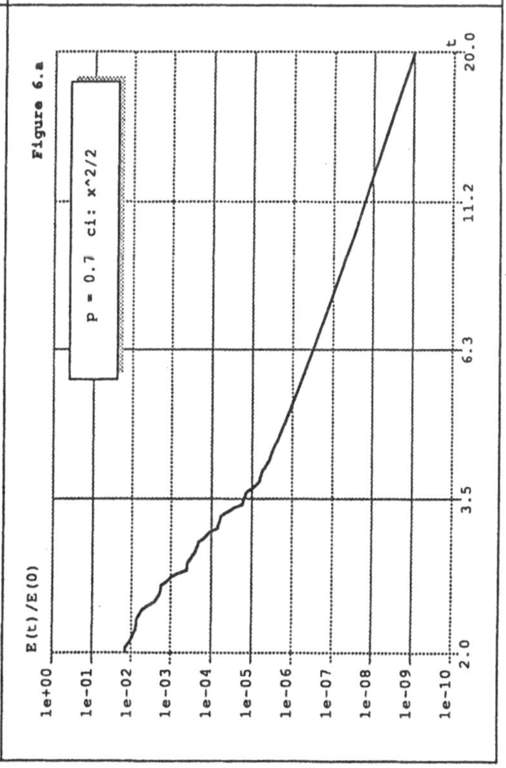

Figure 6.a

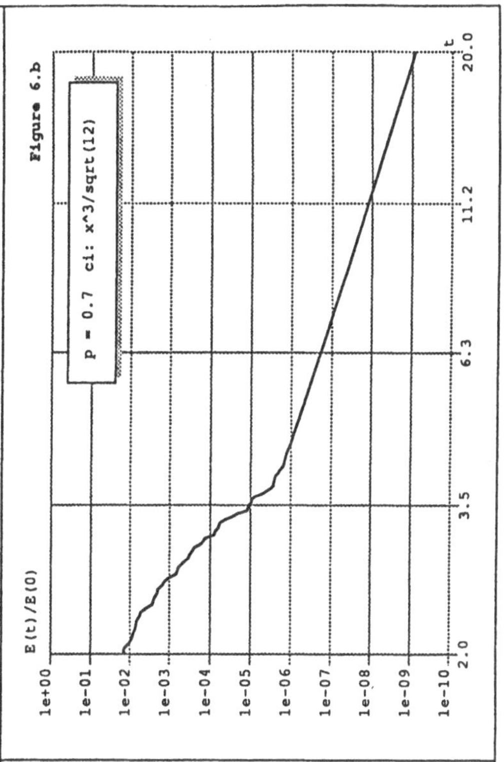

Figure 6.b

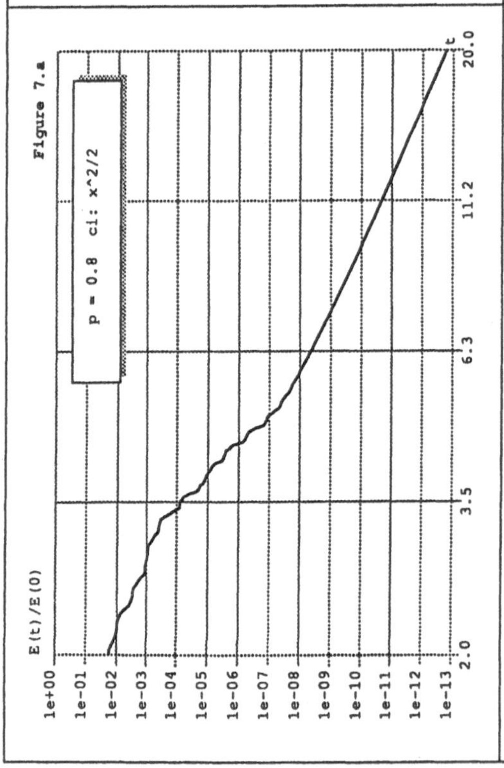
Figure 7.a
p = 0.8 ci: x^2/2
E(t)/E(0)
1e+00
1e-01
1e-02
1e-03
1e-04
1e-05
1e-06
1e-07
1e-08
1e-09
1e-10
1e-11
1e-12
1e-13
2.0
3.5
6.3
11.2
20.0
t

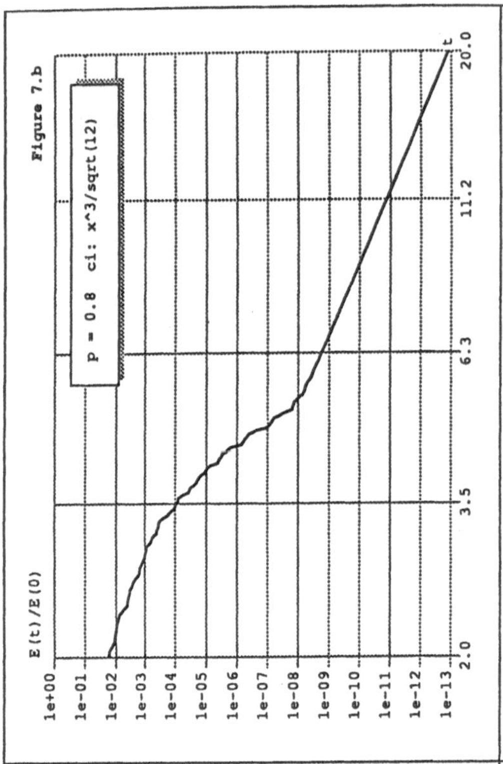
Figure 7.b
p = 0.8 ci: x^3/sqrt(12)
E(t)/E(0)
1e+00
1e-01
1e-02
1e-03
1e-04
1e-05
1e-06
1e-07
1e-08
1e-09
1e-10
1e-11
1e-12
1e-13
2.0
3.5
6.3
11.2
20.0
t

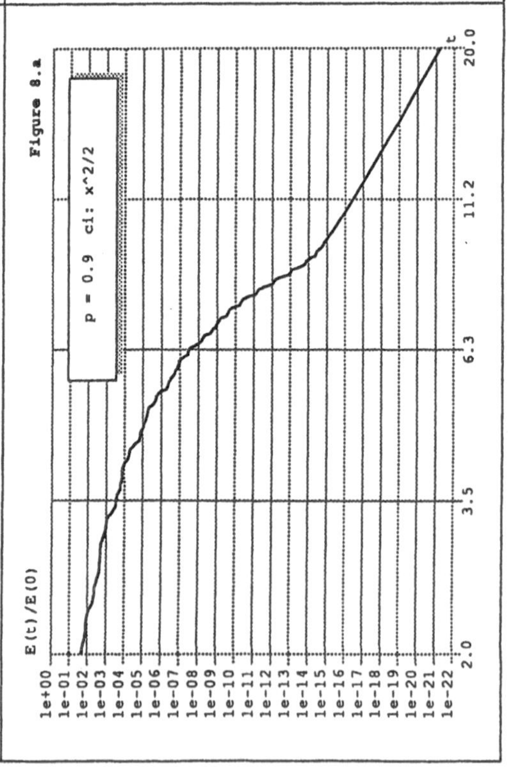
Figure 8.a
p = 0.9 ci: x^2/2
E(t)/E(0)
1e+00
1e-01
1e-02
1e-03
1e-04
1e-05
1e-06
1e-07
1e-08
1e-09
1e-10
1e-11
1e-12
1e-13
1e-14
1e-15
1e-16
1e-17
1e-18
1e-19
1e-20
1e-21
1e-22
2.0
3.5
6.3
11.2
20.0
t

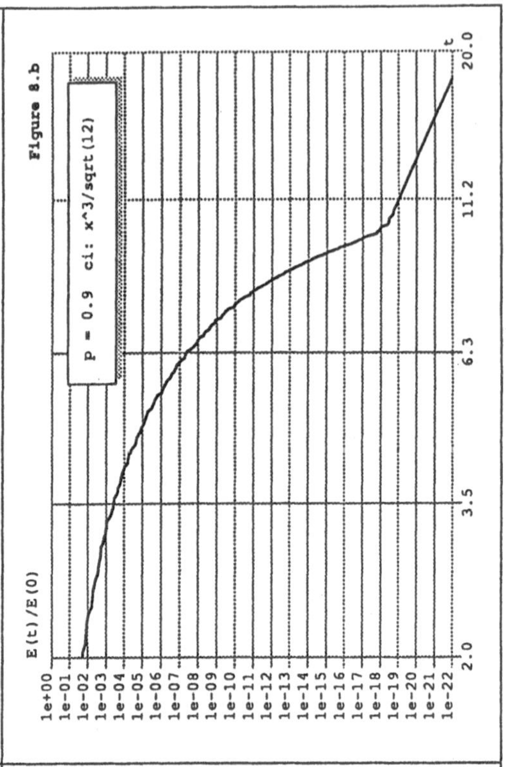
Figure 8.b
p = 0.9 ci: x^3/sqrt(12)
E(t)/E(0)
1e+00
1e-01
1e-02
1e-03
1e-04
1e-05
1e-06
1e-07
1e-08
1e-09
1e-10
1e-11
1e-12
1e-13
1e-14
1e-15
1e-16
1e-17
1e-18
1e-19
1e-20
1e-21
1e-22
2.0
3.5
6.3
11.2
20.0
t

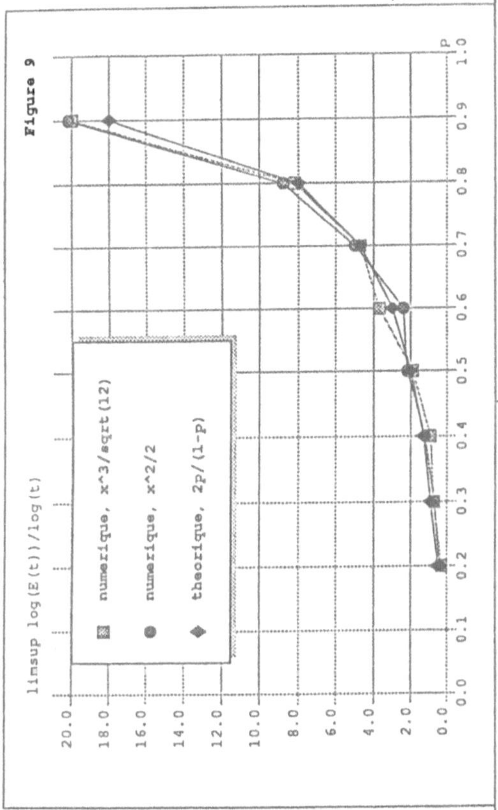

Figure 9

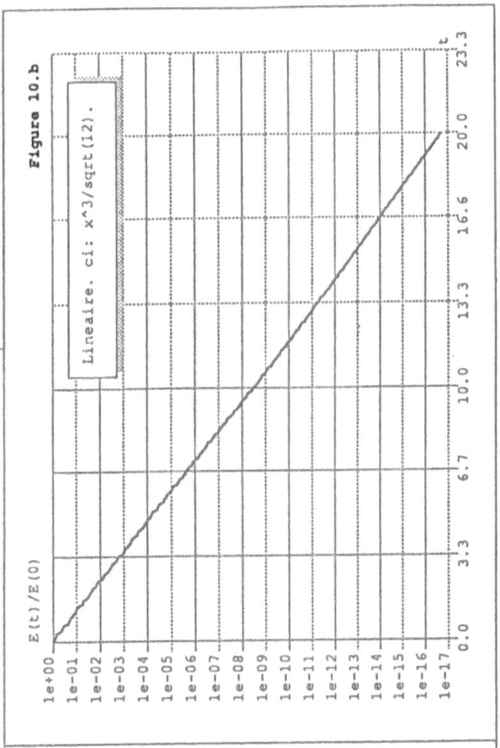

Figure 10.b

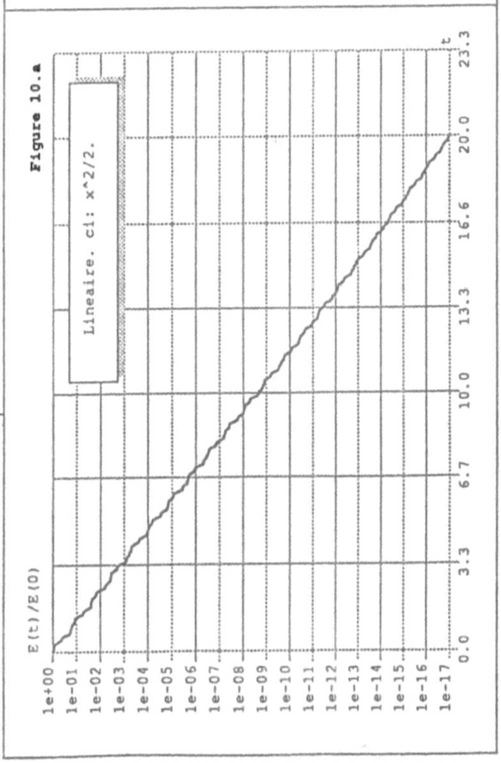

Figure 10.a

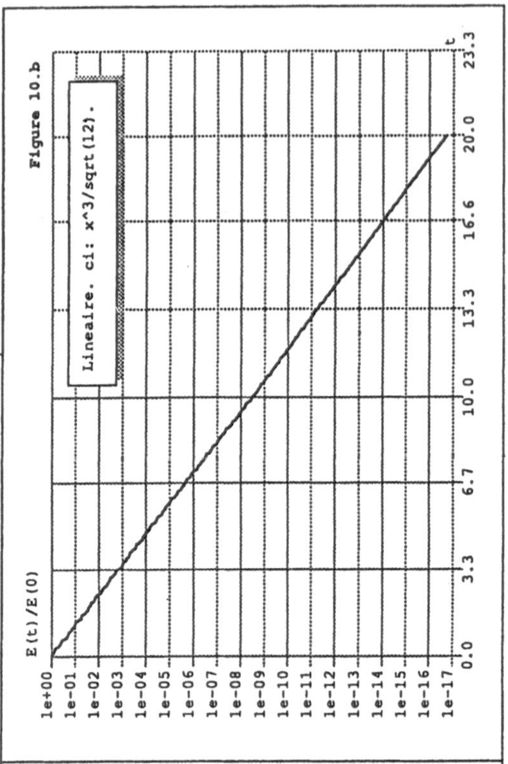
Figure 10.b
Lineaire. ci: x^3/sqrt(12).
E(t)/E(0)
t

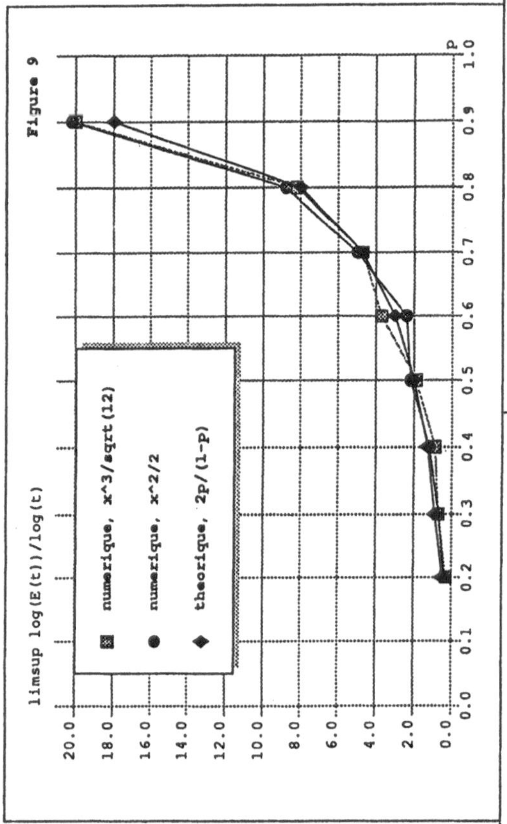
Figure 9
limsup log(E(t))/log(t)
numerique, x^3/sqrt(12)
numerique, x^2/2
theorique, 2p/(1-p)
p

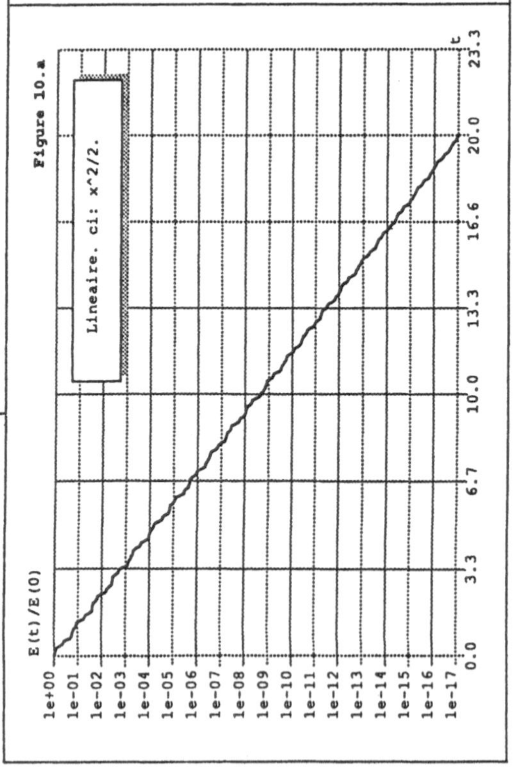
Figure 10.a
Lineaire. ci: x^2/2.
E(t)/E(0)
t

References

[1] J.M. Ball and M. Slemrod. Nonharmonic fourier series and the stabilization of distributed semi-linear control systems. *Communications on Pure and Applied Mathematics*, 32:555–587, 1979.

[2] G. Chen, M. Delfour, A.M. Krall, and G. Payre. Modeling, stabilization and control of serially connected beams. *SIAM Journal on Control and Optimization*, 25(3):526–546, 1987.

[3] G. Chen, S.G. Krantz, D.W. Ma, C.E. Wayne, and H.H. West. The Euler-Bernoulli beam equation with boundary energy dissipation. In *Operator Methods for Optimal Control Problems*, Marcel Dekker Inc., 1988. Edited by Sung J. Lee.

[4] F. Conrad. Stabilization of vibrating beams by a specific feedback. In *Proceedings of the Workshop on Stabilization of flexible structures*, pages 36–51, COMCON, Optimization Software Inc., 1987. Montpellier.

[5] F. Conrad, J. Leblond, and J.P. Marmorat. Stabilization of second order evolution equation by unbounded nonlinear feedback. 1989. Fifth IFAC Symposium on Control of D.P.S, Perpignan.

[6] F. Conrad and M. Pierre. Stabilisation d'une poutre par feedback frontière non linéaire. In preparation.

[7] A. Haraux. Une remarque sur la stabilisation de certains systèmes du deuxième ordre en temps. 1988. A paraître.

[8] A.E. Ingham. Some trigonometrical inequalities with applications to the theory of series. *Mathematische Zeitschrift*, 41:367–379, 1936.

[9] J. Leblond and J.P. Marmorat. Stabilization of a vibrating beam: a regularity result. In *Proceedings of the Workshop on Stabilization of flexible structures*, pages 162–183, COMCON, Optimization Software Inc., 1987. Montpellier.

[10] L. Meirovitch. *Computational methods in structural dynamics.* Sijthoff and Noordhoff, 1980.

[11] A. Pazy. Initial value problems for nonlinear differential equations in Banach spaces. *Séminaire de Mathématiques Appliquées, Collège de France*, 5(84), 1982.

[12] P.A. Raviart and J.M. Thomas. *Introduction à l'analyse numérique des équations aux dérivées partielles.* Masson, 1983.

Uniform Stabilization of the Wave Equation with Dirichlet-Feedback Control without Geometrical Conditions[1]

I. Lasiecka and R. Triggiani

Department of Applied Mathematics
Thornton Hall
University of Virginia
Charlottesville, VA 22903

Table of Contents

---

[1]Research partially supported by National Science Foundation under Grant DMS-8902811 and by Air Force Office of Scientific Research under Grant AFOSR-87-0321.

1. Preliminaries, main statement, literature

Let Ω be an open, bounded domain in R^n, $n \geq 2$, with smooth boundary $\Gamma = \Gamma_0 \cup \Gamma_1$, Γ_i relatively open, Γ_0 possibly empty while Γ_1 is non-empty, with $\bar{\Gamma}_0 \cap \bar{\Gamma}_1 = \varphi$. In Ω we consider the wave equation in $w(t,x)$:

$$w_{tt} = \Delta w \qquad \text{in } (0,T]\times\Omega = Q_T; \tag{1.1}$$

$$w(0,x) = w_0, \quad w_t(0,x) = w_1 \qquad \text{in } \Omega; \tag{1.2}$$

$$w|_{\Sigma_0} = 0; \qquad \text{in } (0,T]\times\Gamma_0 = \Sigma_{0T}; \tag{1.3a}$$

$$w|_{\Sigma_1} = g; \qquad \text{in } (0,T]\times\Gamma_1 = \Sigma_{1T}. \tag{1.3b}$$

Data $\{w_0,w_1,g\}$ in $L_2(\Omega)\times H^{-1}(\Omega)\times L_2(\Sigma_{1T})$ determine the optimal regularity result $\{w,w_t\} \in C([0,T];L_2(\Omega)\times H^{-1}(\Omega))$, any $T < \infty$, for the solution [Lio.1], [L-T.1-2], [LLT.1]. In the present paper, we return to the problem of uniform stabilization of system (1.1)-(1.3) on $L_2(\Omega)\times H^{-1}(\Omega)$, following our previous efforts:

[L-T.4] by an explicit feedback, and [L-T.3, Section 5] (see also the general abstract treatment [FLT.1]) by a Riccati feedback operator.

<u>Uniform stabilization by an explicit dissipative feedback acting on</u> w_t. In [L-T.4] we have introduced the feedback system described by (1.1)-(1.3) with $T = \infty$ and

$$g = \frac{\partial A^{-1} w_t}{\partial \nu} \Big|_{\Sigma_1} \quad \text{in } (0,\infty)\times\Gamma_1 , \tag{1.4}$$

and shown, among others, the following well-posedness and stabilization results. In (1.4) we have set $A\psi = -\Delta\psi$, $\mathcal{D}(A) = H^2(\Omega) \cap H_0^1(\Omega)$, where $H_0^1(\Omega) = \mathcal{D}(A^{1/2})$ (with equivalent norms).

<u>Theorem A</u> [L-T.4]. The following properties hold true for the feedback problem (1.1)-(1.4):

(i) (well-posedness) the map $\{w_0, w_1\} \to \{w(t), w_t(t)\}$ describes an s.c. contraction semigroup on the space $L_2(\Omega)\times H^{-1}(\Omega)$, denoted here by $e^{\mathcal{A}t}$, where $\mathcal{A}$ is the maximal dissipative generator, explicitly defined in [L-T.4; Eq. (1.12)];

(ii) (L_2-property in time of feedback operator) Setting $Z = L_2(\Omega)\times[\mathcal{D}(A^{1/2})]'$, equivalent to $L_2(\Omega)\times H^{-1}(\Omega)$, and defining $\mathcal{E}_w(t) = \|e^{\mathcal{A}t}[w_0, w_1]\|_Z^2 = \|w(t), A^{-1/2} w_t(t)\|_W^2$, $W = L_2(\Omega)\times L_2(\Omega)$, we have

$$\frac{d}{dt}\,\mathcal{E}_w(t) = -2\int_{\Gamma_1} w^2 d\Gamma, \text{ or } \mathcal{E}_w(t)-\mathcal{E}_w(0) = -2\int_0^t\int_{\Gamma_1} w^2 d\Gamma\, dt, \tag{1.5a}$$

$$\int_0^\infty\int_{\Gamma_1} w^2 d\Gamma\, dt \le \tfrac{1}{2}\|\{w_0,w_1\}\|_Z^2 = \tfrac{1}{2}\mathcal{E}_w(0); \tag{1.5b}$$

(iii) (strong stabilization) For any $\{w_0,w_1\} \in Z$, one has

$$\left|\begin{matrix} w(t) \\ w_t(t) \end{matrix}\right| = e^{\mathcal{A}t}\left|\begin{matrix} w_0 \\ w_1 \end{matrix}\right| \to 0 \text{ as } t \to \infty; \text{ in } Z; \tag{1.6}$$

(iv) uniform stabilization: there exist constants $\delta > 0$, $M = M_\delta \ge 1$, such that

$$\left\|\left|\begin{matrix} w(t) \\ w_t(t) \end{matrix}\right|\right\|_Z = \left\|e^{\mathcal{A}t}\left|\begin{matrix} w_0 \\ w_1 \end{matrix}\right|\right\|_Z \le Me^{-\delta t}\left\|\left|\begin{matrix} w_0 \\ w_1 \end{matrix}\right|\right\|_Z, \quad t \ge 0, \tag{1.7}$$

under the following geometrical conditions on $\{\Omega,\Gamma_0,\Gamma_1\}$: there exists a vector field $h(x) = [h_1(x),\dots,h_n(x)] \in C^2(\overline{\Omega})$ such that

(a) $h\cdot\nu \le 0$ on Γ_0; ν = unit outward normal; (1.8)

(b) h is parallel to ν on Γ_1: $h(\sigma) = k(\sigma)\nu(\sigma)$, for a smooth scalar function k, $\sigma \in \Gamma_1$; (1.9)

(c) for some constant $\rho > 0$, and all vectors $y(x) \in [L_2(\Omega)]^n$:

$$\int_\Omega H(x)y(x)\cdot y(x)d\Omega \ge \rho\int_\Omega |y(x)|^2 d\Omega, \tag{1.10}$$

where $H(x)$ is the matrix

$$H(x) = \begin{vmatrix} \frac{\partial h_1}{\partial x_1}, \ldots, \frac{\partial h_1}{\partial x_n} \\ \vdots \\ \frac{\partial h_n}{\partial x_1}, \ldots, \frac{\partial h_n}{\partial x_n} \end{vmatrix}, \tag{1.11}$$

so that (1.10) holds true if $H(x)+H^*(x)$ is strictly positive definite on $\bar{\Omega}$. ∎

If, say, $\Gamma_0 = \phi$, the class of domains Ω which satisfy conditions (a)-(b)-(c) includes strictly convex domains, and set difference of such sets [L-T.4].

The purpose of this paper is, in particular, to remove altogether geometrical conditions on Ω if $\Gamma_0 = \phi$, which derives from assumption (1.9). More precisely, we shall show the following result.

<u>Main Theorem 1.1</u>. Let $h(x) \in C^2(\bar{\Omega})$ be a vector field satisfying assumptions (1.8) and (1.10). Then, conclusion (1.7) on uniform stabilization for problem (1.1)-(1.4) holds true. In particular, if $\Gamma_0 = \phi$, and hence condition (1.8) is empty, then a radial vector field $h(x) = x-x_0$ satisfies (1.10) with $\rho = 1$. Thus, in this case, $\Gamma_0 = \phi$, no geometrical conditions are needed on Ω, except smoothness of Γ. ∎

<u>Orientation on the proof of Theorem 1.1</u>. The proof of uniform stabilization (1.7) in the Dirichlet case (1.3) given in [L-T.4] was inspired by preceding efforts [C.1], [Lag.1] for the

corresponding Neumann problem, which in turn were inspired by corresponding studies for the "exterior problem" [M.1], [S.1], etc. The proof in [L-T.4] is based on two main steps: (i) a change of variable $p = A^{-1}w_t$ to lift the topology $L_2(\Omega)\times H^{-1}(\Omega)$ needed for $\{w,w_t\}$ to the level $H_0^1(\Omega)\times L_2(\Omega)$ for $\{p,p_t\}$; (ii) application of multipliers $h\cdot\nabla p$, etc., to the p-problem.

In this paper we preserve the ideas of [L-T.4] to lift the w-problem (1.1)-(1.4) by one unit in regularity by means of a transformation $w \to v$ to a new variable v, and then apply multipliers to the resulting v-problem. However, unlike before, the transformation $w \to v$ used now is defined by means of pseudo-differential calculus (on the corresponding half-space problem) and is a modification of the lifting transformation of class -1 in both tangential variables used in [L-T.1] in the study of regularity of (1.1)-(1.3).

It should be noted that: (i) the (brief) detour in pseudo-differential calculus is used in this paper only for, and for the sole purpose of, estimating the commutator term ($K\bar{w}$ below), which enters into the v-problem; (ii) instead the v-problem is analyzed by means of multipliers techniques at the level of differential (not pseudo-differential) calculus. More details are given in the orientation at the end of Section 5.2.

Thus, the present proof succeeds to remove any geometrical condition on Ω if $\Gamma_0 = \phi$, while being--we believe--much simpler than the use of geometric optics and micro-local analysis recently employed in the case of uniform stabilization of waves with

Neumann boundary conditions in [B-L-R] where sharp sufficient conditions are obtained. This analysis crucially relies on studies of diffracted rays [M-T.1]; of gliding rays [A-M.1], [M-S.1]; of propagation of singularities [H.2]; etc. Our approach does not require finite speed of propagation. As a result, Theorem 1.1 is now on the same level as known exact controllability results [Lio.2], [H.1], [Tr.1] via multipliers. ∎

For completeness, we provide an overview on uniform stabilization of (1.1)-(1.3) by Riccati feedback operators and related problems.

<u>Uniform stabilization by a Riccati feedback operator acting on</u> $\{w, w_t\}$. Uniform stabilization of (1.1)-(1.4) by means of an 'algebraic' Riccati operator acting on the full pair $\{w, w_t\}$ was obtained in [L-T.3; Section 5] (with an abstract general treatment thereof with further examples given in [FLT.1]). In this approach, the entire theory relies on the natural, preliminary assumption of the 'Finite Cost Condition.' This assumption is automatically fulfilled and guaranteed by either one of the following properties of the dynamics (1.1)-(1.3): that either (1.1)-(1.3) be exactly controllable in finite time T on the space $L_2(\Omega)\times H^{-1}(\Omega)$ by means of $L_2(\Sigma_T)$-controls g, or else (1.1)-(1.3) be uniformly stabilizable on $L_2(\Omega)\times H^{-1}(\Omega)$ by some feedback operator yielding $w|_{\Sigma_1} \in L_2(0,\infty;L_2(\Gamma_1))$. (Indeed, the latter property implies the former, via a well-known result in [R.1], while the

former implies the latter precisely via the algebraic Riccati theory of [L-T.3].) At the time [L-T.3] was written, the results of uniform stabilization, hence of exact controllability, of [L-T.4] provided a sufficiently large class of triples $\{\Omega, \Gamma_0, \Gamma_1\}$ to which the Riccati theory of [L-T.3, Section 5] was applicable with no further assumptions. More recent solutions of the exact controllability problem for (1.1)-(1.3) by direct approaches (either by the Hilbert Uniqueness Method [Lio.2], or by a surjectivity approach [Tr.1], both of which rely on an *a-priori* inequality pointed out explicitly in [H.1], and implicitly contained also in the calculations of [L-T.4]), require no geometrical conditions on Ω, if $g \in L_2(\Sigma_T)$ is applied to all of Γ, and only mild, explicit conditions on the triple $\{\Omega, \Gamma_0, \Gamma_1\}$ if $g \in L_2(\Sigma_{1T})$ is applied only to Γ_1. Even more recently, a geometric optics approach first introduced in [Lit.1] for these problems, has succeeded in obtaining essentially necessary and sufficient conditions for exact controllability of (1.1)-(1.3) in terms of all rays meeting $\Gamma_1 \times (0,T)$ at non-diffractive points.

The case of pathological domains with cracks or corners in two dimensions has been treated in [Gris.1]. Thus, as a result of these efforts on exact controllability, the algebraic Riccati theory, and hence the uniform stabilization theory by a Riccati operator, of [L-T.3, Section 5] has become applicable to a much larger, essentially optimal, class of triples $\{\Omega, \Gamma_0, \Gamma_1\}$ for problem (1.1)-(1.3).

Acknowledgement. We wish to thank C. Bardos for stimulating exchanges.

2. Proof of Main Theorem 1.1. Step 1: Preliminaries

2.1. General strategy

With reference to problem (1.1)-(1.4), we recall the norm equivalence between $\|y\|_{H^{-1}(\Omega)}$ and $\|A^{-1/2}y\|_{L_2(\Omega)}$, and let

$$\mathcal{E}_w(t) \equiv \|w(t)\|^2_{L_2(\Omega)} + \|A^{-1/2}w_t(t)\|^2_{L_2(\Omega)}. \tag{2.1}$$

Our goal is to show, as usual, that there exists a time $0 < T < \infty$ such that with $Z = L_2(\Omega) \times [\mathcal{D}(A^{1/2})]'$, we have

$$\mathcal{E}_w(T) \leq r\mathcal{E}_w(0), \; r < 1; \quad \text{or } \|e^{\mathcal{A}T}\|_{\mathcal{L}(Z)} < 1, \tag{2.2}$$

after which the uniform decay (1.7) is then established. To prove (2.2), it will suffice, as usual, to show that: there is a time $0 < T < \infty$ and a corresponding constant $C_T > 0$ such that

$$\mathcal{E}_w(T) \leq C_T \int_0^T \int_{\Gamma_1} w^2 d\Sigma_1 \tag{2.3}$$

for (2.3), combined with the non-increasing property (1.5) of $\mathcal{E}_w$ will then yield (2.2). Our subsequent effort is aimed at establishing (2.3).

2.2. Operator model for problem (1.1)-(1.4)

We briefly recall the abstract operator model of problem (1.1)-(1.4) introduced by the authors [L-T.1] and used in [L-T.4]. Let $\tilde{D}$ (Dirichlet map) be the operator defined by

$$\tilde{D}g = h \Leftrightarrow \{\Delta h = 0 \text{ in } \Omega;\ h|_{\Gamma_1} = g,\ h|_{\Gamma_0} = 0\}; \tag{2.4}$$

$$\tilde{D}: \text{continuous } H^s(\Gamma) \to H^{s+½}(\Omega),\ s \text{ real}. \tag{2.5}$$

It can be verified by Green's second theorem that [L-T.4]

$$-\tilde{D}^*Ay = \begin{cases} \frac{\partial y}{\partial \nu} & \text{in } \Gamma_1; \\ 0 & \text{in } \Gamma_0, \end{cases} \tag{2.6}$$

where $\tilde{D}^*$ is the adjoint operator of $\tilde{D}$: $(\tilde{D}g,y) = (g,\tilde{D}^*y)_{L_2(\Gamma)}$.

Then the operator model of problems (1.1)-(1.4) is [L-T.1], [L-T.4], etc.

$$w_{tt} = -A(w-\tilde{D}w|_\Gamma) \text{ in } L_2(\Omega); \tag{2.7}$$

$$\begin{cases} g = w|_{\Sigma_1} = -\tilde{D}^*w_t = -\tilde{D}AA^{-1}w_t = \frac{\partial A^{-1}w_t}{\partial \nu}\Big|_{\Sigma_1} & (2.8a) \\ g = 0 \text{ on } \Sigma_0 & (2.8b) \end{cases}$$

2.3. A-priori estimate of $\frac{\partial w}{\partial \nu}$

We let $\Sigma_T = (0,T)\times\Gamma$ and $\Sigma_\infty = (0,\infty)\times\Gamma$, etc.

Lemma 2.1. With reference to problem (1.1)-(1.4)--(or, equivalently, (2.7)--we have the following estimate:

$$\left\|\frac{\partial w}{\partial \nu}\right\|_{H^{-1}(\Sigma_{1\infty})} \leq C\|w|_\Sigma\|_{L_2(\Sigma_{1\infty})}, \tag{2.9}$$

and the same estimate holds for $T < \infty$, with constant C then depending on T. ∎

<u>Proof</u>. The proof is the same for $T < \infty$ or $T = \infty$. Using (2.8) and (2.7), we obtain on Σ:

$$w_t|_\Sigma = \frac{\partial A^{-1}w_{tt}}{\partial \nu} = -\frac{\partial w}{\partial \nu} + \frac{\partial \tilde{D}w}{\partial \nu}. \tag{2.10}$$

By the well-posedness Theorem A(i), we have $w|_\Sigma \in L_2(0,\infty;L_2(\Gamma_1))$, hence $w_t|_\Sigma \in H^{-1}(0,\infty;L_2(\Gamma_1))$, or

$$\|w_t|_\Sigma\|_{H^{-1}(0,\infty;L_2(\Gamma_1))} \leq C\|w|_\Sigma\|_{L_2(0,\infty;L_2(\Gamma_1))}. \tag{2.11}$$

Moreover, by (2.5) with $s = 0$, elliptic theory gives $\tilde{D}w \in L_2(0,\infty;H^{1/2}(\Omega))$, while trace theory for elliptic problems (as in Kellog [K.1, Theorem 3.8.1, p. 71] gives $\frac{\partial \tilde{D}w}{\partial \nu} \in L_2(0,\infty;H^{-1}(\Gamma_1))$ or

$$\left\|\frac{\partial \tilde{D}w}{\partial \nu}\right\|_{L_2(0,\infty;H^{-1}(\Gamma_1))} \leq C\|w|_\Sigma\|_{L_2(0,\infty;L_2(\Gamma_1))}. \tag{2.12}$$

Then, using (2.11) and (2.12) in identity (2.10) yields estimate (2.9), as desired. ∎

3. <u>Proof of Main Theorem 1.1. Step 2: A change of variable from w to v. Statement of a-priori estimates of the v-problem</u>

In Sections 5 and 6, we shall implement the following strategy. Starting with the original variable w of problem

(1.1)-(1.4) on $0 < t < \infty$, we shall define the truncation of w by

$$\bar{w}(t,x,y) = \begin{cases} w(t,x,y), & 0 < t < T; \\ 0, & \text{elsewhere in } R_t^1 = (-\infty,\infty). \end{cases} \tag{3.1}$$

Next, it will be demonstrated in Section 5 (by going over to the half-space problem) that there exists an operator Λ, given explicitly in Section 5, such that the change of variable

$$v = \Lambda\bar{w} \tag{3.2}$$

transforms Eq. (1.1) of the w-problem into the equation

$$v_{tt} = \Delta v + K\bar{w}, \quad (0,T)\times\Omega, \tag{3.3a}$$

and Eq. (1.2) into

$$v|_{\Sigma_0} = 0, \quad R_t^1\times\Omega, \tag{3.3b}$$

K being a commutator operator, possessing the crucial property of Theorem 3.2 and Corollary 3.3 below.

As to the boundary values of the v-problem (3.3), they satisfy the following estimate. In the results below, we let $\Sigma_\infty = R_t^1\times\Gamma$ and $Q_\infty = R_t^1\times\Omega$, $R_t^1 = (-\infty,\infty)$, while $\Sigma_T = (0,T)\times\Gamma$ and $Q_T = (0,T)\times\Omega$.

<u>Lemma 3.1</u>. With reference to problem (3.2), (3.3), we have

$$\|v|_\Sigma\|^2_{H^1(\Sigma_\infty)} + \left\|\frac{\partial w}{\partial \nu}\Big|_\Sigma\right\|^2_{L_2(\Sigma_\infty)} \le C\{\|w|_\Sigma\|^2_{L_2(\Sigma_T)} + \|v\|^2_{L_2(Q_\infty)}\}, \tag{3.4}$$

where the constant C does not depend on T. ∎

Next, the commutator K as acting on the truncation $\bar{w}$, or else the solution w of problem (1.1)-(1.4), equivalently of problem (2.7), satisfies the following basic estimate.

Theorem 3.2. With reference to (3.1) and (3.3), we have

$$\|K\bar{w}\|_{L_2(Q_\infty)} \le C\{\|w|_\Sigma\|^2_{L_2(\Sigma_T)}+\|v\|^2_{L_2(Q_\infty)}+\mathcal{E}_w(T)+\mathcal{E}_w(0)\}, \tag{3.5}$$

where $\mathcal{E}_w(t)$ is defined in (2.1) and the constant C does not depend on T. ∎

Corollary 3.3. With reference to problem (1.1)-(1.4) for w and to v in (3.2), we have

$$\int_0^T\|Kw\|^2_{L_2(\Omega)}dt \le C\{\int_0^T\|w|_{\Gamma_1}\|^2_{L_2(\Gamma_1)}dt+\int_{-\infty}^{\infty}\|v\|^2_{L_2(\Omega)}dt+\mathcal{E}_w(T)+\mathcal{E}_w(0)\}, \tag{3.6}$$

where $\mathcal{E}_w(t)$ is defined in (2.1) and the constant C does not depend on T. ∎

4. Proof of Main Theorem 1.1. Step 3: Analysis of the v-problem by multipliers

4.1. Preliminary energy estimate of the v-problem

We recall the notations $\Sigma_{iT} = (0,T)\times\Gamma_i$; $\Sigma_T = (0,T)\times\Gamma$; $Q_T = (0,T)\times\Omega$; $Q_\infty = (-\infty,\infty)\times\Omega$, to be used throughout this section.

Theorem 4.1. With reference to problem (3.3) for v, if h(x) is the vector field satisfying assumptions (1.8) and (1.10) of the

Main Theorem 1.1, then we have for T sufficiently large (constructively, as below (4.24))

$$\int_0^T E_v(t)dt \le C\{\|w|_{\Sigma_1}\|^2_{L_2(\Sigma_{1T})}+\mathcal{E}_w(T)+\mathcal{E}_w(0)\}$$

$$+ C_T\{\|v\|^2_{C([0,T];L_2(\Omega))}+\|w\|^2_{H^{-1}(Q_T)}\}, \qquad (4.1)$$

where $\mathcal{E}_w(t)$ is defined in (2.1), the constant C does not depend on T (but depends on ρ,h), the constant C_T depends on T, and where we have set

$$E_v(t) = \int_\Omega\{v_t^2(t)+|\nabla v(t)|^2\}d\Omega. \qquad (4.2)$$

<u>Proof</u>. (i) We apply to the v-problem (3.3) the three multipliers $h\cdot\nabla v$; $v \operatorname{div} h$; and v_t. Multiplying (3.3) by $h\cdot\nabla v$ and integrating by parts, we obtain e.g., [Tr.1, Eq. (2.20)]:

$$\int_{Q_T} H\nabla v\cdot\nabla v\, dQ = \int_{\Sigma_T}\frac{\partial v}{\partial\nu}\, h\cdot\nabla v d\Sigma + \tfrac{1}{2}\int_{\Sigma_T} v_t^2 h\cdot\nu d\Sigma - \tfrac{1}{2}\int_{\Sigma_T}|\nabla v|^2 h\cdot\nu d\Sigma$$

$$- \tfrac{1}{2}\int_{Q_T}\{v_t^2-|\nabla v|^2\}\operatorname{div} h\, dQ + \int_{Q_T}(Kw)h\cdot\nabla v\, dQ-[(v_t,h\cdot\nabla v)_\Omega]_0^T, \qquad (4.3)$$

where H(x) is the matrix in (1.11). Multiplying (3.3a) by v div h and integrating by parts yields by (3.3b), e.g., [Tr.1],

$$\int_{Q_T}\{v_t^2-|\nabla v|^2\}\text{div } h\, dQ = [(v_t, v \text{ div } h)_\Omega]_0^T + \int_{Q_T} v\nabla(\text{div } h)\cdot\nabla v\, dQ$$

$$- \int_{Q_T}(Kw)v \text{ div } h\, dQ - \int_{\Sigma_{1T}} \frac{\partial v}{\partial \nu}\, v \text{ div } h\, d\Sigma_1. \tag{4.4}$$

Multiplying (3.2) by v_t and integrating by parts gives by (3.3b),

$$E_v(t) - E_v(0) = 2\int_{Q_T}(Kw)v_t\, dQ + \int_{\Sigma_{1T}} \frac{\partial v}{\partial \nu}\, v_t\, d\Sigma_2. \tag{4.5}$$

Using the boundary condition (3.3b), we have on
Σ_{OT}: $h\cdot\nabla v = \frac{\partial v}{\partial \nu}\, h\cdot\nu$; $\left|\frac{\partial v}{\partial \nu}\right| = |\nabla v|$; and $v_t\big|_{\Sigma_{OT}} = 0$, so that

$$\int_{\Sigma_{OT}} \frac{\partial v}{\partial \nu}\, h\cdot\nabla v\, d\Sigma_0 + ½\int_{\Sigma_{OT}} v_t^2 h\cdot\nu\, d\Sigma_0 - ½\int_{\Sigma_{OT}}|\nabla v|^2 h\cdot\nu\, d\Sigma_0$$

$$= ½\int_{\Sigma_{OT}} \left(\frac{\partial v}{\partial \nu}\right)^2 h\cdot\nu\, d\Sigma_0 \leq 0, \tag{4.6}$$

by assumption (1.8) on h. We return to (4.3): Here on the left-hand side of (4.3), we use assumption (1.10) on H; and on the right-hand side of (4.3), we use (4.6) for the Σ_{OT}-terms of v, estimate (3.4) for the Σ_{1T}-terms of v, and finally estimate (3.6) of Corollary 3.3 on the right-hand side of

$$\left|\int_{Q_T}(Kw)h\cdot\nabla v\, dQ\right| \leq \varepsilon\int_{Q_T}|\nabla v|^2 dQ + \frac{C_h}{\varepsilon}\int_{Q_T}|Kw|^2 dQ \tag{4.7}$$

with $1 \geq \varepsilon > 0$ arbitrary. We obtain

$$(\rho-\varepsilon)\int_{Q_T}|\nabla v|^2 dQ \le C\{\|w|_{\Sigma_1}\|^2_{L_2(\Sigma_{1T})}+\|v\|^2_{L_2(Q_\infty)}\}$$

$$-\tfrac{1}{2}\int_{Q_T}\{v_t^2-|\nabla v|^2\}\,\mathrm{div}\,h\,dQ + b_{0,T}. \tag{4.8}$$

$$b_{0,T} = -[(v_t,h\cdot\nabla v)_{L_2(\Omega)}]_0^T + C\{\mathcal{E}_w(T)+\mathcal{E}_w(0)\}. \tag{4.9}$$

The constant C in (4.8) does not depend on T, but depends on ε, h. We shall only indicate dependence on ε which is crucial. The constant C in (4.9) does not depend on T. Dependence on T will always be noted.

(ii) With reference to identity (4.4), we use again inequality (3.4) on the Σ_{1T}-terms of v, and inequality (3.6) for Kw to obtain for any $1 \ge \varepsilon > 0$,

$$\left|\int_{Q_T}\{v_t^2-|\nabla v|^2\}\mathrm{div}\,h\,dQ\right| \le C\{\|w|_{\Sigma_1}\|^2_{L_2(\Sigma_{1T})}+\|v\|^2_{L_2(Q_\infty)}\}$$

$$+\varepsilon\int_{Q_T}|\nabla v|^2 dQ + \tilde{b}_{0,T}. \tag{4.10}$$

$$\tilde{b}_{0,T} = [(v_t,v\,\mathrm{div}\,h)_{L_2(\Omega)}]_0^T + C\{\mathcal{E}_w(T)+\mathcal{E}_w(0)\}, \tag{4.11}$$

with constants C in (4.10) and (4.11) independent of T. (The C in (4.10) depends on ε, h.) For future reference we note that if we use the version of (4.4) with div h = 1 (i.e., obtained by multiplying (3.3) simply by v this time), then (4.10) holds true with div h = 1.

(iii) Inserting (4.10) into (4.8) yields

$$(\rho-2\varepsilon)\int_{Q_T}|\nabla v|^2 dQ \le C\{\|w|_{\Sigma_1}\|^2_{L_2(\Sigma_{1T})}+\|v\|^2_{L_2(Q_\infty)}\}+\beta_{0,T}. \tag{4.12}$$

$$\beta_{0,T} = b_{0,T} + ½\,\tilde{b}_{0,T} = [½(v_t, v \operatorname{div} h)_\Omega-(v_t,h\cdot\nabla v)_\Omega]_0^T$$
$$+ C\{\mathcal{E}_w(T)+\mathcal{E}_w(0)\}. \tag{4.13}$$

(iv) We next estimate $\beta_{0,T}$. Recalling (4.2), we first obtain from (4.13)

$$|\beta_{0,T}| \le C\{E_v(T)+E_v(0)+\|v\|^2_{C([0,T];L_2(\Omega))}+\mathcal{E}_w(T)+\mathcal{E}_w(0)\}, \tag{4.14}$$

and next we want to express $E_v(T)$ in terms of $E_v(0)$. To this end, we invoke identity (4.5), where we use inequality (3.4) for the Σ_{1T}-terms of v and inequality (3.6) for Kw. We obtain

$$|E_v(T)-E_v(0)| \le \varepsilon\int_{Q_T} v_t^2 dQ+C\{\|w|_{\Sigma_1}\|^2_{L_2(\Sigma_{1T})}+\|v\|^2_{L_2(Q_\infty)}+\mathcal{E}_w(T)+\mathcal{E}_w(0)\}. \tag{4.15}$$

We now set, for convenience of notation,

$$r_T \equiv \|v\|^2_{L_2(Q_\infty)}+\|v\|^2_{C([0,T];L_2(\Omega))}+\mathcal{E}_w(T)+\mathcal{E}_w(0) \tag{4.16}$$

We now extract from (4.15) a bound for $E_v(T)$ in terms of $E_v(0)$ and insert this bound in (4.14). Recalling (4.16), we obtain

$$\|\beta_{0,T}| \le C\ E_v(0)+C\{\|w|_{\Sigma_1}\|^2_{L_2(\Sigma_{1T})}+r_T\}$$

$$+\ \varepsilon\ C\int_{Q_T} v_t^2 dQ. \tag{4.17}$$

(v) Finally, inserting (4.17) into (4.12) yields

$$(\rho-2\varepsilon)\int_{Q_T} |\nabla v|^2 dQ \le C\{\|w|_{\Sigma_1}\|^2_{L_2(\Sigma_{1T})}+r_T\}$$

$$+\ \varepsilon\ C\int_{Q_T} v_t^2 dQ + C\ E_v(0). \tag{4.18}$$

(vi) We now seek an estimate for $\int_{Q_T} v_t^2 dQ$ similar to that for $\int_{Q_T} |\nabla v|^2 dQ$ in (4.18). To this end, we return to (4.10) with div $h \equiv 1$ (as noted below (4.11)). From (4.11) and (4.2),

$$|\tilde{b}_{0,T}| \le C\{E_v(T)+E_v(0)+\|v(T)\|^2+\|v(0)\|^2+\mathcal{E}_w(T)+\mathcal{E}_w(0)\}. \tag{4.19}$$

We now extract from (4.15) a bound for $E_v(T)$ in terms of $E_v(0)$ which we then insert in (4.19), and use the resulting expression in (4.10). We obtain, recalling (4.16),

$$|\int_{Q_T}\{v_t^2-|\nabla v|^2\}dQ \le C\{\|w|_{\Sigma_1}\|^2_{L_2(\Sigma_{1T})}+r_T\}$$

$$(\varepsilon+\varepsilon C)\int_{Q_T} v_t^2+|\nabla v|^2 dQ + C\ E_v(0). \tag{4.20}$$

Using now (4.20) and (4.18), we obtain the sought-after counterpart of (4.18), this time for $\int_{Q_T} v_t^2 dQ$:

$$(\rho-2\varepsilon)\int_{Q_T} v_t^2 dQ \le C\{\|w|_{\Sigma_1}\|^2_{L_2(\Sigma_{1T})}+r_T\}$$

$$\varepsilon\; C\int_{Q_T} v_t^2+|\nabla v|^2 dQ + C\; E_v(0), \qquad (4.21)$$

where we note only crucial dependence of the quantities in terms of ε and T but omit dependence on ρ, h.

(vii) Summing up (4.18) and (4.21), we obtain

$$(\rho-2\varepsilon-\varepsilon C)\int_{Q_T} v_t^2+|\nabla v|^2 dQ = (\rho-2\varepsilon-\varepsilon C)\int_0^T E_v(t)dt$$

$$\le C\{\|w|_{\Sigma_1}\|^2_{L_2(\Sigma_{1T})}+r_T\} + C\; E_v(0). \qquad (4.22)$$

(viii) It remains to estimate $E_v(0)$. To this end, we return to (4.15) with T replaced by a general t, $0 \le t \le T$, and integrate in t over [0,T]. We obtain

$$T\; E_v(0) \le \int_0^T E_v(t)dt+CT\{\|w|_{\Sigma_1}\|^2_{L_2(\Sigma_{1T})}+\|v\|^2_{L_2(Q_\infty)}\}$$

$$+\; \varepsilon T\int_{Q_T} v_t^2 dQ+C\int_0^T \mathcal{E}_w(t)dt+CT\; \mathcal{E}_w(0). \qquad (4.23)$$

Finally, we use estimate (4.22) on the right of (4.23) and obtain for ε small (say $\rho/2 \leq \rho-2\varepsilon-\varepsilon C$):

$$(T-\frac{2C}{\rho})E_v(0) \leq (\frac{2C}{\rho}+CT)\{\|w|_{\Sigma_1}\|^2_{L_2(\Sigma_{1T})}+\|v\|^2_{L_2(Q_\infty)}\}$$

$$+\frac{2C}{\rho}\{\|v\|^2_{C([0,T];L_2(\Omega))}+\mathcal{E}_w(T)+\mathcal{E}_w(0)\}$$

$$+\varepsilon T\int_{Q_T} v_t^2 dQ + C\int_0^T \mathcal{E}_w(t)dt + CT\ \mathcal{E}_w(0), \tag{4.24}$$

which for $T > 2C/\rho$ provides the desired estimate of $E_v(0)$. (We have noted all along that the constant C in front of $E_v(0)$ in (4.22) does not depend on T, as a result of the constant C in (3.6) being independent of T.)

We now recall the dissipativity property (1.5a) for $\mathcal{E}_w(t)$: $\mathcal{E}_w(t) \leq \mathcal{E}_w(0)+2\int_0^t\int_{\Gamma_1} w^2 d\Gamma_1 dt$, and hence use

$$\int_0^T \mathcal{E}_w(t)dt \leq T\ \mathcal{E}_w(0)+T\|w|_{\Sigma_1}\|^2_{L_2(\Sigma_{1T})}$$

in (4.24) and divide (4.24) through by $(T-2C/\rho) > 0$, for T large. We obtain, recalling r_T in (4.16),

$$E_v(0) \leq C_1\{\|w|_{\Sigma_1}\|^2_{L_2(\Sigma_{1T})}+r_T\} + \varepsilon C_2\int_{Q_T} v_t^2 dQ, \tag{4.25}$$

with constants C_1 and C_2 independent of T for T sufficiently large. (What is critical is that the constant in (4.25) in front of $\mathcal{E}_w(T)+\mathcal{E}_w(0)$ be independent of T.)

(ix) Inserting (4.25) into (4.22) results into the estimate

$$\int_0^T E_v(t)dt \leq C\{\|w|_{\Sigma_1}\|^2_{L_2(\Sigma_1)}+r_T\}. \tag{4.26}$$

(x) Next, as a consequence of the definition (5.22) for v, we have: sing supp v ⊂ sing supp $\bar{w}$ = [0,T], and, moreover, the following estimate holds true, say for any $s \geq 0$:

$$\int_{-\infty}^{0} \|v\|^2_{H^s(\Omega)} dt+\int_T^{\infty}\|v\|^2_{H^s(\Omega)} dt \leq C_T\|\bar{w}\|^2_{H^{-1}(Q_T)} = C_T\|w\|^2_{H^{-1}(Q_T)} . \tag{4.27}$$

Finally, recalling the definition of r_T in (4.16) and using (4.27) in (4.26), we arrive at the sought-after estimate (4.1). Theorem 4.1 is proved. ■

4.2. Return from variable v to original variable w

Starting from (4.1) of Theorem 4.1, we shall establish the desired estimate (2.3), as stated in the following theorem.

Theorem 4.2. With reference to problem (1.1)-(1.3) for w, let h(x) be the vector field satisfying the assumptions (1.8) and

(1.10) of the Main Theorem 1.1. Then, there is $C_T > 0$ such that inequality (2.3) holds true; i.e., recalling (2.1):

$$\mathcal{E}_w(T) = \int_\Omega \{|w(T)|^2 + |A^{-\frac{1}{2}} w_t(T)|^2\} d\Omega \le C_T \int_0^T \int_{\Gamma_1} w^2 d\Sigma_1 . \tag{4.28}$$

<u>Proof</u>. (i) From the definitions (5.21), (5.22) of the function $\bar{w}$ and of the transformation $\bar{w} = \Lambda^{-1} v$ in Section 5, it follows readily by (4.27) with $s = 1$ that

$$\|\bar{w}\|^2_{L_2(Q_T)} = \|w\|^2_{L_2(Q_T)} \le C\|v\|^2_{H^1(Q_\infty)}$$

$$\le C\|v\|^2_{H^1(Q_T)} + C_T\|w\|^2_{H^{-1}(Q_T)}, \tag{4.29}$$

with constant C in (4.30) in front of $\|v\|^2_{H^1(Q_T)}$ independent of T.

(ii) Moreover, on $0 < t < T$, equation $w_{tt} = \Delta w$ becomes $\bar{w}_{tt} = \Delta\bar{w} = \Delta\Lambda^{-1}v$ by (5.21), (5.22), where $v \to \Delta\Lambda^{-1}v$ is continuous $H^1(Q_\infty) \to L_2(-\infty,\infty;H^{-2}(\Omega))$, i.e.,

$$\|\bar{w}_{tt}\|^2_{L_2(-\infty,\infty;H^{-2}(\Omega))} = \|w_{tt}\|^2_{L_2(0,T;H^{-2}(\Omega))} \le C\|v\|^2_{H^1(Q_\infty)}$$

$$\le C\|v\|^2_{H^1(Q_T)} + C_T\|w\|^2_{H^{-1}(Q_T)}, \tag{4.30}$$

where in the last step we have used (4.29). The constant C in (4.30) in front of $\|v\|^2_{H^1(Q_T)}$ is independent of T.

(iii) Interpolating between (4.29):

$$\{v,w\} \in H^1(Q_T)\times H^{-1}(Q_T) \to w \in L_2(Q_T),$$

and (4.30):

$$\{v,w\} \in H^1(Q_T)\times H^{-1}(Q_T) \to w_{tt} \in L_2(0,T;H^{-2}(\Omega)),$$

we obtain

$$\|w_t\|^2_{L_2(0,T;H^{-1}(\Omega))} \le C\|v\|^2_{H^1(Q_T)} + C_T\|w\|^2_{H^{-1}(Q_T)}, \qquad (4.31)$$

again with constant C in (4.31) in front of $\|v\|^2_{H^1(Q_T)}$ independent of T.

(iv) Summing up (4.29) and (4.31), we obtain

$$\int_0^T \|w\|^2_{L_2(\Omega)} + \|w_t\|^2_{H^{-1}(\Omega)}\, dt \le C\int_0^T\int_\Omega v_t^2 + |\nabla v|^2 d\Omega\, dt$$

$$+ C_T\|w\|^2_{H^{-1}(Q_T)}, \qquad (4.32)$$

or, recalling the norm-equivalence $\mathcal{D}(A^{1/2}) = H_0^1(\Omega)$ and the definitions (2.1) and (4.2) of $\mathcal{E}_w(t)$ and $E_v(t)$:

$$\int_0^T \mathcal{E}_w(t)dt \le C\int_0^T E_v(t)dt + C_T\|w\|^2_{H^{-1}(Q_T)} \qquad (4.33)$$

with constant C in (4.34) independent of T.

(v) We now invoke inequality (4.1) of Theorem 4.1 on the right of (4.33) and obtain

$$\int_0^T \mathcal{E}_w(t)dt \le C\int_0^T\int_{\Gamma_1} w^2 d\Sigma_1 + C\{\mathcal{E}_w(T)+\mathcal{E}_w(0)\}$$

$$+ C_T\{\|w\|^2_{H^{-1}(Q_T)} + \|v\|^2_{C([0,T];L_2(\Omega))}\}, \quad (4.34)$$

with constant C in front of { } in (4.35) independent of T.

(vi) We now recall the dissipativity property (1.5) of $\mathcal{E}_w$ so that

$$T\ \mathcal{E}_w(T) \le \int_0^T \mathcal{E}_w(t)dt \quad \text{and} \quad \mathcal{E}_w(0) = \mathcal{E}_w(T)+2\int_0^T\int_{\Gamma_1} w^2 d\Sigma_1. \quad (4.35)$$

Then (4.35) used in (4.34) yields

$$(T-C)\mathcal{E}_w(T) \le C_T\int_0^T\int_{\Gamma_1} w^2 d\Sigma_1 + C_T\{\|w\|^2_{H^{-1}(Q_T)} + \|v\|^2_{C([0,T];L_2(\Omega))}\}, \quad (4.36)$$

where we have noted all along that the constant C in (4.37) does not depend on T.

(vii) From (4.36) with T > C, we see that in order to prove Theorem 4.2, it remains to perform an "absorption" of the lower order terms $\|w\|^2_{H^{-1}((Q_T)}$, $\|v\|^2_{C([0,T];L_2(\Omega))}$. This is provided by the following:

<u>Lemma 4.3</u>. Inequality (4.36) with T sufficiently large implies that: there is $C_T > 0$ such that

$$\|v\|^2_{C([0,T];L_2(\Omega))} + \|w\|^2_{H^{-1}((Q_T)} \le C_T \int_0^T \int_{\Gamma_1} w^2 d\Sigma_1, \tag{4.37}$$

so that (4.36) becomes (4.28) and Theorem 4.2 is proved.

Proof. (i) As usual, by contradiction, suppose there exists a sequence $\{w_n\}$ of solutions to (1.1)-(1.4) such that with $v_n = \Lambda\bar{w}_n$, we have

$$\|w_n\|_{H^{-1}(Q_T)} + \|v_n\|_{C([0,T];L_2(\Omega))} = 1; \tag{4.38}$$

$$\int_0^T \int_{\Gamma_1} w_n^2 d\Sigma_1 \to 0, \quad \text{as } n \to \infty. \tag{4.39}$$

Then by (4.36), we have $\mathcal{E}_{w_n}(T) \le$ const, and by (4.35), $\mathcal{E}_{w_n}(0) \le$ const, uniformly in n. Thus,

$$w_n(0) \to \text{some } \tilde{w}_0 \quad \text{weakly in } L_2(\Omega); \tag{4.40a}$$

$$\dot{w}_n(0) \to \text{some } \tilde{w}_1 \quad \text{weakly in } H^{-1}(\Omega). \tag{4.40b}$$

Denote by $\{\tilde{w},\tilde{w}_t\}$ the solution pair of problem (1.1)-(1.4) corresponding to initial data $\{\tilde{w}_0,\tilde{w}_1\}$ obtained in (4.40). Then

$$\{w_n,\dot{w}_n\} \to \{\tilde{w},\tilde{w}_t\} \text{ in } L^\infty(0,T;L_2(\Omega)\times H^{-1}(\Omega)) \text{ weak star,} \tag{4.41}$$

so that $\{w_n,\dot{w}_n\}$ is uniformly bounded in $L^\infty(0,T;L_2(\Omega)\times H^{-1}(\Omega))$. By Simon's compactness results [S.1, Corollary 4],

$$w_n \to \tilde{w} \text{ strongly in } H^{-1}(Q_T). \tag{4.42}$$

Moreover, if we define $\tilde{v} \equiv \Lambda\tilde{\bar{w}}$, we obtain from (4.42), (4.38), and (4.27),

$$\{v_n\} \text{ uniformly bounded in } L_2(Q_\infty); \tag{4.43a}$$

$$v_n = \Lambda\bar{w}_n \to \tilde{v} = \Lambda\tilde{\bar{w}} \text{ weakly in } L_2(Q_\infty). \tag{4.43b}$$

(ii) Next recall from Eq. (3.3a): $\ddot{v}_n = \Delta v_n + K\bar{w}_n$ in Q_T, which implies

$$\|v_n\|^2_{H^1(Q_T)} \leq C\left\{\|K\bar{w}_n\|^2_{L_2(Q_T)} + \|v_n|_\Sigma\|^2_{H^1(\Sigma_\infty)} + \left\|\frac{\partial v_n}{\partial\nu}\right\|^2_{L_2(\Sigma_\infty)}\right\}$$

$$\leq C\{\|w_n|_{\Sigma_1}\|^2_{L_2(\Sigma_{1T})} + \|v_n\|^2_{L_2(Q_\infty)}\} + C\{\mathcal{E}_{w_n}(T) + \mathcal{E}_{w_n}(0)\}, \tag{4.44}$$

where in the last step we have invoked both Lemma 3.1, Eq. (3.5) of Lemma 3.1 and Eq. (3.5) of Theorem 3.5. From (4.39), (4.43), and $\mathcal{E}_{w_n}(T)$, $\mathcal{E}_{w_n}(0)$ uniformly bounded (as already noted below (4.39), we have that $\|v_n\|_{H^1(Q_T)} \leq$ const, and thus by compactness and the $L_2(Q_\infty)$-limit $\tilde{v}$, we have that

$$v_n \to \tilde{v} \quad \text{strongly in } C([0,T];L_2(\Omega)). \tag{4.45}$$

(iii) The limits (4.42) and (4.45) imply via (4.38) that

$$\|\tilde{w}\|_{H^{-1}(Q_T)} + \|\tilde{v}\|_{C([0,T];L_2(\Omega))} = 1. \tag{4.46}$$

(iv) On the other hand, the limit solution $\tilde{w}$ satisfies problem (1.1)-(1.3), (1.4), as well as

$$\tilde{w}|_{\Sigma_{1T}} \equiv 0 \text{ by (4.39), and of course, } \tilde{w}|_{\Sigma_{0T}} \equiv 0. \tag{4.47}$$

By Eq. (1.4) applied to $\tilde{w}$, we then have also

$$\frac{\partial p}{\partial \nu}\Big|_{\Sigma_{1T}} = 0, \quad \text{where } p \equiv A^{-1}\tilde{w}_t. \tag{4.48}$$

The new variable p was introduced in [L-T.4, Eq. (3.20)]: As shown there it solves the problem [L-T.4, Eq. (3.23)],

$$\begin{cases} p_{tt} = \Delta p + \tilde{D}^* \frac{\partial}{\partial \nu} p_t & \text{in } Q_T; \quad (4.49a) \\ p|_{\Sigma_T} = 0 & \text{in } \Sigma_T; \quad (4.49b) \\ \frac{\partial p}{\partial \nu}\Big|_{\Sigma_{1T}} = 0 & \text{in } \Sigma_{1T}, \quad (4.49c) \end{cases}$$

where the last equation comes from (4.48). But by (4.48) we have $\frac{\partial p_t}{\partial \nu}\Big|_{\Sigma_{1T}} = 0$, and so the right hand side in (4.49a), $\tilde{D}^* \frac{\partial p_t}{\partial \nu} = 0$. Then, for T sufficiently large, a standard Holmgren's uniqueness theorem yields $p \equiv 0$ in Q_T; i.e., by (4.48), $\tilde{w}_t \equiv 0$ in Q_T, hence $\tilde{w}_t \equiv$ const in Q_T. But, by the boundary condition on the right of (4.47), we conclude $\tilde{w} \equiv 0$ in Q_T, and this contradicts (4.46). Lemma 4.3 is proved. ■

Then, finally, Theorem 4.2 is established. To completely establish Theorem 1.1, we need to prove Lemma 3.1 and Theorem 3.2. This will be done in the next sections.

5. Implementation of change of variable $w \to v$. Proof of Lemma 3.1

5.1. Preliminaries

Implementation of the change of variable $w \to v$ will proceed via pseudo-differential calculus. Thus, in this and in the next section, we work on a half-space. Let $x > 0$ be a scalar positive varible, t be a real variable, and let $y = [y_1, \ldots, y_{n-1}]$ be an $(n-1)$-dimensional vector with real components. In symbols: $x \in R^1_{x^+}$; $t \in R^1_t$; $y \in R^{n-1}_y$. Throughout Sections 5 and 6, we shall call $\Omega = R^1_{x^+} \times R^{n-1}_y$ the half-space we work on, and $\Gamma = R^{n-1}_y = \Omega|_{x=0}$ its boundary where $n = \dim \Omega \geq 2$ in the interesting cases: No confusion is likely to arise with the bounded domain $\Omega \subset R^n$ and its boundary Γ of Sections 1-4. In Ω, we consider the second order differential operator,

$$P(x,y;D_t,D_x,D_y) = -aD_t^2 + \sum_{i,j=1}^{n-1} a_{ij} D_{y_i} D_{y_j} + 2\sum_{j=1}^{n-1} a_{n_j} D_{y_j} D_x + D_x^2, \tag{5.1}$$

with space-dependent but time-independent coefficients $a = a(x,y)$, $a_{ij} \leq a_{ij}(x,y)$, $a_{n_j} = a_{n_j}(x,y)$, $(x,y) \in \Omega$, satisfying the symmetricity condition $a_{ij} = a_{ji}$. Here and throughout Sections 5 and 6, we use the notation $D_t = \frac{1}{\sqrt{-1}} \frac{\partial}{\partial t}$; $D_x = \frac{1}{\sqrt{-1}} \frac{\partial}{\partial x}$; $D_{y_j} = \frac{1}{\sqrt{-1}} \frac{\partial}{\partial y_j}$, etc. On the boundary Γ of Ω, we consider the first order operator on Γ:

$$B(y;D_x,D_y) = D_x + \sum_{j=1}^{n-1} b_j D_{y_j}, \tag{5.2}$$

with space-dependent, but time-independent coefficients $b_j = b_j(y)$, $y \in \Gamma$. The operators P and B arise from the original problem (1.1)-(1.3). Indeed, the following equation,

$$Pw = 0 \quad \text{in } \Omega\times(0,\infty), \tag{5.3}$$

corresponds to Eq. (1.1) via partition of unity. Thus the coefficients a, a_{ij}, a_{n_j}, b_j are smooth, and constant outside a compact set $\mathcal{K}_{xy}$ of Ω. Moreover, the boundary Γ is non-characteristic for P, and P is "regularly hyperbolic with respect ot t"; i.e., the characteristic polynomial of P,

$$p(x,y;\tau,\xi,\eta) = -a\tau^2 + \sum_{i,j=1}^{n-1} a_{ij}\eta_i\eta_j + 2\xi\sum_{j=1}^{n-1} a_{n_j}\eta_j + \xi^2 \tag{5.4a}$$

$$= -a\tau^2 + \left(\xi + \sum_{j=1}^{n-1} a_{n_j}\eta_j\right)^2 + \sum_{i,j=1}^{n-1} a_{ij}\eta_i\eta_j - \left(\sum_{j=1}^{n-1} a_{n_j}\eta_j\right)^2 \tag{5.4b}$$

has two real and distinct roots in τ, for $(x,y) \in \Omega$ and for (ξ,η) on the unit sphere $\xi^2+|\eta|^2 = 1$, where $|\eta|^2 = \sum_{j=1}^{n-1} \eta_j^2$. Moreover, we have that: min $a(x,y) > 0$ in Ω, and that the quadratic form in η,

$$d(x,y;\eta) = a^2(x,y)\left\{\sum_{i,j=1}^{n-1} a_{ij}(x,y)\eta_i\eta_j - \left(\sum_{j=1}^{n-1} a_{n_j}(x,y)\eta_j\right)^2\right\} \tag{5.5a}$$

$$\geq c|\eta|^2 \tag{5.5b}$$

independent of ξ and τ, is positive definite, uniformly in $(x,y) \in \Omega$, for some constant $c > 0$. The symbol

$$\tilde{\xi}(x,t;\xi,\eta) = \xi + \sum_{j=1}^{n-1} a_{n_j}(x,y)\eta_j \tag{5.6}$$

gives rise to the pseudo-differential operator $\tilde{D}_x$,

$$\tilde{D}_x \equiv D_x + \sum_{j=1}^{n-1} a_{n_j}(x,y) D_{y_j}. \tag{5.7}$$

Throughout, $\tau = \sigma - i\gamma$, $\gamma > 0$, $\sigma \in R^1$ is the 'Laplace-variable' corresponding to t: $D_t \to \tau$; and $\eta \in R^{n-1}$ is the 'Fourier-variable' corresponding to y: $D_y \to \eta$. Introduce the symbols

$$d_1(x,y;\sigma,\eta) = a(x,y)(\sigma^2 - \gamma^2) - \frac{1}{a^2(x,y)} d(x,y;\eta) \tag{5.8}$$

$$d_2(x,y;\sigma) = a(x,y)\sigma, \tag{5.9}$$

and let D_1 and D_2 denote the corresponding pseudo-differential operators. Thus, from (5.4b), (5.6), (5.8), (5.9),

$$p(x,y;\ \tau = \sigma - i\gamma, \xi, \eta) = \tilde{\xi}^2 - (d_1 - 2i\gamma d_2) \tag{5.10}$$

with corresponding operator,

$$P(x,y;D_t,D_x,D_y) = \tilde{D}_x^2 - (D_1 - 2i\gamma D_2). \tag{5.11}$$

The first order operator $\tilde{D}_x$ in (5.7), restricted on Γ, will coincide with the co-normal operator B in (5.2), i.e., $b_j(y) = a_{n_j}(0,y)$. All the symbols are constant in (x,y) outside a compact set $\mathcal{K}_{xy}$ of Ω. For our purposes below, it will suffice to take $\gamma = 0$.

5.2. Division of the $(x,y;\sigma,\eta)$-space and definition of the operator Λ

Even though we shall specialize to the case $\gamma = 0$, we give a description for $\gamma \geq 0$ at no extra effort.

<u>Division of $(x,y;\sigma,\eta)$-space</u>. As the point (x,y) varies and $\gamma \geq 0$ is fixed, the equation

$$d_1(x,y;\sigma,\eta) = 0, \quad \text{i.e., by (5.8): } \sigma^2 - \frac{d(x,y;\eta)}{a^3(x,y)} = \gamma^2 \tag{5.12}$$

describes a family of hyperboloids in $\gamma > 0$ (cones if $\gamma = 0$) in the space $R^1_\sigma \times R^{n-1}_\eta$ (which reduces to a fixed hyperboloid (cone) for (x,y) outside $\mathcal{K}_{xy}$), all passing through the points $\sigma = \pm\gamma$, $\eta = 0$. Henceforth, because of the symmetry in σ of d_1, we may restrict our analysis to the half-space $\sigma > 0$. Setting

$$\inf_{\substack{x,y\\|\eta|=1}} \frac{d(x,y;\eta)}{a^3(x,y)} \equiv m^2; \quad \sup_{\substack{x,y\\|\eta|=1}} \frac{d(x,y;\eta)}{a^3(x,y)} \equiv M^2, \tag{5.13}$$

we have $m > 0$ and $M < +\infty$. Then, from (5.12)-(5.13),

$$m^2|\eta|^2 \leq \sigma^2 = \gamma^2 + \frac{d(x,y;\eta)}{a^3(x,y)} \leq \gamma^2 + M^2|\eta|^2 \leq 2M^2|\eta|^2$$

for all η outside the η-sphere of radius γ/M centered at the origin. (5.14)

Thus, for $\sigma > 0$, all points of the family of hyperboloids for $\gamma > 0$ (cones for $\gamma = 0$) with $|\eta| \geq \gamma/M$, lie between two equilateral cones: $\sigma = m|\eta|$ and $\sigma = \sqrt{2M}\,|\eta|$, uniformly in (x,y). Because of the symmetry of d_1 also in η, we may restrict our attention only to the quarter space $R^{2n}(+) = \{(x,y;\sigma,\eta):$ $(x,y) \in \Omega = R^1_{x^+} \times R^{n-1}_y;\ \sigma,\ \eta_j > 0,\ j = 1,\dots,n-1\}$. In $R^{2n}(+)$, we define the following mutually disjoint regions (see Fig. 1 for $\gamma = 0$, the case of our interest):

$$\mathcal{R}_1 = \{(x,y;\sigma,\eta) \in R^{2n}(+): \tfrac{3}{4}\, m|\eta| < \sigma\}; \tag{5.15}$$

$$\mathcal{R}_{tr} = \{(x,y;\sigma,\eta) \in R^{2n}(+): \frac{m}{2}\, |\eta| \le \sigma \le \tfrac{3}{4}\, m|\eta|\}; \tag{5.16}$$

$$\mathcal{R}_2 = \{(x,y;\sigma,\eta) \in R^{2n}(+): \sigma < \frac{m}{2}\, |\eta|\}; \tag{5.17}$$

whose union is all of $R^{2n}(+)$. The subscript 'tr' stands for 'transition,' as the region $\mathcal{R}_{tr}$ is a region of transition in the definition of the symbols $f(\sigma,\eta)$ and λ below. We note that:

$$\mathcal{R}_{tr} \cup \mathcal{R}_2 \subset \text{'elliptic' cone} = \{(x,y;\sigma,\eta) \in R^{2n}(+): \sigma < m|\eta|\},$$

where $d_1(x,y;\sigma,\eta) < 0$. In fact, we have that there exists a constant $c > 0$ such that in

$$\mathcal{R}_{tr} \cup \mathcal{R}_2: \ -d_1(x,y;\sigma,\eta) \ge c(\sigma^2+|\eta|^2) \quad \text{uniformly in } (x,y). \tag{5.18}$$

<u>Definition of symbols $f(\sigma,\eta)$ and $\lambda(\sigma,\eta)$</u>. On the unit sphere $\sigma^2+|\eta|^2 = 1$ within $R^{2n}(+)$, we define a function $f(\sigma,\eta)$, which does not depend on (x,y), by setting

$$f(\sigma,\eta) = \begin{cases} \sigma \ \text{in } \mathcal{R}_1 & (5.19a) \\ C^\infty\text{-smooth changing from } \sigma \text{ to } |\eta| \text{ in } \mathcal{R}_{tr} & (5.19b) \\ |\eta| \ \text{in } \mathcal{R}_2 & (5.19c) \end{cases}$$

as to have a C^∞-function in (σ,η). Once $f(\sigma,\eta)$ is defined on the quarter-unit sphere, we complete its definition to all of $R^{2n}(+)$ by extending it by homogeneity of order 1 in (σ,η) so that in $R^{2n}(+)$ we obtain

$$f(\sigma,\eta) = \text{positive, } C^\infty\text{-function, constant in } (x,y), \text{ homogeneous of order 1 in } (\sigma,\eta). \tag{5.19d}$$

By symmetry in σ and η, the function f can in fact be defined on the entire space $R^1_{x^+} \times R^{n-1}_y \times R^1_\sigma \times R^{n-1}_\eta$. Then in $R^{2n}(+)$--which excludes the origin--we introduce the C^∞-symbol

$$\lambda(\sigma,\eta) = \frac{1}{f(\sigma,\eta)} \in S^{-1}(R^1_t \times R^{n-1}_y), \tag{5.20}$$

likewise constant in (x,y), which can be defined on the entire space save for the origin. The symbol λ is homogeneous of order -1 in (σ,η) as indicated by the standard notation in (5.20). We shall then indicate by $\Lambda \in OPS^{-1}(R^1_t \times R^{n-1}_y)$ (uniformly in $x \in R^1_{x^+}$) the pseudo-differential operator corresponding to the symbol λ.

Definition of new variable v. We return to the original w-equation (1.1) which yields Equation (5.3) in the half-space. We next truncate w at $t = T$ and extend it by zero outside $[0,T]$ by setting

$$\bar{w}(t,x,y) = \begin{cases} w(t,x,y) & \text{in } [0,T]; \\ 0 & \text{elsewhere in } (-\infty,\infty). \end{cases} \tag{5.21}$$

Finally, we introduce a new variable $v = v(t,x,y)$ by setting

$$v = \Lambda\bar{w}, \text{ i.e., } \hat{v}(\sigma,x,\eta) = \lambda(\sigma,\eta)\hat{\bar{w}}(\sigma,x,\eta), \tag{5.22}$$

where ^ indicates Fourier transform in $t \to i\sigma$ (same as Laplace transform in $t \to \gamma+i\sigma$, with $\gamma = 0$) and Fourier transform in $y \to i\eta$. If $\{w(T),w_t(T)\}$ is the solution pair determined by the initial condition $\{w_0, w_1\}$ of the w-problem, then we see that the variable $\bar{w}$ satisfies the equation

$$P\bar{w} = F \quad \text{in } \Omega\times R_t^1, \tag{5.23}$$

where $F = F_{0,T}$ is defined by

$$F = -w(T)\otimes\delta_T' + w_0\otimes\delta_0' + w_t(T)\otimes\delta_T - w_1\otimes\delta_0, \tag{5.24}$$

where δ is the Dirac measure at 0 and T and δ' its derivative. Moreover, since Λ and D_t commute, we obtain from (5.23) that the new variable v satisfies the equation

$$Pv = [P,\Lambda]\bar{w}+\Lambda F \quad \text{in } \Omega\times R_t^1, \tag{5.25}$$

where $K \equiv [P,\Lambda] = P\Lambda-\Lambda P$ is the commutator of P and Λ. Since $w|_{\Sigma_0} \equiv 0$ from (1.2), application of the tangential operator Λ on the boundary $x = 0$ yields likewise $v|_{\Sigma_0} \equiv 0$, i.e., (3.3b).

<u>Orientation</u>. The operator Λ defined in (5.22) above (in effect, we have defined a whole class of such operators) 'lifts' the regularity of the original variable w to the new variable v by one unit in the tangential variables t and y (and through the equation in x as well), to the H^1-level, where then multipliers techniques on the v-problem are effective, as in Section 4. While there are many choices of transformations $w \to v$ which accomplish this goal, the operator Λ (the class of operators Λ) introduced in (5.22) has most convenient features when it comes to estimating the action of the commutator K on the solution $\bar{w}$, i.e., the term $K\bar{w}$. In the

region $\mathcal{R}_1$, the action of the commutator K vanishes, since the symbol $\lambda(\sigma,\eta)$ in (5.20) of Λ depends only on σ which corresponds to time (see Definition (5.19a)), while P is time independent. Instead, in the complementary region $\mathcal{R}_{tr} \cup \mathcal{R}_2$ where the commutator K is active, the symbol p of the operator P is elliptic, and thus elliptic estimates for v (or $\bar{w}$) apply. In short: the action $K\bar{w}$ of the commutator is estimated only by elliptic estimates, while the new problem in the lifted variable v preserves, or course, the original hyperbolicity character, which is dealt with in Section 4 via multipliers.

5.3. Proof of Lemma 3.1

We shall use the notation $\Sigma_\infty = R_t^1 \times \Gamma$, $R_t^1 = (-\infty,\infty)$, and $\Sigma_T = (0,T)\times\Gamma$.

(i) Since $\Lambda \in OPS^{-1}(R_t^1 \times R_y^{n-1})$, we have Λ: $H^s(R_t^1\times R_y^{n-1}) \rightarrow H^{s+1}(R_t^1\times R_y^{n-1})$ for any real s. But for $\bar{w}$ solution of (5.23) we have $\bar{w}(t,x=0,y) \in L_2(R_t^1\times R_y^{n-1})$ by (5.21) and Theorem A(ii), so that $v(t,x=0,y) \in H^1(R_t^1\times R_y^{n-1})$ continuously, by (5.22) and the above statement with s = 0, i.e.,

$$\|v|_\Sigma\|_{H^1(\Sigma_\infty)} \leq C\|\bar{w}|_\Sigma\|_{L_2(\Sigma_\infty)} = C\|w|_\Sigma\|_{L_2(\Sigma_T)}. \tag{5.26}$$

(ii) To study $\frac{\partial v}{\partial \nu}\big|_\Gamma$, we consider the first order operator $\tilde{D}_x$ in (5.7) (where we recall that $b_j(y) = a_{n_j}(0,y)$, see below (5.11)) and obtain from (5.22),

$$\tilde{D}_x v = \tilde{D}_x(\Lambda\bar{w}) = \Lambda\tilde{D}_x\bar{w}+[\tilde{D}_x,\Lambda]\bar{w} \tag{5.27}$$

on the boundary Γ, where $x = 0$. As to the first term on the right of (5.27), we have $\tilde{D}_x\bar{w}|_{x=0} \in H^{-1}(\Sigma_\infty)$ by (2.9) and the pseudo-local property sing supp$\{\tilde{D}_x\bar{w}\} \subset$ sing supp$\{\bar{w}\}$. Hence, as in (i) above with $s = -1$, $\Lambda(\tilde{D}_x\bar{w})|_{x=0} \in L_2(\Sigma_\infty)$ continuously

$$\|\Lambda\tilde{D}_x\bar{w}|_\Sigma\|_{L_2(\Sigma)} \leq C\|\bar{w}|_\Sigma\|_{L_2(\Sigma_\infty)} = C\|w|_\Sigma\|_{L_2(\Sigma_T)}. \tag{5.28}$$

As to the second term on the right of (5.27), we shall prove that on $x = 0$:

$$[\tilde{D}_x, \Lambda] \in OPS^{-1}(R_t^1 \times R_y^{n-1}), \tag{5.29}$$

from which it then follows that $[\tilde{D}_x, \Lambda]: H^s(\Sigma_\infty) \to H^{s+1}(\Sigma_\infty)$, in particular for $s = -1$,

$$\|[\tilde{D}_x, \Lambda]\bar{w}|_\Sigma\|_{L_2(\Sigma_\infty)} \leq C\|\bar{w}|_\Sigma\|_{H^{-1}(\Sigma_\infty)} = C\|w|_\Sigma\|_{H^{-1}(\Sigma_T)} \quad . \tag{5.30}$$

Then, estimates (5.28) and (5.30) used in (5.27) yield

$$\|\tilde{D}_x v|_\Sigma\|_{L_2(\Sigma_\infty)} \leq C\|w|_\Sigma\|_{L_2(\Sigma_T)} \tag{5.31}$$

as desired, and Lemma 3.1 is proved. It remains to establish (5.29). To this end, we consider the asymptotic expansion of the symbol of $[\tilde{D}_x, \Lambda]$. We have [Ta.1, p. 46],

$$\text{symb}\{\tilde{D}_x\Lambda\} \sim \sum_{\alpha\geq 0} \frac{i^{|\alpha|}}{\alpha!} \left\{ D_\sigma^\alpha \begin{array}{c} \tilde{\xi} \\ \xi \\ \eta \end{array} \right\} \left\{ D_t^\alpha \begin{array}{c} \lambda \\ x \\ y \end{array} \right\} = \tilde{\xi}\lambda, \tag{5.32}$$

since λ in (5.20) is constant in (t,x,y), while

$$\text{symb}\{\Lambda\tilde{D}_x\} \sim \sum_{\alpha\geq 0} \frac{i^{|\alpha|}}{\alpha!} \begin{Bmatrix} D^\alpha_\sigma & \lambda \\ \xi & \\ \eta & \end{Bmatrix} \begin{Bmatrix} D^\alpha_t & \tilde{\xi} \\ x & \\ y & \end{Bmatrix} = \lambda\tilde{\xi}+i(D^1_\eta\lambda)(D^1_y\tilde{\xi})$$

$$+ \text{ higher derivatives,} \tag{5.33}$$

since $\tilde{\xi}$ does not depend on t, and λ does not depend on ξ. In (5.33), with an abuse of notation that will be used henceforth, D^1_η and D^1_y mean D^α_η and D^α_y for a multi-index α with $|\alpha| = 1$. Moreover, by (5.6) and, respectively, (5.19), (5.20), we have

$$D^1_y\tilde{\xi} = \sum_{j=1}^{n-1} (D^1_y a_{n_j}(x,y))\eta_j \in S^1(R^{n-1}_y); \; D^1_\eta\lambda \in S^{-2}(R^1_t\times R^{n-1}_y) \tag{5.34}$$

uniformly in $x \in R^1_{x^+}$ (Definition 3.2 in [L-T.6]). Hence, (5.33), (5.34) imply

$$\text{symb}\{[\tilde{D}_x,\Lambda]\}\begin{cases} = 0 & \text{in } \mathcal{R}_1 \quad (5.35a) \\ \sim i(D^1_\eta\lambda)(D^1_y\tilde{\xi})+\cdots & \text{in } \mathcal{R}_{tr}\cup\mathcal{R}_2 \quad (5.35b) \end{cases} \in S^{-1}(R^1_t\times R^{n-1}_y)$$

uniformly in $x \in R^1_{x^+}$: the vanishing of the symbol in $\mathcal{R}_1$ as in Eq. (5.25a) holds true, since λ in (5.19), (5.20) does not depend on η in $\mathbb{R}_1$. Thus, (5.35) proves *a fortiori* (5.29), as required. ■

6. Proof of Theorem 3.2

From (5.11) we have with $\gamma = 0$,

$$K\bar{w} = [P,\Lambda]\bar{w} = [\tilde{D}^2_x,\Lambda]\bar{w}-[D_1,\Lambda]\bar{w}. \tag{6.1}$$

In the present section we shall prove the following results. Throughout this section, we shall let $Q_\infty = R^1_t \times R^1_{x^+} \times R^{n-1}_y$, and $\Sigma_\infty = R^1_t \times R^{n-1}_y$ while $Q_T = (0,T)\times\Omega$ and $\Sigma_T = (0,T)\times\Gamma$.

Theorem 6.1. Recalling $\mathcal{E}_w(t)$ from (2.1), we have

$$\|[\tilde{D}^2_x,\Lambda]\bar{w}\|^2_{L_2(Q_\infty)} \le C\{\|w|_\Sigma\|^2_{L_2(\Sigma_T)}+\|v\|^2_{L_2(Q_\infty)}+\mathcal{E}_w(T)+\mathcal{E}_w(0)\}, \quad (6.2)$$

where the constant C does not depend on T. ■

Theorem 6.2. Recalling $\mathcal{E}_w(t)$ from (2.1), we have

$$\|[D_1,\Lambda]\bar{w}\|_{L_2(Q_\infty)} \le C\{\|w|_\Sigma\|^2_{L_2(\Sigma_T)}+\|v\|^2_{L_2(Q_\infty)}+\mathcal{E}_w(T)+\mathcal{E}_w(0)\}. \quad (6.3)$$

where C does not depend on T. ■

Assuming for the moment the validity of these two theorems, we readily obtain from (6.2) and (6.3) used in (6.1),

$$\|K\bar{w}\|^2_{L_2(Q_\infty)} \le C\{\|w|_\Sigma\|^2_{L_2(\Sigma_T)}+\|v\|^2_{L_2(Q_\infty)}+\mathcal{E}_w(T)+\mathcal{E}_w(0)\}, \quad (6.4)$$

from which Theorem 3.2, Eq. (3.4), then follows. Thus, it remains to prove Theorems 6.1 and 6.2.

6.1. Analysis of $[\tilde{D}^2_x,\Lambda]\bar{w}$: Proof of Theorem 6.1

It is based on the following two propositions.

Proposition 6.3. There exist pseudo-differential operators $\Pi^0 \in OPS^0(R^1_t \times R^{n-1}_y)$, $\Pi^{-1} \in OPS^{-1}(R^1_t \times R^{n-1}_y)$, and $L^0 \in OPS^0(R^1_t \times R^{n-1}_y)$ such that

$$[\tilde{D}^2_x, \Lambda]\bar{w} = \tilde{D}_x \Pi^0 v + \tilde{D}_x \Pi^{-1} + L^0 v. \tag{6.5}$$

In particular, Π^0 is explicitly constructed in the proof below and has symbol π^0 given by (6.12). ■

Since the last two terms on the right of (6.5) are lower order, we shall concentrate on the term $\tilde{D}_x \Pi^0 v$ and prove

Proposition 6.4. Recalling $\mathcal{E}_w(t)$ in (2.1) and $\tilde{D}_x$ in (5.7), we have

$$\|\tilde{D}_x \Pi^0 v\|^2_{L_2(Q_\infty)} \le c\|\Pi^0 v\|^2_{H^1(Q_\infty)}$$

$$\le C\{\|w|_\Sigma\|^2_{L_2(\Sigma_T)} + \|v\|^2_{L_2(Q_\infty)} + \mathcal{E}_w(T) + \mathcal{E}_w(0)\}, \tag{6.6}$$

where the generic constant C does not depend on T. ■

Then, estimate (6.6) yields inequality (6.2) by use of (6.5). It remains to prove Propositions 6.3 and 6.4.

6.1.1. Proof of Proposition 6.3

We analyze the symbol of $[\tilde{D}^2_x, \Lambda]$ by proceeding as in (5.32), (5.33). We obtain [Ta.1, p. 46],

$$\operatorname{symb}\{\tilde{D}_x^2\Lambda\} \sim \sum_{\alpha\geq 0} \frac{i^{|\alpha|}}{\alpha!} \begin{Bmatrix} D_\sigma^\alpha & \tilde{\xi}^2 \\ \xi & \\ \eta & \end{Bmatrix} \begin{Bmatrix} D_t^\alpha & \lambda \\ x & \\ y & \end{Bmatrix} = \tilde{\xi}^2\lambda \tag{6.7}$$

$$\operatorname{symb}\{\Lambda\tilde{D}_x^2\} \sim \sum_{\alpha\geq 0} \frac{i^{|\alpha|}}{\alpha!} \begin{Bmatrix} D_\sigma^\alpha & \lambda \\ \xi & \\ \eta & \end{Bmatrix} \begin{Bmatrix} D_t^\alpha & \tilde{\xi}^2 \\ x & \\ y & \end{Bmatrix} = \lambda\tilde{\xi}^2 + i(D_\eta\lambda)(D_y^1\tilde{\xi}^2)$$

$$- \tfrac{1}{2}(D_\eta^2\lambda)(D_y^2\tilde{\xi}^2) + \text{higher derivatives.} \tag{6.8}$$

Moreover, from (5.6) we have

$$D_y^1\tilde{\xi}^2 = 2\tilde{\xi}\left[\sum_{j=1}^{n-1} (D_y^1 a_{n_j}(x,y))\eta_j\right] \tag{6.9}$$

$$D_y^2\tilde{\xi}^2 = 2\tilde{\xi}\left[\sum_{j=1}^{n-1} (D_y^2 a_{n_j})\eta_j\right] + 2\left[\sum_{j=1}^{n-1} (D_y^2 a_{n_j})\eta_j\right]^2. \tag{6.10}$$

Thus, with $\hat{\bar{w}}(\sigma,x,\eta)$ and $\hat{v}(\sigma,x,\eta)$ as in (5.22), we obtain from (6.7)-(6.10),

$$\operatorname{symb}\{[D_x^2,\Lambda]\}\hat{\bar{w}} \sim i(D_\eta^1\lambda)(D_y^1\tilde{\xi}^2)\hat{\bar{w}} - \tfrac{1}{2}(D_\eta^2\lambda)(D_y^2\tilde{\xi}^2)\hat{\bar{w}} + \cdots$$

$$= \pi^0(x,y;\sigma,\eta)\tilde{\xi}\hat{v} + \pi^{-1}(x,y;\sigma,\eta)\tilde{\xi}\hat{v} + \phi^0(x,y;\sigma,\eta)\hat{v} + \cdots, \tag{6.11}$$

where $\cdots$ indicate higher order terms and where the symbols π^0, π^{-1}, and ϕ^0 are defined by (we use $\hat{v} = \lambda\hat{\bar{w}}$ by (5.22)),

$$\pi^0(x,y;\sigma,\eta) \begin{Bmatrix} = 0 & \text{in } \mathcal{R}_1 \quad (6.12a) \\ 2i(D_\eta^1\lambda)\lambda^{-1}[\sum_{j=1}^{n-1} (D_y^1 a_{n_j})\eta_j] & \text{in } \mathcal{R}_{tr}\cup\mathcal{R}_2 \quad (6.12b) \end{Bmatrix}$$

$$\in OPS^0(R_t^1\times R_y^{n-1}) = \text{homogeneous symbol of order 0 in } (\sigma,\eta);$$

$$\pi^{-1}(x,y;\sigma,\eta)\begin{cases} = 0 & \text{in } \mathcal{R}_1 \quad (6.13a)\\ \sim (D^2_\eta\lambda)\lambda^{-1}[\sum_{j=1}^{n-1}(D^2_y a_{n_j})\eta_j] & \text{in } \mathcal{R}_{tr}\cup\mathcal{R}_2 \quad (6.13b)\end{cases}$$

$\in S^{-1}(R^1_t\times R^{n-1}_y)$ = homogeneous symbol of order -1 in (σ,η);

$$\phi^0(x,y;\sigma,\eta)\begin{cases} = 0 & \text{in } \mathcal{R}_1 \quad (6.14a)\\ \sim (D^2_\eta\lambda))\lambda^{-1}[\sum_{j=1}^{n-1}(D^1_y a_{n_j})\eta_j]^2 & \text{in } \mathcal{R}_{tr}\cup\mathcal{R}_2 \quad (6.14b)\end{cases}$$

$\in S^0(R^1_t\times R^{n-1}_y)$ = homogeneous symbol of order 0 in (σ,η),

where the vanishing of the symbols in $\mathcal{R}_1$ is due to λ not being dependent on η in $\mathcal{R}_1$, see (5.19)-(5.20). Since $(D^1_\eta\lambda)$, $(D^2_\eta\lambda)$ and λ^{-1} are homogeneous symbols of order -2, -3, and 1, respectively, in (σ,η), we then have that π^0, π^{-1}, and ϕ^0 are homogeneous symbols of order 0, -1, and 0 in (σ,η), respectively. Let then Π^0 and Π^{-1} be the corresponding pseudo-differential operators arising from π^0 and π^{-1}. Since

$$\text{symb}\{\tilde{D}_x\Pi^0\} \sim \sum_{\alpha\geq 0}\frac{i^{|\alpha|}}{\alpha!}\begin{Bmatrix} D^\alpha_\sigma & \tilde{\xi}\\ & \xi\\ & \eta\end{Bmatrix}\begin{Bmatrix} D^\alpha_t & \pi^0\\ & x\\ & y\end{Bmatrix}$$

$$= \tilde{\xi}\pi^0 + (\text{zero order in } \sigma,\eta) + \cdots\ ; \tag{6.15}$$

$$\text{symb}\{\tilde{D}_x\Pi^{-1}\} \sim \tilde{\xi}\pi^{-1} + (\text{zero order in } \sigma,\eta) + \cdots, \tag{6.16}$$

we see then that (6.11), (6.15), (6.16) yield the desired expansion (6.4), where we combine all zero order terms in L^0. ∎

6.1.2. Proof of Proposition 6.4

(i) By (6.12), $\text{symb}\{\Pi^0\} = \pi^0$ vanishes in $\mathcal{R}_1$. On the other hand, the region $\mathcal{R}_{tr} \cup \mathcal{R}_2$ lies in the elliptic region for p: recalling $p = \tilde{\xi}^2 - d_1$ from (5.10) with $\gamma = 0$ with $\tilde{\xi}$, d_1 real symbols, and recalling the lower bound $-d_1 \geq c(\sigma^2 + |\eta|^2)$ in $\mathcal{R}_{tr} \cup \mathcal{R}_2 = \text{supp}\{\pi^0\}$ from (5.18), we obtain

$$|p| \geq |\tilde{\xi}^2 - d_1| \geq c_0(\tilde{\xi}^2 + \sigma^2 + |\eta|^2) \text{ in } \mathcal{R}_{tr} \cup \mathcal{R}_2,$$
$$\text{some } c_0 > 0, \tag{6.17}$$

so that p is elliptic of order 2 in all variables in $\mathcal{R}_{tr} \cup \mathcal{R}_2$. Thus the solution v of $Pv = K\bar{w} + \Lambda F$, see (5.25), satisfies elliptic estimates in particular in $\mathcal{R}_{tr} \cup \mathcal{R}_2 = \text{supp}\{\pi^0\}$, so that

$$\|\Pi^0 v\|^2_{H^1(Q_\infty)} \leq C\left\{\|K\bar{w}\|^2_{H^{-1}(Q_\infty)} + \|\Lambda F\|^2_{H^{-1}(Q_\infty)} + \|v|_\Sigma\|^2_{H^{1/2}(\Sigma_\infty)}\right\} \tag{6.18}$$

(ii) We return to the term $F = F_{0,T}$ in (5.24) and see that with $\{w_0, w_1\} \in L_2(\Omega) \times H^{-1}(\Omega)$, and hence by Theorem A $\{w(T), w_t(T)\} \in L_2(\Omega) \times H^{-1}(\Omega)$ as well, we have by (5.20),

$$\Lambda(w(T)\otimes\delta'_T), \Lambda(w_0\otimes\delta'_0) \in H^{-1/2-\varepsilon}(R^1_t; L_2(R^1_{x^+} \times R^{n-1}_y)) \tag{6.19a}$$

$$\Lambda(w_t(T)\otimes\delta_T), \Lambda(w_1\otimes\delta_0) \in H^{1/2-\varepsilon}(R^1_t; H^{-1}(R^1_{x^+} \times R^{n-1}_y)) \tag{6.19b}$$

(since $\delta' \in H^{-3/2-\varepsilon}(R^1_t)$, $\delta \in H^{-1/2-\varepsilon}(R^1_t)$, and Λ is of order -1 in t and y), as one verifies by symbol analysis. Hence, with reference to (5.24), we conclude *a fortiori* from (6.19) that

$$\Lambda F \in H^{-1}(Q_\infty) \tag{6.20}$$

continuously, i.e., recalling (2.1),

$$\|\Lambda F\|^2_{H^{-1}(Q_\infty)} \le C\left\{\|w(T)\|^2_{L_2(\Omega)}+\|w_t(T)\|^2_{H^{-1}(\Omega)}+\|w_0\|^2_{L_2(\Omega)}+\|w_1\|^2_{H^{-1}(\Omega)}\right\}$$

$$= C\{\mathcal{E}_w(T)+\mathcal{E}_w(0)\}. \tag{6.21}$$

(iii) Moreover, since by (5.22), $K\bar{w} = [P,\Lambda\}\Lambda^{-1}v$, we have

$$\|K\bar{w}\|_{H^{-1}(Q_\infty)} \le C\|v\|_{L_2(Q_\infty)}, \tag{6.22}$$

where $[P,\Lambda]\Lambda^{-1}$ is of class $OPS^1(R^1_t \times R^{n-1}_y \times R^1_{x^+})$.

(iv) Finally, we use (6.21), (6.22), and (5.26) in (6.18) and obtain (6.6) as desired. ∎

Thus, Theorem 6.1 is established. ∎

6.2. <u>Analysis of</u> $[D_1,\Lambda]\bar{w}$: <u>Proof of Theorem 6.2</u>

(i) From (5.8), (5.5), and from (5.20), we have $[D_1,\Lambda] \in OPS^0(R^1_t \times R^{n-1}_y)$. More precisely, the analysis of the symbols gives the counterpart of (5.32), (5.33), and (5.35):

$$\text{symb}\{D_1\Lambda\} \sim \sum_{\alpha \ge 0} \frac{i^{|\alpha|}}{\alpha!} \begin{Bmatrix} D^\alpha_\sigma & d_1 \\ \xi & \\ \eta & \end{Bmatrix} \begin{Bmatrix} D^\alpha_t & \lambda \\ x & \\ y & \end{Bmatrix} = d_1\lambda \tag{6.23}$$

$$\operatorname{symb}\{\Lambda D_1\} \sim \sum_{\alpha\geq 0} \frac{i^{|\alpha|}}{\alpha!}\left\{\begin{matrix} D_\sigma^\alpha & \lambda \\ \xi & \\ \eta & \end{matrix}\right\}\left\{\begin{matrix} D_t^\alpha & d_1 \\ x & \\ y & \end{matrix}\right\} = d_1\lambda + i(D_\eta^1\lambda)(D_y^1 d_1)+\cdots \tag{6.24}$$

$$\operatorname{symb}\{[D_1,\Lambda]\} \begin{cases} = 0 & \text{in } \mathcal{R}_1 \quad (6.25a) \\ \sim i(D_\eta^1\lambda)(D_y^1 d_1)+\cdots & \text{in } \mathcal{R}_{tr}\cup\mathcal{R}_2 \quad (6.25b) \end{cases} \in S^0(R_t^1\times R_y^{n-1})$$

(ii) We recall from (5.19), (5.20) that the symbol $f(\sigma,\eta)$ of Λ^{-1} is of order 1 in σ, η and independent of η in $\mathcal{R}_1$. This, along with (6.25), and the usual analysis of symbol, yields

$$[D_1,\Lambda]\Lambda^{-1} \in OPS^1(R_t^1\times R_y^{n-1}), \quad \text{uniformly in } x \in R^1_{x^+} \tag{6.26}$$

$$\operatorname{symb}\{[D_1,\Lambda]\Lambda^{-1}\} = \begin{cases} 0 & \text{in } \mathcal{R}_1 \quad (6.27a) \\ \begin{pmatrix}\text{homogeneous of} \\ \text{order 1 in } \sigma,\eta\end{pmatrix}+\cdots & \text{in } \mathcal{R}_{tr}\cup\mathcal{R}_2 \quad (6.27b) \end{cases}$$

so that [L-T.6, Lemma 3.1],

$$[D_1,\Lambda]\Lambda^{-1} \text{ continuous } H^1(Q_\infty) \to L_2(Q_\infty). \tag{6.28}$$

(iii) We write $[D_1,\Lambda]\bar{w} = [D_1,\Lambda]\Lambda^{-1}v$. By the vanishing of the $\operatorname{symb}\{[D_1,\Lambda]\Lambda^{-1}\}$ in $\mathcal{R}_1$, we can apply elliptic estimates (as in step (i) of Proposition 6.4) to the variable v which satisfies $Pv = K\bar{w}+\Lambda F$, see (5.25). Recalling (6.28), we have that v enjoys H^1 estimates in $\operatorname{supp}\{[D_1,\Lambda]\Lambda^{-1}\} = \mathcal{R}_{tr} \cup \mathcal{R}_2$, and hence,

$$\|[D_1,\Lambda]\bar{w}\|^2_{L_2(Q_\infty)} = \|[D_1,\Lambda]\Lambda^{-1}v\|^2_{L_2(Q_\infty)} \leq C\|[D_1,\Lambda]v\|^2_{H^1(Q_\infty)}$$

$$\leq C\left\{\|K\bar{w}\|_{H^{-1}(Q_\infty)} + \|\Lambda F\|_{H^{-1}(Q_\infty)} + \|v|_\Sigma\|^2_{H^{1/2}(\Sigma_\infty)}\right\} \tag{6.29}$$

as on the right hand side of (6.18). Recalling (6.21), (6.22), and (5.26) in (6.29), we finally obtain

$$\|[D_1,\Lambda]\overline{w}\|^2_{L_2(Q_\infty)} = C\{\|\overline{w}|_\Sigma\|^2_{L_2(Q_\infty)}+\|v\|^2_{L_2(Q_\infty)}+\mathcal{E}_w(T)+\mathcal{E}_w(0)\} \quad (6.30)$$

with C independent of T, i.e., (6.3) (by (5.21) Theorem 6.2 is proved. ■

References

[A-M.1] K. Anderson and R. Melrose, The propagation of singularities along gliding rays, Invent. Math. 41 (1977), 197-232.

[B-L-R.1] C. Bardos, G. Lebeau, and R. Rauch, Controle et stabilisation dans des problems hyperboliques, Appendix II in J. L. Lions [Lio.2].

[C.1] G. Chen, Energy decay estimates and exact boundary balue controllability for the wave equation in a bounded domain, J. Math. Pures et Appliques (9) 58 (1979), 249-274.

[F-L-T.1] F. Flandoli, I. Lasiecka, and R. Triggiani, Algebraic Riccati equations with non-smoothing observation arising in hyperbolic and Euler-Bernoulli equations, Annali di Matematica Pura e Applicata.

[G.1] P. Grisvard, Controlabilitè exacte des solutions de certains problemes mixtes pour l'equation des ondes dans un polygone et polyedre, J. Math. Pure et Appliq.

[H.1] F. L. Ho, Observabilitè frontiere de l'equation des ondes, C. R. Acad. Sci. Paris Ser. I Math. 302 (1986), 443-446.

[H.2] L. Hormander, The Analysis of Linear Partial Differential Operators, vol. I, III, Springer Verlag, Berlin-Heidelberg, 1983, 1985.

[Lag.1] J. Lagnese, Decay of solutions of wave equations in a bounded region with boundary dissipation, J. Diff. Eqns. 50 (1983), 163-182.

[Lag.2] J. Lagnese, A note on the boundary stabilization of wave equations, SIAM J. Control and Optimiz. 26 (1988), 1250-1256.

[Lio.1] J. L. Lions, Control of singular distributed systems, Gauthier-Villars, Paris 1983.

[Lio.2] J. L. Lions, Exact controllability, stabilization, and perturbations, SIAM Review 30 (1988), 1-68. Extended version collection RMA, vol. 8, Masson, Paris, 1988.

[Lit.1] W. Littman, Near optimal time boundary controllability for a class of hyperbolic equations, pp. 307-312, Springer-Verlag Lecture Notes LNCIS #97, 1987.

[L-L-T.1] I. Lasiecka, J. L. Lions, and R. Triggiani, Non-homogeneous boundary value problems for second order hyperbolic operators, J. de Mathematiques Pures et Appliques 65 (1986), 149-192.

[L-T.1] I. Lasiecka and R. Triggiani, A cosine operator approach to modeling $L_2(0,T;L_2(\Gamma)$-boundary input hyperbolic equations, Applied Math. and Optimiz. 7 (1981), 35-83.

[L-T.2] I. Lasiecka and R. Triggiani, Regularity of hyperbolic equations under $L_2(0,T;L_2(\Gamma)$-boundary terms, Applied Math. and Optimiz. 10 (1983), 275-286.

[L-T.3] I. Lasiecka and R. Triggiani, Riccati equations for hyperbolic partial differential equations with $L_2(0,T;L_2(\Gamma))$-Dirichlet boundary terms, SIAM J. Control and Optimiz. 24 (1986), 884-926.

[L-T.4] I. Lasiecka and R. Triggiani, Uniform experimental energy decay of the wave equation in a bounded region with $L_2(0,\infty;L_2(\Gamma))$-feedback control in the Dirichlet boundary conditions, J. Diff. Eqns. 66 (1987), 340-390.

[L-T.5] I. Lasiecka and R. Triggiani, Exact controllability for the wave equation with Neumann boundary control, Applied Math. and Optimiz. 19 (1989), 243-290. Preliminary version in Springer Verlag Lecture Notes LNICS, n100 (J. P. Zolesio, ed.), pp. 317-371.

[L-T.6] I. Lasiecka and R. Triggiani, Sharp regularity results for mixed second order hyperbolic equations of Neumann type: the L_2-boundary case, Annali di Matematica Pura e Applicata; to appear.

[M.1] C. S. Morawitz,, Energy identities of the wave equation NYU Courant Institute Mat. Sciences, Res. Rep. No. IMM 346, 1976.

[M-S.1] R. Melrose and J. Sjostrand, Singularities of boundary value problems I, II. Comm. Pure Appl. Math. 31 (1978), 593-617, 35 (1982), 129-168.

[R.1] D. L. Russell, Exact boundary controllability theorems for wave and heat processes in star complemented regionds, in "Differential Games and Control Theory," Roxin-Lin-Sternberg editors, Marcell Dekker, New York, 1974, pp. 291-320.

[S.1] J. Simon, Compact sets in the space $L^p(0,T;B)$, Annali di Matematica Pura e Appl. (iv), vol. CXLVI, pp. 65-96.

[S.2] W. Strauss, Dispersal of waves vanishing on the boundary of an exterior domain, Comm. Pure Appl. Math. 28 (1976), 265-278.

[T.1] M. Taylor, Pseudodifferential operators, Princeton University Press, 1981.

[T.2] M. Taylor, Reflection of singularities of solutions to systems of differential equations, Comm. Pure Appl. Math. 28 (1975), 457-478.

[Tr.1] R. Triggiani, Exact boundary controllability on $L_2(\Omega)\times H^{-1}(\Omega)$ of the wave equation with Dirichlet boundary control acting on a portion of the boundary and related problems, Appl. Math. and Optimiz. 18 (1988), 241-277. Also, Lecture Notes LNICS, Springer Verlag, 102 (1987), 291-332; Proceedings, Workshop on Control for Distributed Parameter Systems, University of Graz, Austria (July 1986).

[Tr.2] R. Triggiani, Wave equation on a bounded domain with boundary dissipation: An operator approach, J. Mathem. Anal. and Appl. 137 (1989), 438-461. Also in Operator methods for optimal control problems (S. J. Lee, ed.), Lecture Notes in Pure and Applied Mathematics, vol. 108, 1988, 283-309.

ACTUATORS AND CONTROLLABILITY

OF DISTRIBUTED SYSTEMS

A. EL JAI

IMP - Automatique
University of Perpignan
50, Av. de Villeneuve
F-66000 PERPIGNAN

Abstract

The purpose of this paper is to show the link which exists between abstract concepts of the systems analysis and concrete notions of input output dynamical systems that is to say actuators and sensors. More precisely we are concerned by controllability and we recall abstract definitions of controllability and make a parallel with concrete definitions via actuators. The considered system is parabolic in the case of weak controllability and hyperbolic for exact controllability.

I INTRODUCTION

The systems approach for Automaticians and Engineers is always considered by use of the three usual following steps :

- Step 1 : Modelling
- Step 2 : Identification
- Step 3 : Control

Each of these steps consists on usual work. First one looks for the model, and then tries to achieve it by good choice of certain coefficients or functions. Finally for given criterion we try to find the optimal control law to be applied.

Nowadays with the help of computers and the huge amount of work on systems theory one can go further. Before identification, it is possible to add a step on system analysis. The purpose of this step is to have a better understanding of the system. It consists of studying some notions which have an obvious link with the considered problem. Moreover after the identification step, one can use some fundamental tools for the best choice of the geometrical design of the system : it is the structure optimization step.

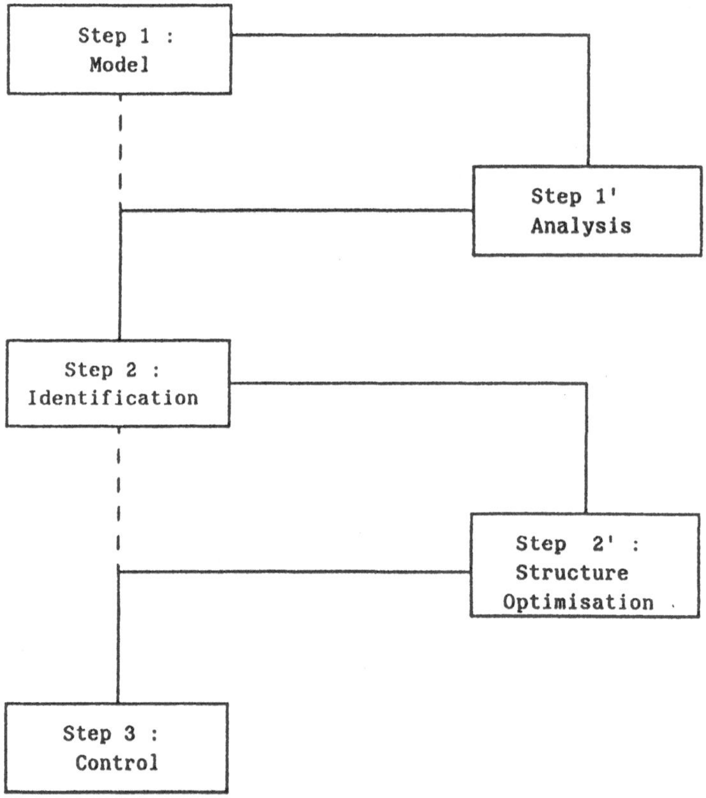

The aim of the analysis step 2' is to obtain best understanding of the system. For this purpose we have to study certain among the concepts of controllability, observability, stability and stabilizability, compensators, detectability, observers, identifiability, ...

Each of these concepts is well defined and many results adapted to numerous situations are known. Of course it is the case of linear dynamical systems which is the best developed and which is now in the standard knowledge of systems analysis.

In the case where the considered system is modelized with a spatial variable and has to be conceived, then the question is : taking into account some technological constraints, is it possible to conceive this system in such a way that the objective becomes easier or cheaper to reach ? This is the purpose of the step 2' and will not be developed in this paper.

In this paper we will be concerned by the step 1'. For any of the listed notions, it is possible to do a theoretical study like for abstract systems or a more concrete work via the input output parameters, that is to say actuators and sensors. So you can link :

- Controllability .... with .... actuators
- Observability .... with .... sensors
- Stabilizability .... with .... actuators
- Compensators .... with .... sensors and actuators
- Detectability .... with .... sensors
- Observers .... with .... sensors and actuators
- Identifiability, ... with .... sensors

etc ......

The controllability concept leads to two major problems which are very often ignored by researchers. First the choice of concrete equivalences making a concept more reachable to Engineers. We shall develop the controllability concept in a concrete way taking into account the geometrical meaning of input parameters, i.e. actuators. Other examples are developed by A. EL JAI -A.J.PRITCHARD in [10].

Secondly one must not ignore numerical and computational aspects making an analysis concept more acceptable and useful. In a last paragraph we shall study the exact controllability problem for hyperbolic systems and we shall show how to compute the control which steers the system to rest.

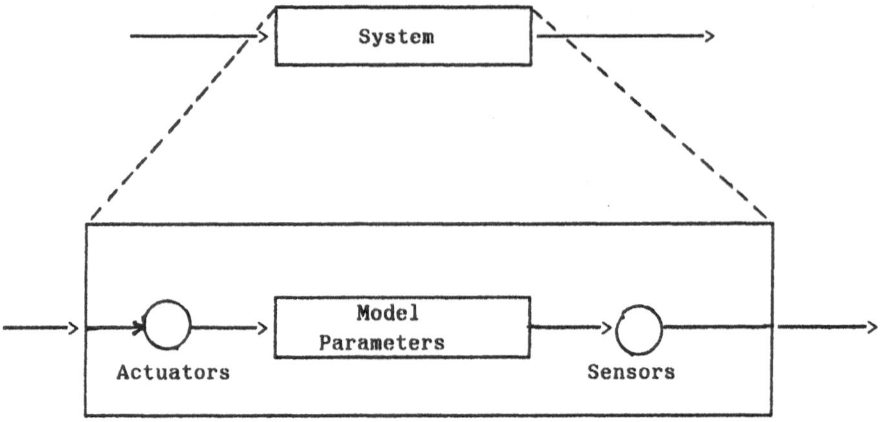

II WEAK CONTROLLABILITY OF PARABOLIC SYSTEMS

Ω is a bounded regular open set of R^n with boundary Γ and $T > 0$. Let us consider the parabolic system described by the following equation :

$$(1) \quad \begin{cases} \frac{\partial y}{\partial t} - \Delta y = g(x)\, u(t) & \Omega \times]0,T[\\ y(x,0) = y_0 & \Omega \\ y = 0 & \Gamma \times]0,T[\end{cases}$$

We suppose :

- $g \in L^2(\Omega_0)$ where $\Omega_0 \subset \Omega$,
- $u \in U = L^2(0,T)$ is the control space,
- $X = L^2(\Omega)$ is the state space.

1. Abstract definition of weak controllability

The system (1) is weakly controllable if :

$\forall\, y_d \in X,\ \forall\, \varepsilon > 0,\ \exists\, u \in L^2(0,T)$ such that

$$(2) \qquad \|y(.,T) - y_d\|_X \leq \varepsilon$$

where $y(.,T)$ is the reached state at time T of the system (1) excited by u.

The choice of the state space X and the control space U is done à priori and may be changed. This occurs when the reached state is not in X. Many examples are given in [1,10,14]. This situation occurs in the case of pointwise or boundary control.

Many equivalent definitions and characterizations properties are given in [4].

2. Concrete definition of weak controllability

For this purpose we recall some basic concepts related to systems input.

* Actuator concept

1. An actuator is the couple (Ω_0, g_0) where
   - $\Omega_0 \subset \Omega$ is the actuator support,
   - $g_0 \in L^2(\Omega_0)$ defines the spatial distribution of the actuator.

The above definition concerns a zone type actuator. This actuator can be located in the domain Ω or the boundary Γ.

For n = 1, one have the following :

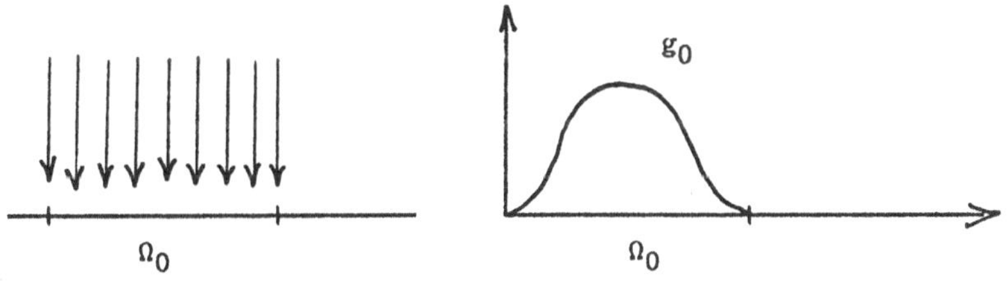

In the case of pointwise actuator Ω_0 becomes $b \in \Omega$ and g_o is the Dirac mass δ_b concentrated in b.

2. An actuator (Ω_0, g_0) is strategic if the excited system is weakly controllable (in the sense of the above definition).

So given state space X and control space U, with these definitions we can classify actuators in strategic ones and non strategic ones.

The strategic actuators are those which can steer the system to desired states. The questions now are : do strategic actuators exist ? and in this case how can they be characterized ?
The main existence result is shown for the chosen state and control spaces X and U.

3. Proposition 1 : existence result

For given $\Omega_0 \subset \Omega$ (resp. $g_0 \in L^2(\Omega)$) / $\exists$ g_0 (resp. $\exists$ $\Omega_0 \subset \Omega$) such that :

- $\operatorname{supp}(g_0) \subset \Omega_0$ (resp. $g_0 \in L^2(\Omega_0)$)
- (Ω_0, g_0) is a strategic actuator

For the proof of these results, see [2,6,10].
The characterization is specific of the considered problem. It depends on the geometry of Ω, the boundary conditions, ...
Suppose that the system (1) is excited by p actuators $(\Omega_i, g_i)_{1 \le i \le p}$. Let φ_{nj} the eigenfunctions of Δ with the Dirichlet boundary conditions and λ_n the associated eigenvalues with r_n the multiplicity of λ_n.

4. Proposition 2 : Characterization of strategic actuators

The actuators $(\Omega_i, g_i)_{1 \le i \le p}$ are strategic if and only if :

- $p \ge \sup(r_n)$
- $\operatorname{rank}(G_n) = r_n$, $\forall$ n where

(3) $$(G_n)_{ij} = \langle g_i, \varphi_{nj} \rangle_{L^2(\Omega_i)}$$

It is obvious that this characterization leads to concrete choice of the actuators structures so that the system is weakly controllable.
This characterization is shown for many types of actuators.

- For zone actuator see [3,17].
- For pointwise and boundary actuators see [2,6]
- For hyperbolic case see [2,5,10].

Many concrete choices of the actuators structures for different geometric situations are developed in [2,10].

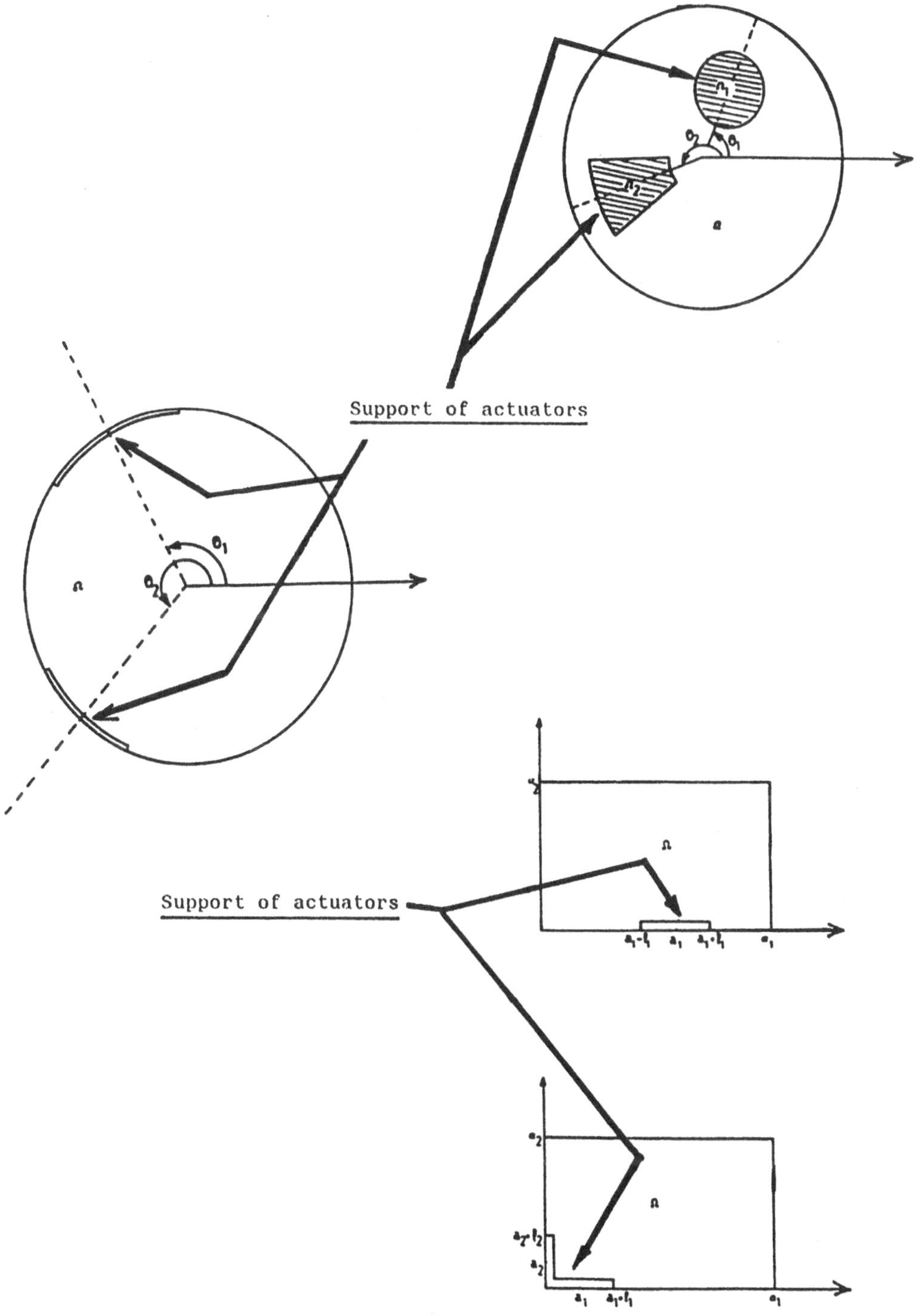

Remark 1 : Choice of the state space X

Suppose that the actuator (Ω_0, g_0) steers the system to final states belonging to a vector space $Y \supset X$. Then one have to change the state space $(X \equiv Y)$ if possible or to make the control more regular. This situation is frequent in the case of pointwise or boundary actuators. Let us consider the following example :

$$(4) \quad \begin{cases} \dfrac{\partial y}{\partial t} - \Delta y = & \text{in } \Omega \times]0,T[\\ y(x,0) = 0 & \text{in } \Omega \\ y(\xi,t) = \sum\limits_{i=1}^{i=p} g_i(\xi) u_i(t) & \text{in } \Gamma \times]0,T[\end{cases}$$

This system is excited by p boundary zone actuators $(\Gamma_i, g_i)_{1 \le i \le p}$, with $\Gamma_i \subset \Gamma$. If the controls are L^2, the reached state at time T is not in $L^2(\Omega) = X$ [14]. To ensure this result the controls must be more regular [6]. We have the following result :

Proposition 3

If $u_i \in L^r(0,T)$, $\forall$ i, $1 \le i \le p$, with $r > 4$ then the state space X can be chosen as $X = L^2(\Omega)$.

For the proof see [2,6].
The characterization result becomes in this case :

The actuators $(\Gamma_i, g_i)_{1 \le i \le p}$ are strategic if and only if :
- $p \ge \sup r_n$
- $\text{rank}(G_n) = r_n$, $\forall$ n , where

$$(5) \qquad (G_n)_{ij} = \langle g_i, \frac{\partial \varphi_{nj}}{\partial \nu} \rangle_{L^2(\Gamma_i)} \qquad 1 \le i \le p \text{ and } 1 \le j \le r_n$$

In the Neumann case, the result is the same with :

(6) $(G_n)_{ij} = \langle g_i, \varphi_{nj} \rangle_{L^2(\Gamma_i)}$ $1 \le i \le p$ and $1 \le j \le r_n$

Remark 2 :

Through the previous result the number of actuators plays a fundamental role (p must be $\ge \sup r_n$). This can be a surprising result for an Engineer who can in fact excite the system with any number of actuators. This difficulty can be rejected because, in any case, it is possible to have $r_n = 1$, $\forall$ n by a good adjustment of the boundary of Ω [15], and then $p = 1$ can be considered.

Remark 3

All the previous results don't lead to any formula for the control which ensure the weak controllability of the system. Usually these controls are obtained by pseudo-inverse techniques or by solving a particular optimal control problem [4,10].

III EXACT CONTROLLABILITY FOR HYPERBOLIC SYSTEMS

In this paragraph we are going to show how to link exact controllability problem of hyperbolic systems to concrete choice of input parameters. Moreover we'll give explicit formulas of the control steering the system to desired states.

Let us consider the system described by the hyperbolic equation :

$$(7)\qquad \begin{cases} \dfrac{\partial^2 y}{\partial t^2} - \Delta y = \sum_{i=1}^{i=p} \delta(x-b_i)\, u_i(t) & \Omega x]0,T[\\ y(x,0) = y^0(x) \quad , \quad \dfrac{\partial y}{\partial t}(x,0) = y^1(x) & \Omega \\ y = 0 & \Gamma x]0,T[\end{cases}$$

with the hypothesis :

- p pointwise actuators located in $b_i \in \Omega$ excite the system,
- the state space is $X = H^1{}_0(\Omega) x L^2(\Omega)$
- $u_i \in L^2(0,T)$, $\forall$ i, $1 \le i \le p$
- y^0 and y^1 are in some convenient spaces
- T > 0 is big enouph

The weak controllability of the system (7) can be defined as for parabolic systems and many results and characterizations of strategic actuators for such systems are developed in [2,10].

Let (w_j) defined by :

$$(8)\qquad \begin{cases} \Delta w_j = \lambda_j w_j & \Omega \\ w_j = 0 & \Gamma \\ \|w_j\|^2 = 1 \end{cases}$$

and suppose the eigenvalues are such that :

$$\ldots < \lambda_n < \ldots < \lambda_2 < \lambda_1 < 0$$

Considering the characterization given in the previous paragraph, the pointwise actuator (b_i, δ_{bi}) is strategic if and only if :

$$(9)\qquad w_j(b_i) \neq 0 \qquad \forall\, j$$

Let us consider now the standard definition of exact controllability.

1. Definition of exact controllability

The system (7) is exactly controllable at time T, if :

$\forall\, z_d \in X,\ \exists\, u = (u_1, u_2, \ldots, u_p)$ with $u_i \in L^2(0,T)$ such that

(10) $$z_d = z(.,T)$$

where

$z(.,T) = (y(.,T), \partial y(.,T)/\partial t)$ and y is the solution of the system (7) excited by u.

This well known definition does not lead to the expression of the control which steers the system to desired states. For this purpose we consider Lions Hilbert Uniqueness Method (HUM) [14].

Suppose the desired state is $z_d = (0,0)$ and consider the autonomous system in φ :

(11) $$\begin{cases} \dfrac{\partial^2\varphi}{\partial t^2} - \Delta\varphi = 0 & \Omega\times]0,T[\\ \varphi(x,0) = \varphi^0(x)\ ,\ \dfrac{\partial\varphi}{\partial t}(x,0) = \varphi^1(x) & \Omega \\ \varphi = 0 & \Gamma\times]0,T[\end{cases}$$

Then we have the result :

2. Proposition 4

For suitable φ^0 and φ^1 and $T > 0$ big enough, the control $u = (u_1, u_2, \ldots, u_p)$ with

(12) $$u_i(t) = -\varphi(b_i,t),\ 1 \le i \le p,$$

steers the system (2) to rest at time T.

The method is based on a convenient choice of the state space X so that the operator Λ defined below is an isomorphism from the closure F of $\mathcal{D}(\Omega)\times\mathcal{D}(\Omega)$ into F'.

(13)

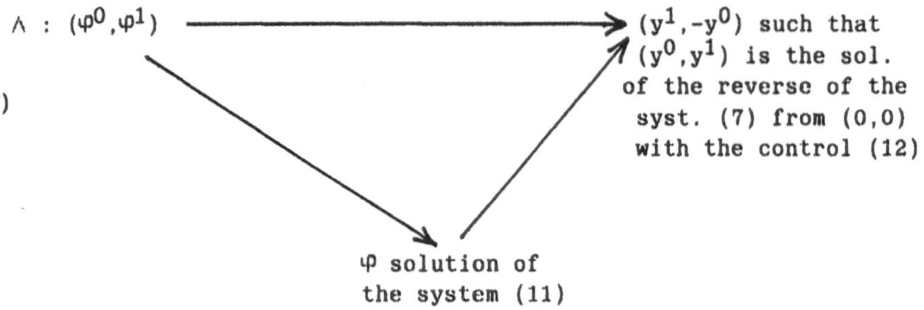

So one must first solve the equation :

$$\Lambda\ (\varphi^0, \varphi^1) = (y^1, -y^0) \tag{14}$$

and then applies the associated control.

Many numerical works has been done for the resolution of (14) and numerous examples can be found in [8,9,11]. It was noticed that it is important to use the asymptotic behaviour of the control. Using this idea it is then possible to give explicit formulas for the solutions φ^0 and φ^1 of (14). This is widely developed in [8] and the most important result is the following one.

3. <u>Proposition 5</u>

Suppose $T > 0$ big enough and at least one of the pointwise actuators $(b_i, \delta_{b_i})_{1 \le i \le p}$ is strategic then the solution of the equation (14) is given by :

$$\varphi^0 = \frac{1}{T} \sum_{j=1}^{\infty} \frac{2}{\sum_{i=1}^{i=p} w_j(b_i)^2} (y^1, w_j)\ w_j \tag{15}$$

and

$$\varphi^1 = \frac{1}{T} \sum_{j=1}^{\infty} \frac{2\lambda_j}{\sum_{i=1}^{i=p} w_j(b_i)^2} (y^0, w_j)\ w_j \tag{16}$$

For the proof see [8].

One can achieve computations by implementing the following approximated formulas with fixed M :

$$(17) \qquad \varphi^0 \approx \frac{1}{T} \sum_{j=1}^{M} \frac{2}{\sum_{i=1}^{i=p} w_j(b_i)^2} (y^1, w_j)\, w_j$$

and

$$(18) \qquad \varphi^1 \approx \frac{1}{T} \sum_{j=1}^{M} \frac{2\lambda_j}{\sum_{i=1}^{i=p} w_j(b_i)^2} (y^0, w_j)\, w_j$$

These formulas are well defined because we supposed at least one of the actuators to be strategic.

Other situations have been developed [8]. For example, in the case of boundary pointwise actuator located in $b \in \Gamma$, the formulas (17) and (18) become :

$$(19) \qquad \varphi^0 \approx \frac{1}{T} \sum_{j=1}^{M} \frac{2}{\frac{\partial}{\partial \nu} w_j(b)^2} (y^1, w_j)\, w_j$$

and

$$(20) \qquad \varphi^1 \approx \frac{1}{T} \sum_{j=1}^{M} \frac{2\lambda_j}{\frac{\partial}{\partial \nu} w_j(b)^2} (y^0, w_j)\, w_j$$

<u>Remark 4</u>

The obtained formulas are satisfactory on the computational level. But for the hyperbolic case the difficulty with this approach comes from the non feedback form for the control (12).

<u>Remark 5</u>

This approach leads to explicit controls steering the system to desired states and was extended for parabolic case and used even for some optimal control problems [7].

Numerical examples

Let us consider the one-dimensional case with $\Omega =]0,1[$

We choose the following numerical values :

- the control time $T = 2$
- The approximated formulas (17)-(18) are implemented with $M = 5$
- the initial conditions of the system are taken to be :

(21) $$y^0(x) = A\, x^2(1 - x^2)$$

(22) $$y^1(x) = (1 + B)\, y^0(x)$$

where A and B are given constants,

The EDP are solved by a reverse Newmark method

When we apply the control given by (12), the reached state is computed and one compare with the desired state by considering the final error :

(23) $$\|\varepsilon\|^2 = \|y(.,T)\|^2 + \|\frac{\partial y}{\partial t}(.,T)\|^2$$

Finally we make the actuator position moving so that one can appreciate how the choice of the actuator location is important.

We give one example with 1 pointwise actuator moving in Ω and an other example with 2 pointwise actuators in Ω.

Case of one actuator

(b is the varying location of the actuator)

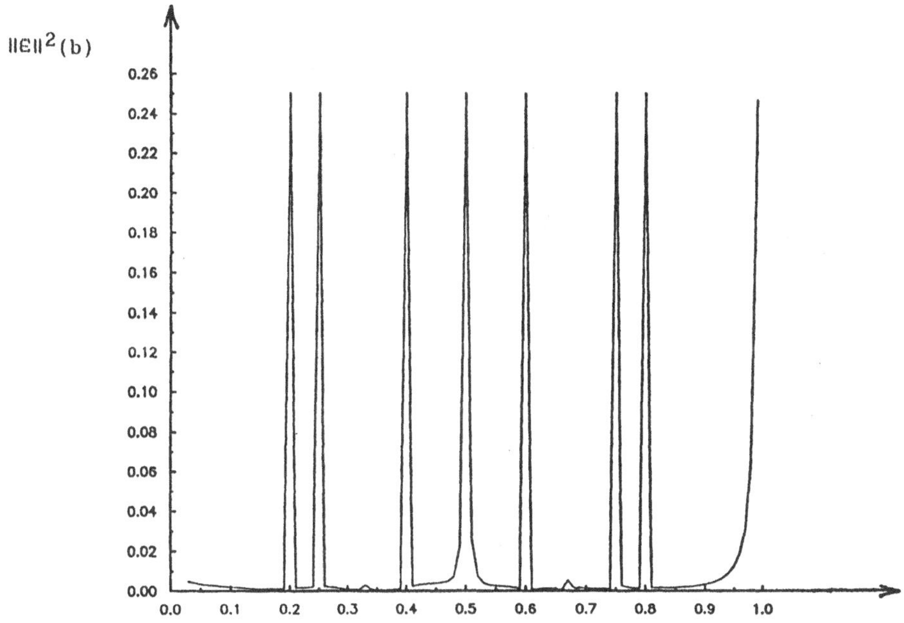

Case of two actuators

(The location of the first actuator is fixed at $b_1 = 0.24$ and the location b_2 of the second actuator is varying in Ω)

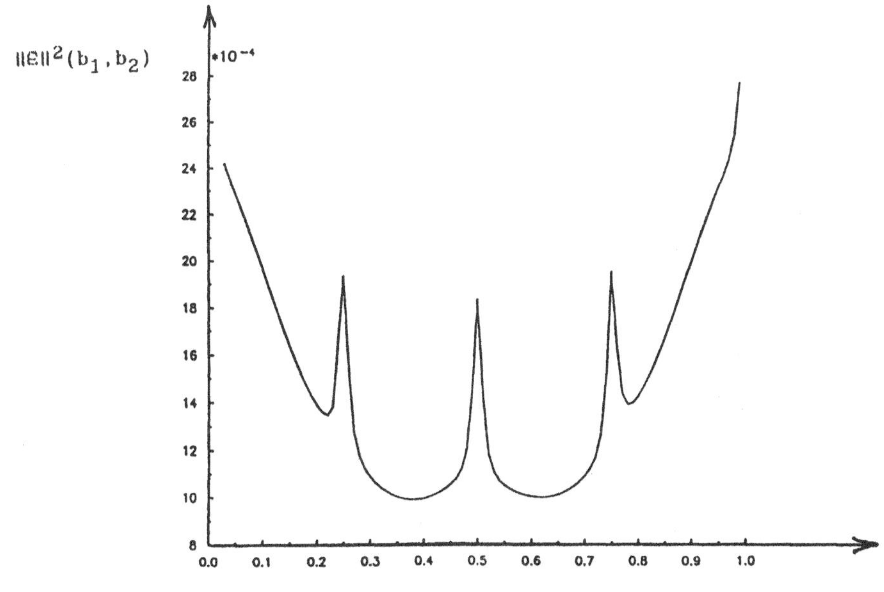

REFERENCES

[1] M. AMOUROUX - A. EL JAI
Des systèmes localisés aux systèmes distribués.
Hermès. A paraître.

[2] L. BERRAHMOUNE
Localisation d'actionneurs pour la contrôlabilité de systèmes paraboliques et hyperboliques. Application par dualité à la localisation de capteurs. Thèse. Avril 1984. Faculté des Sciences. Rabat.

[3] R.F. CURTAIN - A.J. PRITCHARD
Infinite dimensional linear systems theory. Lecture notes in Control and Information Science. Vol. 8. Springer. 1978.

[4] A. EL JAI
Etude d'algorithmes pour la commande de systèmes à paramètres répartis paraboliques. Thèse de Doctorat d'Etat. 1978. Université aul Sabatier. Toulouse.

[5] A. EL JAI
Controllability and actuators for hyperbolic systems.
Fourth IFAC Symposium on Control of Distributed Parameter Systems. Los Angeles. Juin 1986.

[6] A. EL JAI - L. BERRAHMOUNE
- Localisation d'actionneurs zones pour la contrôlabilité de systèmes paraboliques. C.R.Ac.Sc. Paris. T. 297. 1983.
- Localisation d'actionneurs ponctuels pour la contrôlabilité de systèmes paraboliques. C.R.Ac.Sc. Paris. T. 298. N°3. 1984.
- Localisation d'actionneurs frontieres pour la contrôlabilité de systèmes paraboliques. C.R.Ac.Sc. Paris. T. 298. N°8. 1984.

[7] A. EL JAI - A. BELFEKIH
Exacte contrôlabilité et contrôle optimal de systèmes paraboliques. APII. Control/Systems Analysis. 1989. A paraître.

[8] A. EL JAI - J. BOUYAGHROUMNI
Numerical approach for exact pointwise controllability of hyperbolic systems.
Fifth IFAC Symposium on Control of Distributed Parameter Systems. Perpignan, June 26-29, 1989.

[9] A. EL JAI - A. GONZALEZ
Actionneurs frontières pour l'exacte Contrôlabilité des systèmes hyperboliques. Ist International Workshop on sensors and actuators in Distributed systems. Perpignan. 16-18 décembre 1987.

[10] A. EL JAI - A.J. PRITCHARD
Capteurs et actionneurs dans l'analyse des systèmes distribués. Masson. Série RMA. 1986.
Sensors and controls in the analysis of distributed systems. J. Wiley.1988.

[11] R.GLOWINSKI - C. H. LI - J.L. LIONS
A numerical approach to the exact boundary controllability of the wave equation. Dirichlet controls : description of the numerical methods. Research report UH/MD-22. University of Houston. Department of Mathematics. January 1988.
et Jap. Journal of Applied Mathematics. 1989.

[12] A. HARAUX
Séries lacunaires et contrôle semi-interne des vibrations d'une plaque rectangulaire. Journal de Math. pures et Appl. 1989.

[13] S.JAFFARD
Contrôle interne exact des vibrations d'une plaque carrée. C.R.A.S. Paris. 307. 1988.

[14] J.L. LIONS
- Contrôle optimal des systèmes gouvernés par des équations aux dérivées partielles. Dunod. 1968.

- Contrôlabilité exacte des systèmes distribués
Collection R.M.A. Masson. 1988. Vol. 1 : Méthode HUM

[15] A.M. MICHELETTI
Perturbazione dello spettro di un operatore ellitico di tipo variazionale, in relazione ad una variazione del campo
Annali di Matematica pura ed applicata (IV), Vol. XCVII. 1973.

[16] P. RAVIART - J.M. THOMAS
Introduction à l'analyse numérique des équations aux dérivées partielles. Masson. 1983.

[17] Y. SAKAWA
Controllability for partial differential equations of parabolic type. SIAM J. on Control.Vol. 12. 1974.

Linear Quadratic Control Problem without Stabilizability

Giuseppe Da Prato[(1)]
Scuola Normale Superiore
Piazza dei Cavalieri 7, 56126 Pisa, Italy

Michel Delfour[(2)]
Centre de recherches mathématiques et
Département de Mathématique et de Statistique
Université de Montréal
C.P.6128, Succ.A, Montréal (Québec)
Canada, H3C 3J7

1 Introduction.

This paper deals with the linear quadratic optimal control problem over an infinite time horizon for infinite dimensional systems in Hilbert spaces with bounded control and observation operators. In [1] the authors have recently constructed examples where the system is not stabilizable and yet the algebraic Riccati equation has a positive self-adjoint unbounded solution. This phenomenon is intimately related to the fact that stabilizability only occurs for a dense subset of initial conditions.

The object of this paper is to fill up this gap in the theory. Under no stabilizability hypothesis we a priori define the set Σ of initial states which can be stabilized and show that it can be given a natural Hilbert space structure. When Σ is dense in the space of initial conditions, we construct the minimum positive self adjoint unbounded solution to the algebraic Riccati equation. A new technique is introduced to directly obtain the semigroup associated with the closed loop system and the properties of the feedback operator. If the usual detectability hypothesis is added we recover the fact that the closed loop system is exponentially stable. Examples are also included to illustrate the theoretical considerations. Extensions to systems with unbounded control and observation operators are possible and will be reported in a forthcoming paper. We felt that it was more instructive to first illustrate the phenomenon and the main features of the theory for the bounded case.

2 Problem formulation.

Let H (*state space*), U (*control space*) and Y (*observation space*) be three Hilbert spaces. Let A: $D(A)\subset H\rightarrow H$ be the infinitesimal generator of a strongly continuous semigroup e^{tA}. Let $B\in \mathcal{L}(U;H)$ and $C\in \mathcal{L}(H;Y)$ be the control and observation operators. Consider the system

---

(1) Work partially supported by the Italian National Project M.P.I. 40% "Equazioni di Evoluzione e Applicazioni Fisico-Matematiche"

(2) This research has been supported in part by Canada Natural Sciences and Engineering Research Council Grant A-8730 while the author was a Killam fellow from the Canada Council.

(2.1) $$\begin{cases} x'(s) = Ax(s) + Bu(s) \;, \; s\geq 0 \\ x(0) = h \end{cases}$$

and the associated cost function

(2.2) $$J(u,h) = \int_0^\infty \{ |Cx(s)|^2 + |u(s)|^2 \} ds$$

Denote by V the value function

(2.3) $$V(h) = \inf\{ J(u,h) \; ; \; u \in L^2(0,\infty;U) \}$$

with domain

(2.4) $$\mathrm{dom}\, V = \{ h \in H \; ; \; V(h) < \infty \}$$

which will be referred to as the *domain of stabilizability* for the triple (A,B,C).

3 An example of unbounded solution to the Riccati equation.

Let $H = \ell^2$ be the Hilbert space of all sequences $x = \{x_n\}_{n\in N}$, with norm

(3.1) $$|x|^2 = \sum_{k=1}^{\infty} x_k^2$$

Let $\{e_k\}$ be the orthonormal basis in ℓ^2

(3.2) $$(e_k)_n = \delta_{kn} \quad , \quad k \in N$$

Define the bounded operators

(3.3) $$Ae_k = \frac{k}{k+1} e_k \quad , \quad Be_k = \frac{\sqrt{2k+1}}{k+1} e_k \quad , \quad k \in N$$

Notice that their spectra is made up of a point and continuous part

(3.4) $$\sigma_p(A) = \{ \tfrac{k}{k+1} \; ; \; k \in N \} , \; \sigma_c(A) = \{1\}$$

(3.5) $$\sigma_p(B) = \{ \tfrac{\sqrt{2k+1}}{k+1} \; ; \; k \in N \} , \; \sigma_c(B) = \{0\}$$

Associate with A and B the control system

$$(3.6) \qquad \begin{cases} x'(s) = Ax(s) + Bu(s) \;,\; s\geq 0 \\ x(0) = h \end{cases}$$

and the observation x(s) (the observation operator C is the identity). Consider the cost function

$$(3.7) \qquad J(u,h) = \int_0^\infty \{|x(s)|^2+|u(s)|^2\}\,ds$$

It is well known that if the pair (A,B) is stabilizable , then there exists a bounded symmetric nonnegative linear operator P_∞ on H which is the minimum nonnegative solution of the algebraic Riccati equation

$$(3.8) \qquad P_\infty A+A^*P_\infty-P_\infty BB^*P_\infty + I = 0$$

However , it is easy to check that the unbounded operator

$$(3.9) \qquad P_\infty e_k = (k+1)e_k \;,\quad k\in N$$

is the only solution to (3.8). This means that only initial conditions h in the domain $D(P_\infty^{1/2})$ of $P_\infty^{1/2}$

$$(3.10) \qquad \begin{cases} \{x\in \ell^2 \;;\; \sum_{k=1}^\infty (k+1)x_k^2 < \infty\} \\ P_\infty^{1/2}e_k = \sqrt{k+1}\, e_k \end{cases}$$

can be stabilized. For all others

$$(3.11) \qquad J(u,h) = \infty \;,\quad h\notin D(P_\infty^{1/2})$$

Hence dom V = $D(P_\infty^{1/2})$ in this example.

The interpretation of this phenomenon is that , for $h\notin D(P_\infty^{1/2})$, the state x cannot be stabilized with a finite energy control u on $[0,\infty[$ with values in H. Yet the closed loop system is given by the operator

$$(3.12) \qquad A-BB^*P_\infty = -I$$

which is exponentially stable in H. So, for all h in H, the solution x^* of the closed loop system

$$(3.13) \qquad \begin{cases} x'(s) = [A-BB^*P_\infty]x(s) \;,\; s\geq 0 \\ x(0) = h \end{cases}$$

is given by $x^*(s) = e^{-s}h$ and belongs to $L^2(0,\infty;H)$, whereas the optimal control $u^*(s)$ is given by

(3.14) $$u^*(s) = -B^*P_\infty x^*(s) = -B^*P_\infty e^{-s}h$$

and u belongs to $L^2(0,\infty;H)$, if and only if $h \in D(P_\infty^{1/2})$.

Finally it is useful to notice that

(3.15) $$B^*P_\infty e_k = \sqrt{2k+1}\, e_k \ , \quad D(B^*P_\infty) = D(P_\infty^{1/2}).$$

and

(3.16) $$BB^*P_\infty e_k = \frac{2k+1}{k+1} e_k \ , \quad D(BB^*P_\infty) = H$$

4 Asymptotic behaviour of the solution P(t) of the associated Riccati equation.

It is well known that we can associate with the control problem (2.1)-(2.2) the mild solution $P \in C_s([0,\infty[;\Sigma^+(H))$ of the Riccati equation

(4.1) $$\begin{cases} P' = A^*P + PA - PBB^*P + C^*C \\ P(0) = 0 \end{cases}$$

We have denoted by $\Sigma^+(H)$ the set of all positive symmetric operators in H and by $C_s([0,\infty[;\Sigma^+(H))$ the set of all the mappings T: $[0,\infty[\to \Sigma^+(H)$, such that $T(\cdot)x$ is continuous for all $x \in H$.

For each $h \in H$ the function $(P(\cdot)h,h)$ is nondecreasing , moreover the following identity holds :

(4.2) $$(P(t)h,h) + \int_0^t |u(s)+B^*P(t-s)x(s)|^2 ds = \int_0^t \{|Cx(s)|^2 + |u(s)|^2\} ds$$

for all $u \in L^2_{loc}(0,\infty;U)$.

We set

(4.3) $$\phi(h) = \lim_{t\to\infty} (P(t)h,h) \ , \quad \forall h \in H$$

The function ϕ: $H \to [0,\infty]$ is convex, strict and lower semi-continuous, with domain

(4.4) $$\Sigma = \{h \in H \ ; \ \phi(h) < \infty\}$$

Lemma 4.1. *The following statements hold:*

(i) For all h *and* k *in* Σ, $(P(\cdot)h,k)$ *is bounded.*

(ii) Σ *is a vector subspace of* H.

(iii) For all h *and* k *in* Σ, *the following limit exists*

$$(4.5) \qquad \psi(h,k) = \lim_{t\to\infty} (P(t)h,k)$$

Moreover ψ *is a bilinear form on* $\Sigma\times\Sigma$ *and*

$$(4.6) \qquad \psi(h,h) = \phi(h), \quad \forall h\in H$$

Proof.

(i) For all h and k in Σ and $t\geq 0$, we have

$$(4.7) \quad |(P(t)h,k)|^2 \leq (P(t)h,h)\,(P(t)k,k) \leq \phi(h)\phi(k)$$

and the conclusion follows.

(ii) For all h in Σ and λ in **R**, $\phi(\lambda h) = \lambda^2\phi(h)$ and so $\lambda h\in\Sigma$. For all h and k in Σ

$$(P(t)(h+k),h+k) = (P(t)h,h) + (P(t)k,k) + 2(P(t)h,k)$$

and from (i) $\phi(h+k) \leq [\phi(h)^{1/2}+\phi(k)^{1/2}]^2$. Thus Σ is a linear subspace of H. Part (iii) is an immediate consequence of parts (i) and (ii).#

We define the following inner product on Σ

$$(4.8) \qquad (h,k)_\Sigma = (h,k) + \psi(h,k)$$

which makes it a pre-Hilbert space.

Lemma 4.2. *The space* Σ, *endowed with the inner product* (4.8), *is a Hilbert space.*

Proof. It is sufficient to show that Σ is complete with respect to the norm

$$(4.9) \qquad |h|_\Sigma = [|h|^2+\phi(h)]^{1/2}.$$

Let $\{h_n\}$ be a Cauchy sequence in Σ, then there exists $h\in H$ such that $h_n\to h$. Moreover there exists $\lambda\geq 0$ such that

$$(4.10) \qquad |h_n|^2 + \phi(h_n) \to \lambda$$

and

(4.11) $\phi(h_n) \to \lambda - |h|^2$.

By lower semi-continuity of ϕ, we have

(4.12) $\lambda - |h|^2 = \lim_{n\to\infty} \phi(h_n) \geq \phi(h)$

and, by definition of Σ, h belongs to Σ.

Finally, for each $\varepsilon>0$, there exists a positive integer $N(\varepsilon)$ such that

$$|h_n - h_m|_\Sigma^2 = |h_n - h_m|^2 + \phi(h_n - h_m) \leq \varepsilon \quad , \quad \forall m,n \geq N(\varepsilon)$$

As n goes to infinity we get

$$|h - h_m|^2 + \phi(h - h_m) \leq \varepsilon \quad , \quad \forall m \geq N(\varepsilon)$$

by continuity of the norm in H and lower semi-continuity of ϕ. This shows that $h_n \to h$ in Σ and completes the proof.#

We have constructed the space Σ of initial conditions for which the expression $(P(t)h,h)$ has a limit. In general its closure in H will not be dense and it will be natural to decompose H as a direct sum

(4.13) $H = \bar{\Sigma} \oplus \Sigma^\perp$

where $\bar{\Sigma}$ is the closure of Σ in H and $\Sigma^\perp$ is the orthogonal complement to Σ. In the sequel we identify the elements of the dual H' of H with those of H. We shall denote by Σ' the dual of Σ and by $\langle\cdot,\cdot\rangle_\Sigma$ the duality pairing between Σ and Σ'.

Proposition 4.3. *Assume that Σ is dense in* H. *Then there exists a unique linear operator* $P_\infty \in \mathcal{L}(\Sigma; \Sigma')$ *such that*

(4.14) $\langle P_\infty h,k \rangle_\Sigma = \psi(h,k) \quad , \quad \forall h,k \in \Sigma.$

P_∞ *can also be viewed as a closed self adjoint positive operator on H with dense domain*

(4.15) $D(P_\infty) = \{h \in \Sigma \,;\, \psi(h,\cdot)$ *is continuous in* $H\}$.

We have

(4.16) $\phi(h) = (P_\infty h,h), \quad \forall h \in D(P_\infty)$

(4.17) $\psi(h,k) = (P_\infty h,k), \quad \forall h \in D(P_\infty), \forall k \in H$

and the subdifferential of ϕ is given by

(4.18) $$\frac{1}{2}\partial\phi(h) = \begin{cases} P_\infty h\,, & \text{if } h\in D(P_\infty) \\ \emptyset\,, & \text{if } h\notin D(P_\infty) \end{cases}$$

Moreover $P_\infty^{1/2}$ *is well defined and*

(4.19) $$D(P_\infty^{1/2}) = \Sigma = [D(P_\infty),H]_{1/2}$$

where $[X,Y]_{1/2}$ *denotes the interpolation space between* Y *and its dense subspace* X.

Proof. By definition of the inner product on Σ, the symmetrical bilinear form ψ on $\Sigma\times\Sigma$ is continuous and there exists a unique $P_\infty \in L(\Sigma;\Sigma')$ such that (4.14) is verified. Moreover ψ is Σ-H coercive and P_∞ is a self-adjoint operator in H with domain $D(P_\infty)$. Expression (4.18) follows from the fact that ϕ is l.s.c. so that $\partial\phi$ is maximal monotone in H. Finally the positive self-adjoint operator P_∞ has a positive square root $P_\infty^{1/2}$ which is a closed linear operator on H with dense domain $D(P_\infty^{1/2})$ which coincides with Σ .#

Assume now that Σ is not dense in H and denote by $\overline{\Sigma}$ the closure of Σ in H; then we have the similar result

Corollary. *There exists a unique linear operator* $P_\infty \in L(\Sigma;\Sigma')$ *such that*

(4.20) $$< P_\infty h,k >_\Sigma = \psi(h,k) \quad , \ \forall h,k\in\Sigma$$

P_∞ *can also be viewed as a closed self adjoint positive operator on* $\overline{\Sigma}$ *with dense domain*

(4.21) $$D(P_\infty) = \{h\in\Sigma\,;\ \psi(h,\cdot) \text{ is continuous in } \overline{\Sigma}\}$$

We have

(4.22) $$\phi(h) = (P_\infty h,h), \quad \forall h\in D(P_\infty)$$

(4.23) $$\psi(h,k) = (P_\infty h,k), \quad \forall h\in D(P_\infty)\,,\ \forall k\in H$$

and the subdifferential of ϕ is given by

(4.24) $$\frac{1}{2}\partial\phi(h) = \begin{cases} P_\infty h\,, & \text{if } h\in D(P_\infty) \\ \emptyset\,, & \text{if } h\notin D(P_\infty) \end{cases}$$

Moreover $P_\infty^{1/2}$ *is well defined and*

(4.25) $$D(P_\infty^{1/2}) = \Sigma$$

where Σ *coincides with the real interpolation space between* $D(P_\infty)$ *and* $\bar{\Sigma}$.

5 Existence of the optimal control and optimal closed loop system.

In this Section we use the asymptotic properties obtained in Section 4, to solve the optimal control problem (2.1)-(2.2). In addition we study the mapping between the initial conditions and the optimal state and control.

Theorem 5.1. *The following statements hold:*

(i) Given any h *in* H, *either* $h \notin \Sigma$ *and*

(5.1) $$J(u,h) = +\infty, \quad \forall u \in L^2(0,\infty;U) \text{ and } V(h) = \phi(h) = +\infty$$

or $h \in \Sigma$ *and there exists a unique optimal control* $\hat{u}(\cdot,h)$ *in* $L^2(0,\infty;U)$ *such that*

(5.2) $$J(\hat{u}(\cdot,h),h) = V(h) = \phi(h)$$

(ii) The mapping

(5.3) $$\Sigma \to L^2(0,\infty;U), \ h \to \hat{u}(\cdot,h)$$

is linear and continuous.

(iii) Denote by $\hat{x}(\cdot,h)$ *the optimal state corresponding to the optimal control* $\hat{u}(\cdot,h)$ *and set*

(5.4) $$S_\Sigma(t)h = \hat{x}(t,h), \quad t \geq 0, \quad h \in \Sigma,$$

Then $S_\Sigma(\cdot)$ *is a strongly continuous semigroup in* Σ.

(iv) Let A_Σ *be the infinitesimal generator of* $S_\Sigma(\cdot)$. *For all* $h \in D(A_\Sigma)$, *we have*

(5.5) $$\hat{u}(\cdot,h) \in H^1(0,\infty,U) \text{ and } \hat{u}'(\cdot,h) = \hat{u}(\cdot,A_\Sigma h)$$

(5.6) $$C\hat{x}(\cdot,h) \in H^1(0,\infty;Y), \ \hat{x}'(\cdot,h) = \hat{x}(\cdot,A_\Sigma h) \text{ and } D(A_\Sigma) \subset D(A).$$

(v) For all h *in* $D(A_\Sigma)$ *the map*

(5.7) $$h \to \hat{u}(\cdot,h) : D(A_\Sigma) \to U$$

is linear and continuous. Its closure in Σ *generates an unbounded linear operator*

(5.8) $K : D(K) \subset \Sigma \to U$ *such that* $D(A_\Sigma) \subset D(K)$

and

(5.9) $D(A_\Sigma) = D(A) \cap D(K), \ A_\Sigma h = Ah + BKh$

Moreover, for all h *in* $D(A_\Sigma)$ *and* $t \in [0,\infty[$, $\hat{x}(t,h) \in D(A_\Sigma)$,

(5.10) $A_\Sigma \hat{x}(t,h) = A\hat{x}(t,h) + B\hat{u}(t,h) = [A+BK]\hat{x}(t,h)$

(5.11) $\hat{u}(t,h) = K\hat{x}(t,h)$

(vi) *For all* h *in* $D(A_\Sigma)$

(5.12) $Kh = \lim_{t\to\infty} -B^*P(t)h,$

and for all h *in* Σ *and almost all* t *in* $[0,\infty[$

$$\hat{u}(t,h) = K\hat{x}(t,h), \ \hat{x}(t,h) \in D(K)$$

When $\bar{\Sigma} = H$ *the closure* K_∞ *of the operator* $-B^*P_\infty$ *in* Σ *coincides with* K *on* $D(A_\Sigma)$.

Proof .

(i). By definition of Σ, for all $h \notin \Sigma$ $\lim_{t\to\infty} (P(t)h,h) = \infty$, and in view of identity (4.2)

$$(P(t)h,h) \le J(u,h), \quad \forall u \in L^2(0,\infty;U), \ \forall t \ge 0$$

By letting t go to infinity we obtain (5.1).

When $h \in \Sigma$ identity (4.2) yields

$$\phi(h) \le J(u,h), \ \forall u \in L^2(0,\infty;U)$$

For each $t>0$, let (x_t,u_t) be defined by

$$x_t'(s) = Ax_t(s) - BB^*P(t-s)x_t(s), \text{ in } [0,t] \ , \ x_t(0) = h$$

(5.13) $u_t(s) = -B^*P(t-s)x_t(s)$, in $[0,t]$.

The pair (x_t,u_t) is the optimal solution on the interval $[0,t]$.

Consider the extension $\hat{u}_t$ of u_t from $[0,t]$ to $[0,\infty[$

(5.14) $$\hat{u}_t(s) = \begin{cases} u_t(s) \ , \text{ if } 0 \le s \le t \\ 0 \ , \text{ if } s > t \end{cases}$$

and let $\hat{x}_t$ be the corresponding extension of the solution x_t of the state equation on [0,t]

$$\hat{x}_t(s) = \begin{cases} x_t(s), & \text{if } 0\le s\le t \\ 0, & \text{if } s>t \end{cases}$$

Again by (4.2) and (5.14)

$$(P(t)h,h) = \int_0^t \{|Cx_t(s)|^2+|u_t(s)|^2\}ds \ge \int_0^\infty |\hat{u}_t(s)|^2 ds$$

Hence, for any sequence $\{t_n\}$, $t_n\to\infty$, $\{\hat{u}_{t_n}\}$ is bounded in $L^2(0,\infty;U)$ and there exists $\hat{u}$ in $L^2(0,\infty;U)$ and a subsequence of $\{t_n\}$ (still denoted $\{t_n\}$) such that

(5.16) $$\hat{u}_{t_n} \to \hat{u} \text{ in } L^2(0,\infty;U)\text{-weak}$$

Denote by $\hat{x}$ the solution of

(5.17) $$\hat{x}'(s) = A\hat{x}(s) + B\hat{u}(s), \text{ in } t\ge 0,\ \hat{x}(0) = h$$

Then for any fixed T>0 and $t_n>T$

$$\hat{u}_{t_n} \to \hat{u} \text{ in } L^2(0,T;U)\text{-weak},\ \hat{x}_{t_n} \to \hat{x} \text{ in } L^2(0,T;H)\text{-weak}$$

But for $t_n>T$

$$(P(t_n)h,h) \ge \int_0^T \{|Cx_{t_n}(s)|^2+|u_{t_n}(s)|^2\}ds$$

and by weak lower semicontinuity

$$\phi(h) \ge \int_0^T \{|C\hat{x}(s)|^2+|\hat{u}(s)|^2\}ds$$

As T goes to infinity

(5.18) $$\phi(h) \ge \int_0^\infty \{|C\hat{x}(s)|^2+|\hat{u}(s)|^2\}ds = J(\hat{u},h)$$

Combining (5.18) and (5.13) it follows that there exists $\hat{u} = \hat{u}(\cdot,h)\in L^2(0,\infty;U)$ such that

$$J(\hat{u},h) \le \phi(h) \le J(u,h) \quad ,\ \forall u\in L^2(0,\infty;U)$$

It follows $V(h) \le J(\hat{u},h) \le \phi(h) \le V(h)$. This establishes (5.2).

As for the uniqueness of $\hat{u}$, assume that $\hat{u}_1$ and $\hat{u}_2$ are two optimal controls in $L^2(0,\infty;U)$. Then

$$J(\hat{u}_1,h) = J(\hat{u}_2,h) = V(h).$$

So for $\hat{u}_1 \neq \hat{u}_2$

$$J((\hat{u}_1 + \hat{u}_2)/2,h) = \frac{1}{2}[J(\hat{u}_1,h)+J((\hat{u}_2,h)] - J((\hat{u}_1 - \hat{u}_2)/2,h) =$$

$$= V(h)-J((\hat{u}_1 - \hat{u}_2)/2,h) \leq V(h) - \frac{1}{4}\|\hat{u}_1 - \hat{u}_2\|^2 < V(h)$$

which contradicts the optimality of $\hat{u}_1$ and $\hat{u}_2$.

(ii) Let $\hat{u}_t$ be defined by (5.13), then

(5.19) $$\|\hat{u}_t\|^2_{L^2(0,\infty;U)} \leq (P(t)h,h) \leq |h|^2_\Sigma$$

Moreover, since the optimal control is unique, in the step (i) we have proved that

$$\lim_{t\to\infty} \hat{u}_t = \hat{u}, \quad \text{in } L^2(0,\infty;U) \text{ weak, for any } h \in \Sigma$$

We now prove that $\hat{u}_t \to \hat{u}$ in $L^2(0,\infty;U)$-strong. By optimality of the pair (x_t,u_t) on $[0,t]$

$$J^t(u_t,h) = \operatorname{Inf}\{J^t(v,h)\ ;\ v \in L^2(0,\infty;U)\}$$

where

$$J^t(v,h) = \int_0^t \{|Cx(s;v)|^2 + |v(s)|^2\}ds$$

We want to prove that

$$\lim_{t\to\infty} J^t(u_t,h) = J(\hat{u},h)$$

By definition of the minimizing element u_t on $[0,t]$

$$J^t(u_t,h) \leq J^t(\hat{u}(\cdot,h),h) = \int_0^t \{|C\hat{x}(s,\hat{u}(\cdot,h))|^2 + |\hat{u}(s,h)|^2\}ds$$

and necessarily

$$\limsup_{t\to\infty} J^t(u_t,h) \leq \int_0^\infty \{|C\hat{x}(s,\hat{u}(\cdot,h))|^2 + |\hat{u}(s,h)|^2\}ds = J(\hat{u}(\cdot,h),h)$$

We have shown in Section (i) that

$$\hat{u}_t \to \hat{u}, \quad \text{in } L^2(0,\infty;U) \text{ weak}$$

and we can show by the same technique that $\{C\hat{x}_t\}$ is bounded in $L^2(0,\infty;Y)$ and that weak subsequences $\{C\hat{x}_{t_n}\}$ converging to some y in $L^2(0,\infty;Y)$ can be extracted :

$$C\hat{x}_{t_n} \to y, \text{ in } L^2(0,\infty;Y) \text{ weak}$$

By continuity of the state $x(\cdot;u)$ with respect to the control u on a finite time interval [0,T], T>0, the map $u \to x(\cdot;u) : L^2(0,T;U) \to L^2(0,T;H)$ is weakly continuous and finally

$$u \to Cx(\cdot;u) : L^2(0,T;U) \to L^2(0,T;Y)$$

is also weakly continuous . This implies that for all T>0, $y = C\hat{x}(\hat{u},h)$ in $L^2(0,T;Y)$ and hence in $L^2(0,\infty;Y)$. As a result

$$\hat{u}_t \to \hat{u}, \text{ in } L^2(0,\infty;U) \text{ weak and } C\hat{x}_t \to C\hat{x}, \text{ in } L^2(0,\infty;Y) \text{ weak.}$$

But the functional

$$(v,y) \to \int_0^\infty \{|y(s)|^2+|v(s)|^2\}ds) : L^2(0,\infty;U)\times L^2(0,\infty;Y) \to R$$

is lower weakly continuous and necessarily

$$\liminf_{t\to\infty} \int_0^\infty [\,|C\hat{x}_t|^2 + |\hat{u}_t|^2\,]ds \geq \int_0^\infty [\,|C\hat{x}|^2 + |\hat{u}|^2\,]ds$$

that is

$$\liminf_{t\to\infty} J^t(u_t,h) \geq J(\hat{u},h)$$

Finally

$$J(\hat{u},h) \leq \liminf_{t\to\infty} J^t(u_t,h) \leq \limsup_{t\to\infty} J^t(u_t,h) \leq J(\hat{u},h)$$

and this proves that

$$\lim_{t\to\infty} J^t(u_t,h) = J(\hat{u},h) \quad .$$

The strong continuity will now be obtained by the following simple computation

$$\|C\hat{x}_t - C\hat{x}\|^2 + \|\hat{u}_t - u\|^2 = \|C\hat{x}_t\|^2 + \|\hat{u}_t\|^2 + \|C\hat{x}\|^2 + \|\hat{u}\|^2 - 2(C\hat{x}_t, C\hat{x}) - 2(\hat{u}_t, \hat{u}) =$$

$$= J^t(u_t,h) + J(\hat{u},h) - 2(C\hat{x}_t, C\hat{x}) - 2(\hat{u}_t, \hat{u}).$$

As t goes to ∞ $J^t(u_t,h) \to J(u,h)$ and by weak convergence

$$(C\hat{x}_t, C\hat{x}) \to (C\hat{x}, C\hat{x}) = \|C\hat{x}\|^2 .$$

$$(\hat{u}_t,\hat{u}) \to (\hat{u},\hat{u}) = \|\hat{u}\|^2 .$$

So we conclude that

$$\lim_{t\to\infty} \{\|C\hat{x}_t - C\hat{x}\|^2 + \|\hat{u}_t - u\|^2\} = 2J(\hat{u},h) - 2[\|C\hat{x}\|^2 + \|\hat{u}\|^2] = 0$$

and

$$\hat{u}_t \to \hat{u}, \text{ in } L^2(0,\infty;U)\text{-strong and } C\hat{x}_t \to C\hat{x}, \text{ in } L^2(0,\infty;Y)\text{-strong.}$$

By (5.18) and by the Uniform Boundnedess Theorem it follows that the mapping

$$\Sigma \to L^2(0,\infty;U),\ h \to \hat{u}\,(\cdot,h)$$

is linear and continuous.

(iii) We first remark that, by Bellman's Optimality Principle we have $\hat{x}\,(t,h) \in \Sigma$ for all $h \in \Sigma$ and

(5.20) $$\hat{x}(t+s,h) = \hat{x}(t;\hat{x}(s,h)) \ , \quad \forall t\geq 0,\ \forall s\geq 0$$

(5.21) $$V(\hat{x}\,(t,h)) = \int_t^\infty \{|C\hat{x}(s,h)|^2 + |\hat{u}(s,h)|^2\}\,ds$$

Thus $S_\Sigma(t)$ is a linear operator in Σ for all $t\geq 0$. We prove now that $S_\Sigma(t)$ is bounded in Σ. By (5.17) we have

(5.22) $$\hat{x}\,(t,h) = e^{tA}h + \int_0^t e^{(t-s)A}B\hat{u}(s,h)ds$$

It follows that for any $T>0$ there exists $C_T>0$ such that

(5.23) $$|\hat{x}\,(t,h)|_H^2 \leq |h|_H^2$$

moreover, by (5.21) $\phi(\hat{x}(t,h)) \leq \phi(h)$, and the continuity of $S_\Sigma(t)$ follows. We prove now

$$\lim_{t\to 0} \hat{x}(t,h) = h \ , \quad \forall h \in \Sigma \text{ in } \Sigma$$

By (5.21) we have $\lim_{t\to 0} \hat{x}(t,h) = h$ in H. It remains to show that $\hat{x}(t)$ is continuous at $t = 0$ with respect to the seminorm $\phi(h)^{1/2}$. We have, by the linearity of $\hat{x}(\cdot,h)$ and $\hat{u}(\cdot,h)$ in h

$$\phi(\hat{x}(t,h)-h) = \int_0^\infty \{|C\hat{x}(t+s,h) - C\hat{x}(s,h)|^2 + |\hat{u}(t+s,h) - \hat{u}(s,h)|^2\}\,ds$$

Since $C\hat{x}(\cdot,h) \in L^2(0,\infty;Y)$ and $\hat{u}(\cdot,h) \in L^2(0,\infty;U)$, we have $\lim_{t\to 0} \phi(\hat{x}(t,h)-h) = 0$, and (iii) is proved.

(iv) Let $h \in D(A_\Sigma)$; then $\hat{x}(\cdot,h) \in C^1([0,\infty[;\Sigma)$ and $\hat{x}'(0,h) = A_\Sigma h$. Denote by $\hat{w}(\cdot) = \hat{u}(\cdot,A_\Sigma h)$ the optimal control corresponding to $A_\Sigma h$. So, for all $t>0$

$$\phi\left[\frac{\hat{x}(t,h)-h}{t} - A_\Sigma h\right] =$$

$$= \int_0^\infty \left\{ \left|\frac{C\hat{x}(t+s,h)-C\hat{x}(s,h)}{t} - C\hat{x}'(s,h)\right|^2 + \left|\frac{\hat{u}(t+s,h)-\hat{u}(s,h)}{t} - \hat{w}(s)\right|^2 \right\} ds$$

As t goes to zero the first two terms go to zero and necessarily

$$\lim_{t\to 0} \int_0^\infty \left|\frac{\hat{u}(t+s)-\hat{u}(s)}{t} - \hat{w}(s)\right|^2 ds = 0$$

which implies $\hat{w} = \hat{u}'$ and $\hat{u} \in H^1(0,\infty;U)$, $\forall h \in D(A_\Sigma)$. By (5.22) it follows that $h \in D(A)$ and (5.6) follows.

(v) We have shown in (ii) that the map

$$h \to \hat{u}(\cdot,h) : \Sigma \to L^2(0,\infty;U),$$

is linear and continuous. In particular

$$h \to \hat{u}'(\cdot,h) := \hat{u}(\cdot, A_\Sigma h): D(A_\Sigma) \to L^2(0,\infty;U)$$

is also continuous. Hence

$$h \to \hat{u}(\cdot,h) : D(A_\Sigma) \to H^1(0,\infty;U)$$

is linear and continuous when $D(A_\Sigma)$ is endowed with the graph norm topology :

$$\|h\|^2_{D(A_\Sigma)} = \|h\|^2_\Sigma + \|A_\Sigma h\|^2 .$$

In particular, $\hat{u}(\infty) = 0$, $\hat{u} \in C([0,\infty];U)$ and the map

$$h \to \hat{u}(0,h) : D(A_\Sigma) \to U$$

is linear and continuous . We denote it by K. Equivalently K is a closed linear unbounded operator from Σ to U with domain

$$D(K) = \{ h \in \Sigma : Kh \in U\} \supset D(A_\Sigma)$$

In view of this and identity (5.6)

$$\forall h \in D(A_\Sigma),\ A_\Sigma h = Ah + B\hat{u}(0,h) = [A+BK]h$$

Conversely, if $h\in D(A)\cap D(K)$,

$$A_\Sigma h = Ah + BKh \Rightarrow h\in D(A_\Sigma)$$

anfd $D(A_\Sigma) = D(A)\cap D(K)$.

(vi) To relate K and the limit of P(t), we go back to formula (4.2) with $h\in\Sigma$, $u = \hat{u}(\cdot,h)$ and $x = \hat{x}(\cdot,h)$:

$$(5.24)\qquad \langle P(t)h,h\rangle_\Sigma + \int_0^t |\hat{u}(s,h)+B^*P(t-s)\hat{x}(s,h)|^2 ds = \int_0^t \{|C\hat{x}(s,h)|^2+|\hat{u}(s,h)|^2\}\, ds$$

As t goes to infinity we obtain

$$\lim_{t\to\infty} \int_0^t |\hat{u}(s,h)+B^*P(t-s)\hat{x}(s,h)|^2 ds = 0$$

Setting $P(r) = 0$ for $r\le 0$, then

$$(5.25)\qquad \lim_{t\to\infty} \int_0^\infty |\hat{u}(s,h)+B^*P(t-s)\hat{x}(s,h)|^2 ds = 0$$

since

$$\lim_{t\to\infty} \int_t^\infty |\hat{u}(s,h)|^2 ds = 0.$$

Now repeat the same estimate with $A_\Sigma h$ instead of h and $\hat{x}(\cdot,A_\Sigma h) = \hat{x}'(\cdot,h)$, $\hat{u}(\cdot,A_\Sigma h) = \hat{u}'(\cdot,h)$. Then by the same argument

$$(5.26)\qquad \lim_{t\to\infty} \int_0^\infty |\hat{u}'(s,h)+B^*P(t-s)\hat{x}'(s,h)|^2 ds = 0$$

Introduce the notation and use (5.25) and (5.26) :

$$v_t(s) = \hat{u}(s,h)+B^*P(t-s)\hat{x}(s,h),\quad v_t\to 0 \text{ in } L^2(0,\infty;U)$$

$$w_t(s) = \hat{u}'(s,h)+B^*P(t-s)\hat{x}'(s,h),\quad w_t\to 0 \text{ in } L^2(0,\infty;U)$$

For h in $D(A_\Sigma)$, differentiate (5.24) with respect to t

$$\frac{d}{dt}\langle P(t)h,h\rangle + |\hat{u}(0,h)+B^*P(t)h|^2+2\int_0^t (v_t(s),w_t(s))ds = |C\hat{x}(t,h)|^2+|\hat{u}(t,h)|^2$$

But for $t'\ge t$

$$<P(t')h,h> - <P(t)h,h> \geq 0 \Rightarrow \frac{d}{dt}<P(t)h,h> \geq 0$$

and notice that

$$\lim_{t\to\infty}\int_0^t (v_t(s),w_t(s))ds = \lim_{t\to\infty}\int_0^\infty (v_t(s),w_t(s))ds \to 0 \text{ as } t\to\infty$$

Hence

$$0 \leq \liminf_{t\to\infty} \frac{d}{dt}<P(t)h,h> \leq \liminf_{t\to\infty}\left\{|C\hat{x}(s,h)|^2+|\hat{u}(s,h)|^2\right\}$$

$$0 \leq \lim_{t\to\infty} |\hat{u}(0,h)+B^*P(t)h|^2 \leq \liminf_{t\to\infty}\left\{|C\hat{x}(s,h)|^2+|\hat{u}(s,h)|^2\right\}$$

But recall that $C\hat{x}(\cdot,h)\in H^1(0,\infty;Y)$ and $\hat{u}(\cdot,h)\in H^1(0,\infty;U)$ and this implies that

$$\lim_{t\to\infty} C\hat{x}(\cdot,h) = 0 \text{ and } \lim_{t\to\infty}\hat{u}(\cdot,h) = 0$$

Finally for all h in $D(A_\Sigma)$

$$Kh = \lim_{t\to\infty}[-B^*P(t)h]. \ \#$$

Remark 5.1 Theorem 5.1 shows that

$$V(h) = \phi(h) = V(\hat{x}(t,h)) = \int_0^\infty \{|C\hat{x}(s,h)|^2+|\hat{u}(s,h)|^2\}ds. \tag{5.27}$$

Hence $\operatorname{dom} V = \operatorname{dom}\phi = \Sigma$ and Σ coincides with the domain of stabilization of the triple (A,B,C).

Moreover, by the linearity of $\hat{x}(s,h)$ and $\hat{u}(s,h)$ in h it follows that

$$\psi(h,k) = \int_0^\infty \{(C\hat{x}(s,h),C\hat{x}(s,k))+(\hat{u}(s,h),\hat{u}(s,k))|^2\}ds \quad ; \forall h,k\in\Sigma.\# \tag{5.28}$$

6 The algebraic Riccati equation.

In this section we study the relationship between the control problem (2.1)-(2.2) and the *algebraic Riccati equation*

$$<QA_\Sigma h,k>_\Sigma + <Qh,A_\Sigma k>_\Sigma + (Kh,Kk)_U + (Ch,Ck)_Y = 0 \ , \ \forall h,k\in D(A_\Sigma) \tag{6.1}$$

where K is the closure of $-B^*P_\infty$ on Σ and

$$A_\Sigma = A+BK \text{ on } D(A_\Sigma) \tag{6.2}$$

Definition 6.1. We shall say that the triplet (A,B,C) is *approximatively stabilizable* if $\overline{\Sigma} = H$.

Proposition 6.1. The self adjoint operator P_∞ on H defined by (4.15), is a solution of the Algebraic Riccati Equation (6.1).

Proof. For all h and k in $D(A_\Sigma)$, we have, recalling (5.28)

$$\langle P_\infty A_\Sigma h,k\rangle = \int_0^\infty \{(C\hat{x}'(s,h),C\hat{x}(s,k))_Y+(\hat{u}'(s,h),\hat{u}(s,k))_U\}ds$$

$$\langle P_\infty h,A_\Sigma k\rangle = \int_0^\infty \{(C\hat{x}(s,h),C\hat{x}'(s,k))_Y+(\hat{u}(s,h),\hat{u}'(s,k))_U\}ds$$

Now $C\hat{x}(\cdot,h)$ and $C\hat{x}(\cdot,k)$ belong to $H^1(0,\infty;Y)$,$\hat{u}(\cdot,h)$ and $\hat{u}(\cdot,k)$ belong to $H^1(0,\infty;Y)$ and their limits as t goes to infinity are 0. Therefore

$$\langle P_\infty A_\Sigma h,k\rangle + \langle P_\infty h,A_\Sigma k\rangle = \int_0^\infty \frac{d}{dt}\{(C\hat{x}(s,h),C\hat{x}(s,k))_Y+(\hat{u}(s,h),\hat{u}(s,k))_U\}ds =$$

$$= -(Ch,Ck)_Y - (\hat{u}(0,h),\hat{u}(0,k))_U$$

But in view of expression (5.11) and (5.12) in Theorem 5.1 we readily obtain (6.1) and (6.2).#

Example 6.1. Recall the example in Section 3. We have seen that

(6.3) $$H = D(A) = \ell^2 \;,\; \Sigma = \{h\in \ell^2 \,;\, \sum_{k=1}^{\infty}(k+1)h_k^2 < \infty\}, \overline{\Sigma} = H$$

(6.4) $$D(P_\infty) = \{h\in \ell^2 \,;\, \sum_{k=1}^{\infty}(k+1)^2h_k^2 < \infty\};\; P_\infty c_k = (k+1)c_k \,,\; k\in N$$

We have moreover $D(A_\Sigma) = \Sigma$ and K is the closed operator in H

(6.5) $$D(K) = \Sigma,\; Kc_k = \sqrt{2k+1}\; c_k \,,\; k\in N$$

The space Σ is the set of all initial conditions which can be stabilized with a finite energy. However for all h in H

(6.6) $$\int_0^\infty |x(s)|_H^2 ds < \infty$$

and for all h in Σ

(6.7) $$\int_0^\infty |x(s)|_\Sigma^2 ds = \int_0^\infty \{|x(s)|_H^2 + \langle P_\infty x(s),x(s)\rangle_\Sigma\} ds \le c|h|^2 .\#$$

We remark that in general the closed loop system is not exponentially stable, as the following example shows.

Example 6.2. Let $H = D(A) = \ell^2$,

(6.8) $$Ae_k = 0 \ , \ Be_k = \frac{1}{k} e_k .$$

then

(6.9) $$P_\infty e_k = k\, e_k \ , \quad k \in N.$$

(6.10) $$\Sigma = \{h \in \ell^2 ; \sum_{k=1}^\infty k h_k^2 < \infty \ \}, \bar{\Sigma} = H$$

(6.11) $$Fe_k = (A-BB^*P_\infty)e_k = -\frac{1}{k} e_k .$$

Thus F is stable but not exponentially stable both in H and in Σ.#

Proposition 6.2. *If the triplet* (A,B,C) *is approximatively stabilizable and the pair* (A^*,C^*) *is stabilizable, then :*

(6.12) $$\int_0^\infty |\hat{x}(t,h)|_H^2 dt < \infty \quad , \text{ for all } h \in \Sigma.$$

Proof. If (A^*,C^*) is stabilizable, then there exists a minimal positive bounded solution to the Riccati equation :

(6.13) $$AQ + QA^* - QC^*CQ + I = 0$$

and the closed loop system

(6.14) $$y'(t) = [A^*-C^*CQ]y(t) \ , \ y(0) = k$$

is L^2-stable. Denote by T(t) the semigroup associated with the above system. For all h in Σ consider the optimal state $\hat{x}(\cdot,h)$ and control $\hat{u}(\cdot,h)$; then

(6.15) $$\hat{x}'(t,h) = [A^*-C^*CQ]^*\hat{x}(t,h) + QC^*C\hat{x}(t,h) + B\hat{u}(t,h) \ , \ \hat{x}(0,h) = h$$

and

$$\hat{x}(t,h) = T^*(t)h + \int_0^t T^*(t-s)[QC^*C\hat{x}(s,h)+B\hat{u}(s,h)]ds \tag{6.16}$$

It follows that

$$\|\hat{x}(t;h)\|_{L^2(0,\infty;H)} \leq \|T^*(\cdot)h\|_{L^2(0,\infty;H)} + \|T^*(\cdot)h\|_{L^2(0,\infty;H)}\|QC^*C\hat{x}(\cdot,h)+B\hat{u}(\cdot,h)\|_{L^2(0,\infty H)}$$

The right hand side is finite since T^* is exponentially decreasing, QC^* and B are bounded and $C\hat{x}(\cdot,h)$ and $\hat{u}(\cdot,h)$ are $L^2(0,\infty;H)$ functions.#

Remark 6.1. To show the L^2-stability with respect to the Σ norm, one would have to prove that

$$\int_0^\infty <P_\infty\hat{x}(t;h),\hat{x}(t,h)>_\Sigma dt = \int_0^\infty [|C\hat{x}(t,h)|^2+|\hat{u}(t,h)|^2 tdt < \infty.\# \tag{6.17}$$

7 A condition for approximated stabilizability.

We assume here that A is the infinitesimal generator of an analytic semigroup in H and denote by $\sigma(A)$ the spectrum of A and by $\rho(A)$ the resolvent set of A. If $\sigma(A)$ is point then it is possible to generalize the well known Hautus condition. In some applications (see Example 7.1, below), the following hypothesis is fulfilled

(i) $\sigma(A)$ *consists of a convergent sequence* $\{\lambda_i\}$ *of semi-simple eigenvalues plus the limit* $\lambda_\infty = \lim_{i\to\infty}\lambda_i$.

(ii) $\sigma(A)=\sigma^-(A)\cup\sigma^+(A)$ *with* $Re\lambda<0$ *if* $\lambda\in\sigma^-(A)$ *and* $Re\lambda>0$ *if* $\lambda\in\sigma^+(A)$.

(7.1) *We set* $P_\pm = \frac{1}{2\pi i}\int_{\gamma_\pm}(\lambda-A)^{-1}d\lambda$, *where* $\gamma_\pm$ *is a suitable curve around* $\sigma^\pm(A)$. *We have* $I = P_+ + P_-$.

(iii) Setting $P_i = \frac{1}{2\pi i}\int_{C(\lambda_i,\varepsilon_i)}(\lambda-A)^{-1}d\lambda$, *where* $C(\lambda_i,\varepsilon_i)$ *is a circle in* $\rho(A)$, *we have* $e^{tA}P_+x = \sum_{i=1}^\infty e^{t\lambda_i}P_i x$.

Proposition 7.1. *Assume* (7.1) *and let* $B\in \mathcal{L}(U;H)$. *Then the following statements are equivalent :*

(i) The triple (A,B,I) *is approximatively stabilizable.*

(ii) $\mathrm{Ker}(B^*)\cap\mathrm{Ker}(A^*-\lambda_i I) = \{0\}$ *for all* $\lambda_i\in\sigma^+(A)$.

Proof.(i)⇒(ii). Assume, by contradiction, that (A,B,I) is approximatively stabilizable and there exists $\lambda_i\in\sigma^+(A)$ and $|h|=1$ in H such that $A^*h=\bar{\lambda}_i h$, $B^*h = 0$. Now the solution x of (2.1) is given by $x(t) = e^{t\lambda_i}h$ so that h does not belong to $\bar{\Sigma}$.

ii)⇒i). Let $h\in H$ and $u\in L^2(0,\infty;U)$; we can write the solution of problem (2.1) as

(7.2) $$x(t) = e^{tA}P_-h + \int_0^t e^{(t-s)A}P_-Bu(s)ds + \int_t^\infty e^{(t-s)A}P_+Bu(s)ds$$
$$+ e^{tA}\left\{P_+h + \int_0^\infty e^{-sA}P_+Bu(s)ds\right\}$$

thus the control u is admissible if and only if $P_+x + \int_0^\infty e^{-sA}P_+Bu(s)ds = 0$. Consider now the mapping

(7.3) $$\gamma: L^2(0,\infty;U)\to P_+H=H_+,\ u\to x = \int_0^\infty e^{-sA}P_+Bu(s)ds$$

and its adjoint

(7.4) $$\gamma^* ;\ H_+^*\to L^2(0,\infty;U),\ h\to B^*e^{-sA^*}H_+^*h$$

Clearly the triple (A,B,I) is approximatively stabilizable if and only if $\mathrm{Ker}(\gamma^*) = \{0\}$. Now, assume that (ii) holds and, by contradiction, that $\mathrm{Ker}(\gamma^*) \neq \{0\}$. Let $h\in\mathrm{Ker}(\gamma^*)$; by (7.1) we have

(7.5) $$B^*e^{-sA^*}H_+^*h = B^*\sum_{i=1}^{\infty} e^{-t\lambda_i}P_i^*h.= 0$$

which implies $P_i^*h\in\mathrm{Ker}(B^*)$. Since $P_i^*h\in\mathrm{Ker}(A^*-\lambda_i I)$ (because λ_i is semi-simple) we have found a contradiction with (ii).#

Example 7.1.(the nerve axon system).

Let Ω be an open bounded set in **R**; consider the system (introduced in [2])

$$
(7.6)\quad \begin{cases} \frac{\partial x_1}{\partial t}(t,\xi) = \alpha\Delta x_1(t,) + b_{11}x_1(t,\xi) + b_{12}x_2(t,x) + \sum_{j=1}^{J} f_j(t)\phi_j(t,x) \\ t>0,\ \xi\in\Omega \\ \frac{\partial x_2}{\partial t}(t,\xi) = b_{21}x_1(t,x) + b_{22}x_2(t,x) + \sum_{j=1}^{J} g_j(t)\psi_j(t,x), \quad t>0,\ \xi\in\Omega \\ x_1(0,x) = h_1(x),\ \xi\in\Omega \\ x_2(0,x) = h_2(x),\ \xi\in\Omega \\ x_1(t,\xi) = 0,\ x_2(t,\xi) = 0,\quad t>0,\ \xi\in\partial\Omega \end{cases}
$$

where we assume that α, $b_{ij} : \mathbf{R}\to\mathbf{R}$ are given real numbers, with $\alpha > 0$, and $\phi_1,..,\phi_J$; $\psi_1,..,\psi_J \in C(\bar{\Omega})$ are linearly independent functions.

Choose $H= L^2(\Omega)\times L^2(\Omega)$, $U = \mathbf{R}^J\times\mathbf{R}^J$; setting

$$
(7.7)\quad x = \begin{bmatrix} x_1 \\ x_2 \end{bmatrix},\ h = \begin{bmatrix} h_1 \\ h_2 \end{bmatrix} \qquad u = \begin{bmatrix} (f_1,..,f_J) \\ (g_1,..,g_J) \end{bmatrix},\quad Bu = \begin{bmatrix} \sum_{j=1}^{J} f_j\phi_j(t,\cdot) \\ \sum_{j=1}^{J} g_j\psi_j(t,\cdot) \end{bmatrix}
$$

and

$$
(7.8)\qquad b = \begin{bmatrix} b_{11} & b_{12} \\ b_{21} & b_{22} \end{bmatrix}
$$

we can write system (7.6) in the abstract form (2.1). The spectrum $\sigma(A)$ of A consists in two sequences of semi-simple eigenvalues $\{\lambda_\pm(k)\}_{k\in N}$ and the accumulation point

$$
(7.10)\qquad \lambda_\infty = -b_{22}
$$

The eigenvalues $\lambda_\pm(k)$ are defined by

$$
(7.11)\qquad \lambda_\pm(k) = \frac{1}{2}\left\{-\alpha\mu_k + Tr(b) \pm [(-\alpha\mu_k + Tr(b))^2 + 4(\alpha\mu_k b_{22} - det(b)]\right\}
$$

where are the eigenvalues of the Laplacian with Dirichlet boundary conditions.

Now it is easy to check hypothesis 7.1, so that we can apply Proposition 7.1.#

References

[1] G.Da Prato and M.C.Delfour, Stabilization and unbounded solutions of the Riccati equation, Proc. 27th IEEE Conference on Decision and Control, pp.352-357, IEEE Publications, N.Y., 1988

[2] J.Evans , The Stability of Nerve Impulses, I: Linear Approximations, Indiana Univ. Math. Journal, 21 ,pp. 877-885, 1972.

RICCATI EQUATIONS IN NONCYLINDRICAL DOMAINS

PIERMARCO CANNARSA*
Dipartimento di Matematica
Università di Pisa
Via F. Buonarroti, 2. 56127 Pisa, Italy

GIUSEPPE DA PRATO*
Scuola Normale Superiore
56126 Pisa, Italy

JEAN-PAUL ZOLESIO**
CNRS
Laboratoire de Physique Mathématique, U.S.T.L.
34060, Montpellier Cedex, France

1 Introduction

We are here concerened with a dynamical system governed by the following state equation :

$$(1.1)\qquad \begin{cases} u_t(t,x) = \Delta u(t,x) + f(t,x) & ;\ t\in\,]0,T],\ x\in \Omega_t \\ u(t,x) = 0 & ;\ t\in\,]0,T],\ x\in \Gamma_t \\ u(0,x) = u_0(x) & ;\ x\in \Omega_0 \end{cases}$$

where Ω_0 is a bounded domain in $\mathbf{R}^N$ with smooth boundary Γ_0, $\Omega_t = T_t(\Omega_0)$, $\Gamma_t = T_t(\Gamma_0)$ and T_t denote the flow associated to a given vector field $V : [0,T]\times\mathbf{R}^N \to \mathbf{R}^N$, i.e.

$$(1.2)\qquad \frac{\partial}{\partial t} T_t(x) = V(t,T_t(x)),\ \ t\in [0,T],\ x\in \mathbf{R}^N \qquad\qquad T_0(x) = x$$

---

* The author is a member of GNAFA (Consiglio Nazionale delle Ricerche)

** Part of this research was completed while the author was visiting Scuola Normale Superiore, Pisa (Italy).

We assume that $V(\cdot,x)$ is continuous uniformly in x and $V(t,\cdot)$ is twice continuously differentiable with bounded second order derivatives. Moreover $u_0 \in L^2(\Omega_0)$ and the control f belongs to $L^2(Q_T)$,
where $Q_T = \bigcup_{0<t<T} \{t\}\times\Omega_t$.

We want to minimize the cost functional :

(1.3) $$J(f) = \int_0^T dt \int_{\Omega_t} \{|u(t,x)|^2 + |f(t,x)|^2\}dx + \int_{\Omega_T} |u(T,x)|^2 dx$$

over all $f \in L^2(Q_T)$.

This problem has been studied in [8], by reducing (1.1)-(1.3) to a problem in a fixed domain, by a suitable change of variables. With this procedure, one can solve the Riccati equation for the transformed problem, and then perform the inverse change of variables. However it seems simpler to work directly in the moving domain.

We present here an approach which is based on the reduction of problem (1.1) to an abstract evolution equation in a fixed Banach space, which has been introduced in [6]. More precisely, we consider the Hilbert space $L^2(\mathbf{R}^N) = H$ (norm $\|\cdot\|$), and introduce a family of linear operators $\{A(t)\}_{t\in[0,T]}$ in $L^2(\mathbf{R}^N) = H$ as follows:

(1.4) $$\begin{cases} D_{A(t)} = \{u \in L^2(\mathbf{R}^N) \,;\, u_{|\Omega_t} \in H^2(\Omega_t)\cap H_0^1(\Omega_t) \text{ and } u_{|\Omega_t^c} \in H^2(\Omega_t^c)\cap H_0^1(\Omega_t^c)\ \} \\ \\ A(t)y = z \in H \text{ where } \int_{\mathbf{R}^N} z\phi\, dx = \int_{\mathbf{R}^N} y\ \Delta\phi\, dx \end{cases}$$

for all $\phi \in C^\infty(\mathbf{R}^N)$ satisfying $\phi=0$ on Γ_t.

Next we consider the function :

(1.5) $$\beta(t,x) = \begin{cases} 1 \text{ if } x \in \Omega_t \text{ and } t \in [0,T] \\ \\ 0 \text{ otherwise} \end{cases}$$

and introduce the linear operator B(t) in $\mathcal{L}(H)$ as

(1.6) $$(B(t)z)(x) = \beta(t,x)z(x) \quad , \forall z \in H$$

Thus, we transform problem (1.1)-(1.3) in the following :

Minimize :

$$(1.7) \qquad J_1(f) = \int_0^T \{ \|B(t)v(t)\|^2 + \|B(t)f(t)\|^2 \} dt + \|B(T)v(T)\|^2 .$$

over all controls $f \in L^2(0,T;H)$, subject to state equation

$$(1.8) \qquad v'(t) = A(t)v(t) + B(t)f(t) \, , \; v(0) = v_0 \in H$$

Problem (1.1)-(1.3) is related to problem (1.7)-(1.8) in the following way. If (f_1^*, v^*) is an optimal pair for (1.7)-(1.8) and if $\beta(0,\cdot)v_0 = u_0$, then, setting :

$$(1.9) \qquad f^*(t) = \beta(t,\cdot)f_1^*(t)|_{Q_T} \, , \; u^*(t) = \beta(t,\cdot)v^*(t)|_{Q_T}$$

(f^*,u^*) is an optimal pair for (1.1)-(1.3).

Therefore it is sufficient to solve problem (1.7)-(1.8). In order to do this we first solve state equation (1.8) by using Theorem 1.1 below (proved in [6]), then we solve the control problem by using Dynamic Programming. Remark, however that the quadratic form $J_1(f)$ is not coercive, thus we have to generalize the usual theory to cover this case.

Theorem 1.1.
For any $v_0 \in H$ and any $f \in L^2(0,T;H)$, there exists a unique strong solution of (1.8) $v \in C([0,T];H)$.

2 Dynamic Programming

We study here problem (1.7)-(1.8). As previously remarked the cost $J_1(f)$ is not coercive, however Riccati equation can be formally stated as follows :

$$(2.1) \qquad Q' + A^*Q + QA - QBQ + B = 0 \; ; \; Q(T) = B(T)$$

Remark that the term QBQ stays for the meaningless term $QBB^{-1}BQ$.

Let us introduce some notation. We denote by $\Sigma(H)$ (resp. $\Sigma^+(H)$) the cone of symmetric (resp. symmetric and positive) linear bounded operators in H and by $C_s([0,T];\Sigma(H))$ the Banach space of all strongly continuous mappings $F: [0,T]\to\Sigma(H)$, endowed with the norm :

$$(2.2) \qquad \|F\| = \sup\{\|F(t)\| \,;\, t\in[0,T]\}$$

Next, we say that $Q\in C_s([0,T];\Sigma(H))$ is a *mild solution* of problem (2.1) if it is a solution of the following integral equation :

$$(2.3) \quad Q(t)x = U^*(T,t)B(T)U(T,t)x + \int_t^T U^*(s,t)[B(s) - Q(s)B(s)Q(s)]U(s,t)x\,ds \;;\; x\in H$$

where $U(t,s)$ is the evolution operator associated to the family $\{A(t)\}_{t\in[0,T]}$, whose existence is proved in [9]. [3] and [6].

The following result can be proved as in [7] and [2].

Proposition 2.1
Equation (2.1) has a unique mild solution $Q\in C_s([0,T];\Sigma^+(H))$. Moreover, for any $x\in H$ we have :

$$(2.4) \qquad Q_n(\cdot)x \to Q(\cdot)x \text{ in } C([0,T];H) \text{ as } n\to\infty$$

where Q_n is the solution to the approximating Riccati equation :

$$(2.5) \qquad Q_n' + A_n^*Q_n + Q_nA_n - Q_nBQ_n + B = 0 \;;\; Q_n(T) = B(T)$$

and $A_n = nA(n-A)^{-1}$.

We can state now our main result.

Theorem 2.2
There exists a unique optimal pair (f_1^*, v^*) for problem (1.7)-(1.8). Moreover the following statements hold.

i) $f_1^*(t) = - B(t)Q(t)v^*(t)$; $t\in[0,T]$
ii) $v^*(t)$ is the solution to the closed loop equation :

(2.6) $z' = A(t)z - B(t)z \ ; \ z(0) = v_0$

iii) If (g,w) is another optimal pair, we have :

(2.7) $f_1^* = g$ a.e. on Q_T

Proof.

Let $f \in L^2(0,T;H)$ and let v be the corresponding solution of (1.8). Moreover denote by v_n the solution of

(2.8) $$v_n'(t) = A_n(t)v_n(t) + B(t)f(t) \ , \ v(0) = v_0 \in H$$

and by Q_n the solution of (2.5). By computing $\frac{d}{dt}<Q_n(t)v_n(t),v_n(t)>$ and then, by integrating in [0,T], we find :

(2.9) $$<Q_n(0)v_0,v_0> + \int_0^T \|B(s)Q_n(s)v_n(s)+B(s)f(s)\|^2 ds =$$

$$= \int_0^T \{\|B(s)v_n(s)\|^2+\|B(s)f(s)\|^2\} ds \ + \|B(T)v_n((T)\|^2$$

As $n \to \infty$, we have :

(2.10) $$<Q(0)v_0,v_0> + \int_0^T \|B(s)Q(s)v(s)+B(s)f(s)\|^2 ds = J_1(f)$$

Now, let $v^*(t)$ be the solution of (2.6), and set $f_1^*(t) = - B(t)Q(t)v^*(t)$,then by (2.10) it follows that the pair (f_1^*, v^*) is optimal.

Let finally (g,w) be is another optimal pair; then by (2.10) it follows :

(2.11) $$B(t)Q(t)w(t)+B(t)g(t) = 0$$

which implies

(2.12) $$g(t) = -B(t)w(t) \text{ on } Q_T$$

and the last statement holds.

Remark 2.3.

We can go back to the original problem (1.1)-(1.3). Let us define a linear operator F on $L^2(Q_T)$, by setting :

$$(2.13) \qquad (Fu)(t,x) = [B(t)Q(t)B(t)u(t,\cdot)](x) \quad \forall (t,x)\in Q_T$$

Then, by Theorem (2.2) it follows that there exists a unique optimal pair (f^*,u^*) for problem (1.1)-(1.3), where

$$(2.14) \qquad f^* = -Fu^*$$

and u^* is the solution to the closed loop equation

$$(2.15) \qquad \begin{cases} u_t(t,x) = \Delta u(t,x) - F(u(t,\cdot))(x) \ ; \ t\in]0,T], \ x\in \Omega_t \\ u(t,x) = 0 \ ; \ t\in]0,T], \ x\in \Gamma_t \\ u(0,x) = u_0(x) \ ; \ x\in \Omega_0 \end{cases}$$

3. Boundary Control. In almost all examples when a Dirichlet Boundary Control f is applied, the control occurs on a fixed part, Σ, of the boundary. Then, we will study the following modelling problem:

$$(3.1) \qquad \begin{cases} u_t(t,x) = \Delta u(t,x) \ ; \ t\in]0,T], \ x\in \Omega_t \\ u(t,x) = f(t,x) \ ; \ t\in]0,T], \ x\in \Gamma_t \\ u(0,x) = u_0(x) \ ; \ x\in \Omega_0 \end{cases}$$

assuming that there exists a subset Σ contained in Γ_t for all $t\in [0,T]$, of finite superficial measure, such that $\Sigma\supset spt(f)$. Notice that this assumption can be equivalently restated requiring the vector field V in (1.2) to vanish on Σ.

Now, using the well-known procedure introduced by [5] (and developed by many authors such as [10] and [4]), we write problem (3.1) as

$$(3.2) \qquad u(t,\cdot) = U(t,0)u_0 - \int_0^t \frac{\partial}{\partial s}U(t,s)D_s f(s,\cdot)ds$$

where $u(t,\cdot)(x) = u(t,x)$, $f(t,\cdot)(x) = f(t,x)$ and $U(t,s)$ is the evolution operator associated to the family $\{A(t)\}_{t\in[0,T]}$ of linear operators in the Hilbert space $H= L^2(\mathbf{R}^N)$ which is defined in (1.4). Moreover, D_s is the Dirichlet mapping given by $D_s z = w$, where

$$(3.3) \qquad \begin{cases} \Delta w = 0 \quad \text{in} \quad \Omega_t \cup \Omega_t^c \\ w(x) = z(x) \text{ on } \Gamma_t \end{cases}$$

As is well-known, D_s is a continuous map of $L^2(\Gamma_t)$ into $H^{1/2-\varepsilon}(\mathbf{R}^N)$ for all $\varepsilon>0$.

We recall (see [2] and [1]) that there exists $k>0$ such that

$$(3.4) \qquad \left\|\frac{\partial}{\partial s} U(t,s)\right\|_{L(H^{1/2-\varepsilon}(\mathbf{R}^N);L^2(\mathbf{R}^N))} \leq \frac{k}{|t-s|^{1-2\varepsilon}}$$

Therefore, equation (3.2) has a rigorous interpretation.

We want to minimize the cost functional

$$K(f) = \int_0^T [\int_{\Omega_t} |u(t,x)|^2 dx + \int_{\Sigma} |f(t,x)|^2 d\sigma(x)]dt + \int_{\Omega_T} |u(T,x)|^2 dx$$

over all controls $f\in L^2([0,T]\times\Sigma)$.

By introducing the family of operators $\{B(t)\}_{t\in[0,T]}$ as in (1.5)-(1.6) and the control space $U=L^2(\Sigma)$, we transform the above minimization problem into the following:

$$(3.5) \qquad \text{Minimize} \qquad K_1(f) = \int_0^T \left\{\|B(t)u(t)\|_H^2 + \|f(t)\|_U^2\right\} dt + \|B(T)u(T)\|_H^2 ,$$

over all controls $f\in L^2(0,T;U)$ and states u subject to equation (3.2).

Notice that, unlike problem (1.7), problem (3.5) is now coercive. So, we can apply the results of [2] (see also [1]) to solve the Riccati Equation associated to (3.5), and derive a solution of problem (3.5) by Dynamic Programming.

References

[1] P. Acquistapace: to appear this volume.

[2] P. Acquistapace, F. Flandoli, B. Terreni: *Boundary control of nonautonomous parabolic systems*, preprint

[3] P. Acquistapace and B. Terreni: Some existence and regularity results for abstract non-autonomous parabolic equations, *J. Math. Anal. Appl. 99 (1984), 9-61.*

[4] H. Amann: *Feedback Stabilization of Linear and Semilinear Parabolic Systems*, Proceedings of the Meeting "Trends in Semigroup Theory and Applications", Trieste, Sept. 28- Oct. 2, 1987, Lecture Notes in Pure and Applied Mathematics, M. Dekker (to appear)

[5] V. A. Balakrishnan: *Applied Functional Analysis* (Second Edition), Springer-Verlag, New York-Heidelberg-Berlin (1981)

[6] P.Cannarsa,G.Da Prato and J.P.Zolésio : The Damped wave Equation in Moving Domains, (preprint)

[7], G.Da Prato: Quelques résultat d'existence, unicité et régularité pour un problème de la théorie du contrôle.J. Maths. Pures Appl. **52**, (1973), 353-375.

[8], G. Da Prato and J. P. Zolesio: An optimal control problem for a parabolic equation in non-cylindrical domains, *S. & C. Letters* **11** (1988), 73-77.

[9], T.Kato and H.Tanabe: On the abstract evolution equation, *Osaka Math. J.* **14** (1962), 107-133.

[10], I. Lasiecka, R. Triggiani: *Stabilization and structural assignement of Dirichlet boundary feedback parabolic equations*, SIAM J. Control & Optim. **21** (1983), 766-803

BOUNDARY CONTROL PROBLEMS FOR NON-AUTONOMOUS PARABOLIC SYSTEMS

Paolo Acquistapace

Dipartimento di Metodi e Modelli Matematici
per le Scienze Applicate
Università di Roma "La Sapienza", via A. Scarpa 10, Roma

&

Brunello Terreni

Dipartimento di Matematica "F. Enriques"
Università Statale di Milano, via C. Saldini 50, Milano

0. Goal

This paper is concerned with boundary control, over finite time horizon, of some linear non-autonomous systems of parabolic type. We continue the analysis of [1] where we studied , from an abstract point of view, a class of parabolic systems obeying the abstract assumptions of [3,4] which, essentially, force the system to have a variational structure. We consider here a different class of not necessarily variational systems, whose "parabolicity" is somewhat less stringent (for instance, the order of some boundary operators may reduce by one, or more, for some values of t), provided we pay a price in terms of higher time regularity assumptions. The abstract hypotheses corresponding to this class are those of [8,2], which are in fact independent of those of [3,4], as remarked in [3, §7].

1. The initial-boundary value problem

Let Ω be a bounded open set of $\mathbb{R}^n$, $n\geq 1$, with sufficiently smooth $\partial\Omega$. Consider a pair $\big(\mathcal{A}(t,x,D),\mathcal{B}(t,x,D)\big)$ of N-ples of differential operators, acting on functions $u:\Omega\to\mathbb{C}^N$, $N\geq 1$, defined respectively in $[0,T]\times\Omega$ and in $[0,T]\times\partial\Omega$. We assume that $(\mathcal{A},\mathcal{B})$ is an elliptic system of order 2m, $m\geq 1$, in the sense of [7, §5], and that all related properties hold uniformly with respect to $t\in[0,T]$; moreover we require that all coefficients of $(\mathcal{A},\mathcal{B})$ are $C^{1+\alpha}$ in t for some $\alpha\in]0,1[$.

Here is the initial-boundary value problem we are going to control:

$$\begin{cases} y_t(t,x)-\mathcal{A}(t,x,D)y = f(t,x), & (t,x)\in[0,T]\times\Omega, \\ y(0,x) = y_0(x), & x\in\Omega, \\ \mathcal{B}(t,x,D)y = u(t,x), & (t,x)\in[0,T]\times\partial\Omega; \end{cases} \tag{1.1}$$

as we are interested to boundary control, we will take $f\equiv0$.

2. The abstract initial-boundary value problem

Set $H:=[L^2(\Omega)]^N$, and define the linear unbounded operator $A(t):D_{A(t)}\subseteq H\to H$ by:

$$\begin{cases} D_{A(t)}:=\{u\in[W^{2,2}(\Omega)]^N:\ \mathcal{B}(t,\cdot,D)u=0 \text{ on } \partial\Omega\} \\ A(t)u:=\mathcal{A}(t,\cdot,D)u. \end{cases} \tag{2.1}$$

It is known that $\{A(t)\}_{t\in[0,T]}$ is a family of closed, densely defined generators of analytic semigroups in H; in particular it is not restrictive to assume:

$$\begin{cases} \text{there exists } \omega\in]\pi/2,\pi[\text{ such that the resolvent set of} \\ A(t) \text{ contains the sector } S_\omega:=\{0\}\cup\{z\in\mathbb{C}:\ |\arg z|<\omega\}, \text{ and} \\ \qquad \|[\lambda-A(t)]^{-1}\|_{\mathcal{L}(H)} \le \dfrac{M}{1+|\lambda|} \quad \forall\lambda\in S_\omega,\ \forall t\in[0,T]. \end{cases} \tag{2.2}$$

Next, arguing as in [10, § 5.3], it is easy to obtain:

$$\begin{cases} \text{there exists } \alpha\in]0,1[\text{ such that } t\to[\lambda-A(t)]^{-1}\in C^{1+\alpha}([0,T],\mathcal{L}(H)) \\ \text{for each } \lambda\in S_\omega, \text{ and} \\ \qquad \left\|\dfrac{d}{dt}[\lambda-A(t)]^{-1}\right\|_{\mathcal{L}(H)} \le \dfrac{K}{1+|\lambda|^\alpha} \quad \forall\lambda\in S_\omega,\ \forall t\in[0,T], \\ \qquad \left\|\dfrac{d}{dt}A(t)^{-1}-\dfrac{d}{ds}A(s)^{-1}\right\|_{\mathcal{L}(H)} \le N\ |t-s|^\alpha \quad \forall s,t\in[0,T]. \end{cases} \tag{2.3}$$

Conditions (2.2) and (2.3) are the assumptions of [8]. The abstract Cauchy problem

$$\begin{cases} y'(t)-A(t)y(t) = f(t), & t\in[0,T] \\ y(0) = y_0 \end{cases} \tag{2.4}$$

corresponds to the concrete problem (1.1) with $u\equiv0$.

To give an abstract version of the non-homogeneous boundary conditions of (1.1) we introduce the elliptic "Green map" $G(t)$, defined by:

$$v:=G(t)g \Leftrightarrow \begin{cases} \mathcal{A}(t,\cdot,D)v = 0 & \text{in } \Omega, \\ \mathcal{B}(t,\cdot,D)v = 0 & \text{on } \partial\Omega. \end{cases} \tag{2.5}$$

The existence of $v=G(t)g$, for smooth g, is guaranteed by the results of [7]. However we need a stronger property: namely, setting $U:=[L^2(\partial\Omega)]^N$, we assume:

$$\begin{cases} \text{there exists } \vartheta\in]0,1[\text{ such that } G(t)\in\mathcal{L}\left(U, D_{[-A(t)]^{\vartheta}}\right) \text{ for} \\ \text{each } t\in[0,T] \text{ and } t\to[-A(t)]^{\vartheta}G(t)\in C\left([0,T],\mathcal{L}(U,H)\right). \end{cases} \tag{2.6}$$

This requirement is natural in variational problems, but in our situation it is not easy to get it; we will see later some examples where (2.6) is satisfied.

Using the operator $G(t)$, we may rewrite problem (1.1) as:

$$\begin{cases} y'(t)-A(t)[y(t)-G(t)u(t)] = 0, & t\in[0,T] \\ y(0) = y_0 . \end{cases} \tag{2.7}$$

3. The evolution operator

The results of [8] and [2] show that we can construct the evolution operator $\{U(t,s)\}_{0\le s<t\le T}$ associated to $\{A(t)\}_{t\in[0,T]}$; thus if $y_0\in H$ and $f\in L^2(0,T;H)$ the solution of problem (2.4) is

$$y(t) = U(t,0)y_0+\int_0^t U(t,s)f(s)ds, \quad t\in[0,T].$$

Consequently, at least formally, if $u\in L^2(0,T;U)$ the solution of problem (2.7) is

$$y(t) = U(t,0)y_0+\int_0^t U(t,s)A(s)G(s)u(s)ds, \quad t\in[0,T]; \tag{3.1}$$

but (3.1) is not meaningful because the range of $G(s)$ is never contained in $D_{A(s)}$. In order to give sense to (3.1) let us recall some properties of $\{U(t,s)\}_{0\le s<t\le T}$, proved in [5]:

PROPOSITION 3.1 *Under assumptions* (2.1), (2.2), (2.3) *let*

$\{U(t,s)\}$ *be the evolution operator associated to* $\{A(t)\}$. *Then:*

(i) $\|[-A(t)]^{\gamma}U(t,s)[-A(s)]^{\beta}\|_{\mathcal{L}(H)} \le M_{\gamma\beta}[1+(t-s)^{\beta-\gamma}]$

$$\forall 0\le s<t\le T,\ \forall \gamma,\beta\in[0,1];$$

(ii) *for each* $0\le s<t\le T$ *and* $x\in H$ *there exists* $\frac{d}{ds}U(t,s)=:V(t,s)$, *and*

$$\|V(t,s)\|_{\mathcal{L}(D_{[-A(s)]^{\vartheta}},H)} \le B(t-s)^{\vartheta-1} \quad \forall 0\le s<t\le T,\ \forall\vartheta\in[0,1],$$

$$\|V(t,s)+A(s)e^{(t-s)A(s)}\|_{\mathcal{L}(D_{[-A(s)]^{\vartheta}},H)} \le B(t-s)^{\vartheta+\alpha-1}$$

$$\forall 0\le s<t\le T,\ \forall\vartheta\in[0,1];$$

(iii) $[-A(t)]^{-\beta}U(t,s)[-A(s)]^{\gamma}$ *has an extension to* $\mathcal{L}(H)$ *bounded by* $M_{\gamma\beta}[1+(t-s)^{\beta-\gamma}]$ $\forall 0\le s<t\le T,\ \forall\gamma,\beta\in[0,1]$. □

Using Proposition 3.1(ii) and the obvious fact that $\frac{d}{ds}U(t,s)x = -U(t,s)A(s)x$ $\forall x\in D_{A(s)}$, we may rewrite (3.1) as:

$$y(t) = U(t,0)y_0 - \int_0^t \frac{d}{ds}U(t,s)G(s)u(s)ds, \quad t\in[0,T], \tag{3.2}$$

and this expression is obviously meaningful in view of (2.6).

4. The control problem

Equation (3.2) will be considered as the state equation of the following L.-Q.-R. problem:

$$\left\{\begin{array}{l} \text{minimize} \\ J(u):=\int_0^T [\left(M(t)y(t)|y(t)\right)_H+\left(N(t)u(t)|u(t)\right)_U]dt + \\ \qquad\qquad +\left(P_T y(T)|y(T)\right)_H \\ \text{over all controls } u\in L^2(0,T;U), \text{ where } y \text{ is the solution of} \\ \text{problem (2.7), i.e. } y \text{ is given by (3.2).} \end{array}\right. \tag{4.1}$$

Here we assume (besides (2.1), (2.2), (2.3), (2.5) and (2.6)):

$$M\in L^{\infty}\left(0,T;\Sigma^{+}(H)\right), \tag{4.2}$$

$$N\in C_s\big([0,T],\Sigma^+(U)\big) \quad \text{and} \quad N(t)\ge\nu>0 \quad \forall t\in[0,T], \tag{4.3}$$

$$\begin{cases} P_T\in\Sigma^+(H) \text{ and } [-A(T)^*]^{2\beta}P_T \text{ can be continuously extended} \\ \text{to } \mathcal{L}(H) \text{ for some } \beta\in]\frac{1}{2}-\alpha,\frac{1}{2}]\cap[0,\frac{1}{2}]. \end{cases} \tag{4.4}$$

We remark that by (4.4) the term $\big(P_Ty(T)|y(T)\big)_H$ is well defined and in addition, thinking of it as a function of u (via (3.2)), it turns out that such a function is continuous on $L^2(0,T;U)$ [1, Lemma 3.5].

We solve the control problem exactly as in [1], by a dynamic programming technique.

5. The Riccati equation

The Riccati equation associated to the control problem (4.1) is, formally,

$$\begin{aligned} P(t) &= U(T,t)^*P_TU(T,t) + \\ &+ \int_t^T U(s,t)^*[M(s)-P(s)A(s)G(s)N(s)^{-1}G(s)^*A(s)^*P(s)]U(s,t)ds, \end{aligned} \tag{5.1}$$

but it is not meaningful since the range of G(s) is not contained in $D_{A(s)}$. However, recalling (2.6), we rewrite (5.1) as

$$\begin{aligned} P(t) &= U(T,t)^*P_TU(T,t) + \\ &+ \int_t^T U(s,t)^*\Big\{M(s)-[[-A(s)^*]^{1-\vartheta}P(s)]^*K(s)[-A(s)^*]^{1-\vartheta}P(s)\Big\}U(s,t)ds \end{aligned} \tag{5.2}$$

where

$$K(s):=[-A(s)^{\vartheta}]G(s)N(s)^{-1}[[-A(s)]^{\vartheta}G(s)]^*; \tag{5.3}$$

by (2.6) and (4.3) we have $K\in C\big([0,T],\Sigma^+(H)\big)$.

The same argument of [1] leads to:

PROPOSITION 5.1 *Equation* (5.2) *has a unique solution* P *in* [0,T]. *Moreover:*

(i) $P(t)\ge0 \quad \forall t\in[0,T]$;

(ii) P *satisfies the integral equation*

$$P(t) = U(T,t)^*P_T\Phi(T,t) + \int_t^T U(\sigma,t)^*M(\sigma)\Phi(\sigma,t)d\sigma, \quad t\in[0,T], \qquad (5.4)$$

where $\{\Phi(t,s)\}_{0\le s\le t\le T}$ *is an evolution operator which is the unique solution of this further integral equation:*

$$\Phi(t,s)x = U(t,s)x +$$

$$+ \int_s^t [[-A(r)^*]^{1-\vartheta}U(t,r)^*]^*K(r)[[-A(r)^*]^{1-\vartheta}P(r)]^*\Phi(r,s)xdr, \qquad (5.5)$$

$$t\in[0,T], \quad x\in H;$$

(iii) *for each* $\eta\in[0,1]$ *(resp.* $\eta\in[0,\vartheta]$*),* $t\to[-A(t)^*]^{\eta}P(t)\in C\big([0,T[,\mathcal{L}(H)\big)$ *(resp.* $C\big([0,T],\mathcal{L}(H)\big)$*).* □

6. Synthesis

Following again [1] we get:

THEOREM 6.1 *Let* $y_0\in H$ *be given. Then:*

(i) *there exists a unique optimal control* $u_*\in L^2(0,T;U)$ *for problem* (4.1);

(ii) *the optimal cost* $J(u_*)$ *is given by*

$$J(u_*) = \big(P(0)y_0|y_0\big)_H ,$$

where P *is the solution of the Riccati equation* (5.2);

(iii) *The optimal state* $y_*\in L^2(0,T;H)$ *is given by*

$$y_*(t) = \Phi(t,0)y_0 , \quad t\in[0,T],$$

where Φ *is the solution of* (5.5);

(iv) *the optimal control* u_* *is given by the feedback formula*

$$u_*(t) = - N(t)^{-1}G(t)^*A(t)^*P(t)y_*(t), \quad t\in[0,T[,$$

where we have written $G(t)^*A(t)^*$ *instead of* $-[[-A(t)]^{\vartheta}G(t)]^*[-A(t)^*]^{1-\vartheta}$;

(v) *the optimal pair* (u_*,y_*) *is characterized by the optimality system*

$$\begin{cases} y_*(t) = U(t,0)y_0 - \int_0^t \frac{d}{ds}U(t,s)G(s)u(s)ds, \\ u_*(t) = -N(t)^{-1}G(t)^*A(t)^*p(t), & t\in[0,T[. \\ p(t) := U(T,t)^*P_Ty_*(T) + \int_t^T U(s,t)^*M(s)y_*(s)ds, \end{cases}$$

□

7. Examples

EXAMPLE 7.1 Take $N=1$, $\partial\Omega\in C^3$, $\lambda_0>0$ and

$$\mathcal{A}(t,x,D)u := \sum_{s,j=1}^{n} D_s\left(A_{sj}(t,x)D_ju\right) + \lambda_0 u \;, \quad (t,x)\in[0,T]\times\overline{\Omega}, \tag{7.1}$$

$$\mathcal{B}(t,x,D)u := \sum_{s=1}^{n} \beta_s(t,x)D_su + \alpha(t,x)u \;, \quad (t,x)\in[0,T]\times\partial\Omega, \tag{7.2}$$

where

$$A_{sj}, \beta_s, \alpha \text{ are } C^2 \text{ in } x \text{ and } C^{1+\alpha} \text{ in } t; \tag{7.3}$$

for each $x\in\partial\Omega$ and $t\in[0,T]$ the polynomial

$$\sum_{s,j=1}^{n} A_{sj}(t,x)\left(\xi_j+\tau\nu_j(x)\right)\left(\xi_s+\tau\nu_s(x)\right) \tag{7.4}$$

has 1 root $\tau_+(t,x;\xi)$ with positive imaginary part for each $\xi\in\mathbb{R}^n$ tangent to $\partial\Omega$ at x;

$$\sum_{s=1}^{n} \beta_s(t,x)\left(\xi_s+\tau_+(t,x;\xi)\nu_s(x)\right)\neq 0 \quad \forall\xi\in\mathbb{R}^n \text{ tangent to } \partial\Omega \text{ at } x \tag{7.5}$$

(in particular, β may be a real, non-tangential vector).

It is easy to see that the abstract assumptions (2.2) and (2.3) hold for the operator $A(t)$ given by (2.1), provided λ_0 is sufficiently large. Let us verify that $G(t)$, given by (2.5), fulfills (2.6). First of all we have Green's formula:

$$\int_\Omega [\mathcal{A}(t,x,D)u\,\overline{v} - u\,\overline{\mathcal{A}^*(t,x,D)v}]dx = \int_{\partial\Omega} \left[\frac{\partial u}{\partial\nu_A}\,\overline{v} - u\,\overline{\frac{\partial v}{\partial\nu_{A^*}}}\right]d\sigma \tag{7.6}$$

$$\forall u,v\in W^{2,2}(\Omega),$$

where

$$\mathcal{A}^*(t,x,D)v := \sum_{s,j=1}^{n} D_s\left(\overline{{}^tA_{sj}(t,x)D_j v}\right) + \lambda_0 v\ , \qquad (t,x)\in[0,T]\times\overline{\Omega}$$

and $\frac{\partial}{\partial\nu_A}$, $\frac{\partial}{\partial\nu_{A^*}}$ are the conormal derivatives relative to $\mathcal{A}$, $\mathcal{A}^*$.

Let us denote by $\tau^1(x)$, ..., $\tau^{n-1}(x)$ an orthonormal system of vectors tangent to $\partial\Omega$ at x; then

$$\frac{\partial u}{\partial\nu_A} = \sum_{j=1}^{n}\left\{-(\nu_A\cdot\nu)(\beta\cdot\nu)^{-1}(\beta\cdot\tau^j)+(\nu_A\cdot\tau^j)\right\}\frac{\partial u}{\partial\tau^j} +$$

$$+ (\nu_A\cdot\nu)(\beta\cdot\nu)^{-1}\mathcal{B}(t,\cdot,D)u - \alpha(\nu_A\cdot\nu)(\beta\cdot\nu)^{-1}u,$$

so that by (7.6) it is easy to get (compare with [9, Ch.2, Th.2.1])

$$\int_\Omega [\mathcal{A}(t,x,D)u\ \bar{v} - \overline{\mathcal{A}^*(t,x,D)v}]dx =$$

$$= \int_{\partial\Omega} [\mathcal{B}(t,x,D)u\ (\nu_A\cdot\nu)(\beta\cdot\nu)^{-1}\bar{v} - u\ \overline{\mathscr{C}(t,x,D)v}]d\sigma \quad \forall u,v\in W^{2,2}(\Omega), \tag{7.7}$$

where

$$\begin{aligned}\mathscr{C}(t,x,D)v := &-\frac{\partial v}{\partial\nu_{A^*}} + \sum_{j=1}^{n-1}\overline{[(\nu_A\cdot\nu)(\beta\cdot\nu)^{-1}(\beta\cdot\tau^j)-(\nu_A\cdot\tau^j)]}\frac{\partial v}{\partial\tau^j} + \\ &+\left\{\sum_{j=1}^{n-1}\frac{\partial}{\partial\tau^j}\overline{[(\nu_A\cdot\nu)(\beta\cdot\nu)^{-1}(\beta\cdot\tau^j)-(\nu_A\cdot\tau^j)]} - \alpha(\nu_A\cdot\nu)(\beta\cdot\nu)^{-1}\right\}v;\end{aligned} \tag{7.8}$$

in particular we have

$$\begin{cases} D_{A(t)^*} := \{v\in[W^{2,2}(\Omega)]^N:\ \mathscr{C}(t,\cdot,D)v=0 \text{ on } \partial\Omega\} \\ A(t)^*v := \mathcal{A}^*(t,\cdot,D)v. \end{cases} \tag{7.9}$$

Now let $g\in W^{2,2}(\partial\Omega)$ and set $u:=G(t)g$, i.e., by (2.5),

$$\begin{cases} \mathcal{A}(t,\cdot,D)u = 0 & \text{in } \Omega, \\ \mathcal{B}(t,\cdot,D)u = g & \text{on } \partial\Omega. \end{cases}$$

By classical results [7,9]

$$\|u\|_{W^{2,2}(\Omega)} \le c\ \|g\|_{W^{1/2,2}(\partial\Omega)} \qquad \forall g\in W^{1/2,2}(\partial\Omega). \tag{7.10}$$

On the other hand, let $\psi:=[A(t)^*]^{-1}u$, i.e.

$$\begin{cases} \mathcal{A}^*(t,\cdot,D)\psi = u & \text{in } \Omega, \\ \mathcal{C}(t,\cdot,D)\psi = 0 & \text{on } \partial\Omega; \end{cases} \tag{7.11}$$

as $(\mathcal{A}^*,\mathcal{C})$ satisfies the assumptions of [7], again by classical results we have

$$\|\psi\|_{W^{2,2}(\Omega)} \le c\,\|u\|_{L^2(\Omega)}\,. \tag{7.12}$$

If we multiply the equation $\mathcal{A}(t,x,D)u=0$ by ψ, and integrate by parts in Ω, by (7.7) we get:

$$\|u\|^2_{L^2(\Omega)} \le -\int_{\partial\Omega} g\,(\nu_A\cdot\nu)(\beta\cdot\nu)^{-1}\bar{\psi}\,d\sigma =$$

$$\le \left|\left\langle g, \overline{(\nu_A\cdot\nu)(\beta\cdot\nu)^{-1}\psi}\right\rangle_{W^{-3/2,2}(\partial\Omega),W^{3/2,2}(\partial\Omega)}\right| \le$$

$$\le c\,\|g\|_{W^{-3/2,2}(\partial\Omega)}\|\psi\|_{W^{3/2,2}(\partial\Omega)}\,,$$

and by (7.12) we obtain

$$\|u\|_{L^2(\Omega)} \le c\,\|g\|_{W^{-3/2,2}(\partial\Omega)} \qquad \forall g\in W^{1/2,2}(\partial\Omega). \tag{7.13}$$

Finally, interpolating between (7.10) and (7.13) we deduce in a standard way

$$\|u\|_{W^{3/2,2}(\Omega)} \le c\,\|g\|_{L^2(\partial\Omega)} \qquad \forall g\in W^{1/2,2}(\partial\Omega) \tag{7.14}$$

This estimate shows that $G(t)$ has a bounded extension from $L^2(\partial\Omega)$ into $W^{3/2,2}(\Omega)$, and consequently we get assumption (2.6) for each $\vartheta\in]0,3/4[$.

REMARK 7.2 By (7.7) it follows that

$$\left(u|G(t)^*A(t)^*v\right)_H = \int_{\partial\Omega} u\,(\nu_A\cdot\nu)(\beta\cdot\nu)^{-1}\bar{v}\,d\sigma \quad \forall u\in W^{1/2,2}(\partial\Omega),\ \forall v\in D_{A(t)^*},$$

i.e.

$$G(t)^*A(t)^*v = \overline{(\nu_A\cdot\nu)(\beta\cdot\nu)^{-1}}\; v_{|\partial\Omega} \quad \forall v\in D_{A(t)^*} . \tag{7.15}$$

If, in particular, $\beta=\nu_A$ (the case of $A_1(t)$ in [1]), we get $G(t)^*A(t)^*v = v_{|\partial\Omega}$ $\forall v\in D_{A(t)^*}$.

REMARK 7.3 If $\beta\equiv0$ and $\alpha\equiv1$ in (7.2), then we have, as in [1], that (2.6) holds for each $\vartheta\in]0,1/4[$ and that $G(t)^*A(t)^*v = \partial v/\partial\nu_A$ $\forall v\in D_{A(t)^*}$.

REMARK 7.4 In Example 7.1 the assumptions of [1] are also fulfilled, even assuming only C^α regularity in t (but in this case, of course, (2.3) is no longer true).

EXAMPLE 7.5 Take n=1, $\Omega=]0,1[$, $\lambda_0>0$ and

$$\mathcal{A}(t,x,D):= A(t,x)\cdot u'' + B(t,x)\cdot u' + C(t,x)\cdot u + \lambda_0 u , \tag{7.16}$$

$$\mathcal{B}_j(t,D):= \beta_j(t)\cdot u'(j) + \alpha_j(t)\cdot u(j) , \quad j=0,1, \tag{7.17}$$

where $A(t,x)$, $B(t,x)$, $C(t,x)$, $\beta_j(t)$, $\alpha_j(t)$ are N×N matrices whose coefficients are continuous in x and $C^{1+\alpha}$ in t. We assume

$$\det A(t,x) \neq 0 \quad \forall(t,x)\in[0,T]\times[0,1]. \tag{7.18}$$

Consider the boundary value problem (for fixed t)

$$\begin{cases} \mathcal{A}(t,\cdot,D)u = f\in L^2(0,1;\mathbb{C}^N) \\ \mathcal{B}_j(t,D)u = z_j\in\mathbb{C}^N, \quad j=0,1; \end{cases} \tag{7.19}$$

setting v:=u', by (7.18) we may rewrite (7.19) as a first order system of 2N equations:

$$\begin{cases} \begin{pmatrix} u \\ v \end{pmatrix}' = \begin{pmatrix} 0 & I \\ Q(t,x) & R(t,x) \end{pmatrix}\begin{pmatrix} u \\ v \end{pmatrix} + \begin{pmatrix} 0 \\ F(t,x) \end{pmatrix} , \quad x\in]0,1[, \\ \alpha_j(t)\cdot u(j) + \beta_j(t)\cdot v(j) = z_j , \quad j=0,1 , \end{cases} \tag{7.20}$$

where $Q:=-A^{-1}B$, $R:=-A^{-1}(C+\lambda_0 I)$, $F:=-A^{-1}f$. An easy check shows that (7.20) is uniquely solvable if and only if

$$\det \begin{pmatrix} \alpha_0(t)\cdot U(0) & \beta_0(t)\cdot V(0) \\ \alpha_1(t)\cdot U(1) & \beta_1(t)\cdot V(1) \end{pmatrix} \neq 0 , \tag{7.21}$$

where the 2N×2N matrix

$$\begin{pmatrix} U(x) \\ V(x) \end{pmatrix}, \quad x\in[0,1]$$

is any Wronskian relative to the homogeneous system associated to (7.20). Thus the columns of U(x) give 2N linearly independent solutions in $C^1([0,1],\mathbb{C}^N)$ of the homogeneous system $\mathcal{A}(t,\cdot,D)u=0$, and the columns of V(x) give the derivatives of such solutions.

Under assumptions (7.18) and (7.21) (which is obviously intrinsic, i.e. does not depend on the particular Wronskian) it is now easy to verify that the abstract assumptions (2.2), (2.3) and (2.6) are fulfilled.

EXAMPLE 7.6 Our abstract theory also applies to certain dynamic systems acting on non-cylindrical domains, which are studied in [6].

REFERENCES

[1] - P. Acquistapace, F Flandoli, B. Terreni: *Boundary control for non-autonomous parabolic systems*, preprint Dip. di Mat. Univ. di Pisa, n° 258 (1988).

[2] - P. Acquistapace, B. Terreni: *Some existence and regularity results for abstract non-autonomous parabolic equations*, J. Math. Anal. Appl. **99** (1984) 9-64.

[3] - P. Acquistapace, B. Terreni: *A unified approach to abstract linear non-autonomous parabolic equations*, Rend. Sem. Mat. Univ. Padova **78** (1987) 47-107.

[4] - P. Acquistapace, B. Terreni: *On fundamental solutions for abstract parabolic equations*, in *"Differential equations in Banach spaces"*, Proceedings, Bologna 1985, A: Favini & E. Obracht editors, Lect. Notes n° 1223, Springer-Verlag, Berlin/Heidelberg 1986, 1-11.

[5] - P. Acquistapace, B. Terreni: *Regularity properties of evolution operators of abstract parabolic equations*, preprint.

[6] - P. Cannarsa, G. Da Prato, J.-P. Zolesio: *Riccati equations in non-cylindrical domains*, this volume.

[7] - G. Geymonat, P. Grisvard: *Alcuni risultati di teoria spettrale per i problemi ai limiti lineari ellittici*, Rend. Sem. Mat. Univ. Padova **38** (1967) 121-173.

[8] - T. Kato, H. Tanabe: *On the abstract evolution equation*, Osaka Math. J. **14** (1962) 107-133.

[9] - J.-L. Lions, E. Magenes: *"Problémes aux limites non homogènes et applications"*, vol. I, Dunod, Paris 1968.

[10] - H. Tanabe: *"Equations of evolution"*, Pitman, London 1969.

EXISTENCE AND OPTIMAL CONTROL FOR WAVE EQUATION IN MOVING DOMAIN

G.Da Prato, Scuola Normale Superiore, 56126,Pisa, Italy
J.P.Zolésio, Laboratoire de Physique Mathématique,U.S.T.L., Montpellier,France.

Key words: Evolution Equation, Moving domain,Optimal Control,Wave equation.

Abstract .The technique of change of variables is used to solve wave equation in a moving domain and to study a Linear Quadratic Optimal Control problem .

1 Introduction.

We consider the wave equation on a moving domain

$$(1.1)\quad \begin{cases} \phi_{tt}(t,x) = \operatorname{div}(K(t,x)\nabla\phi(t,x)) + u(t,x) & t\in\,]0,T],\ x\in\Omega_t \\ \phi(0,x) = \phi_0(x) \quad \phi_t(0,x) = \phi_1(x) & x\in\Omega_0 \\ \phi(t,x) = 0 \quad t\in\,]0,T],\ x\in\Gamma_t & \end{cases}$$

where Ω_0 is a bounded domain in $\mathbf{R}^N$ with smooth boundary Γ_0, $\Omega_t = T(t,\Omega_0)$, $\Gamma_t = T(t,\Gamma_0)$ and $T(t,\cdot)$ denotes the flow associated to a given vector field $V : [0,T]\times\mathbf{R}^N \to \mathbf{R}^N$, i.e.

(1.2) $$\frac{\partial}{\partial t}T(t,x) = V(t,T(t,x)),\quad t\in[0,T],\ x\in\mathbf{R}^N \qquad T(0,x) = x$$

We shall assume that there exists an open bounded set of $\mathbf{R}^N$ with smooth boundary ∂D such that $D\supset\Omega_t$ for all $t\in[0,T]$ and $V(t,\cdot) = 0$ on ∂D. We assume that $V(\cdot,x)$ and $\frac{\partial}{\partial t}V(\cdot,x)$ are continuous uniformly in x and $V(t,\cdot)$ is twice continuously differentiable with bounded second order derivatives ; then the transformation $T(\cdot,\cdot)$ belongs to $C^2([0,T];C^2(D;D))$, we shall denote by $J(t,x) = \det(DT(t,x))$ the Jacobian of the transformation

Moreover $\phi_0\in H_0^1(\Omega_0)$, $\phi_1\in L^2(\Omega_0)$ and the control u belongs to $L^2(Q_T)$, where $Q_T = \bigcup_{0<t<T}\{t\}\times\Omega_t$. Finally $K(t,x) = (K_{ij}(t,x))$ is a $N\times N$ real symmetric matrix such that :

(1.3) $$\langle K(t,x)\xi,\xi\rangle \geq \alpha_0|\xi|^2 \quad \text{for all } \xi\in\mathbf{R}^N \text{ and some } \alpha_0>0.$$

We consider the optimal control problem :

Minimize :

(1.4) $$J(u) = \int_0^T\int_{\Omega_t}\{|\phi_t(t,x)|^2+|\nabla\phi(t,x)|^2+|u(t,x)|^2\}dx + \int_{\Omega_T}\{|\phi_t(T,x)|^2+|\nabla\phi(T,x)|^2\}dx$$

over all $u\in L^2(Q_T)$, subject to equations (1.1).

In Section 2 , we use the change of variables technique (introduced in [9] and [10]) to reduce problem (1.1) in a cylindrical domain. More precisely, we set

(1.5) $z(t,x) = \phi(t,T(t,x)), \;\; p(t,x) = [J(t,x)]^{1/2}u(t,T(t,x)),$

(The presence of the factor $J^{1/2}$ in the definition of p is to preserve the cost functional and Riccati equation structure in the change of variables at Section 3). Then problem (1.1) reduces to

$$(1.6)\quad \begin{cases} z_{tt}(t,x) = J^{-1}(t,x)\mathrm{div}(G(t,x)\cdot\nabla z(t,x)) + <\gamma(t,x),\nabla z(t,x)> + 2<M(t,x),\nabla z_t(t,x)> \\ \qquad + J^{-1/2}(t,x)p(t,x) \\ z(0,x) = z_0(x) \quad , \; z_t(0,x) = z_1(x) \quad \text{in } \Omega_0 \\ z(t,x) = 0 \text{ in }]0,T]\times\partial\Omega_0 \end{cases}$$

(where G,γ and M are defined in (2.4)-(2.6) below). We prove (Proposition 2.1) that the matrix G(t,x) is strictly positive, provided $|V(t,x)|<\alpha_0$ (if K(t,x) = I this assumption reduces to $|V(t,x)|<1$). Then we are able to solve problem (1.6) by using the Theory of Abstract Hyperbolic equations (whose main results are recalled in the Appendix at the end of the paper). Finally, caming back to problem (1.1) ,we can prove the existence and uniqueness of a solution ϕ. We remark that some existence results, for problem (1.1) have been previously proved in [7] and [8] by using Faedo-Galerkin approximation and in [11], by using a penalization method.However the results we present here are stronger and allow us the construction of a continuous evolution operator, which is useful to study the Riccati equation corresponding to the optimal control problem (1.4).

Section 3 is devoted to the Riccati equation which can be solved as in [1], by using the regularity properties of the evolution operator recalled in the Appendix. Finally, by using Dynamic Programming, we prove the existence of an optimal feedback control.

We end with the remark that the assumption $|V(t,x)|<\alpha_0$ is not restrictive. Assume for simplicity, $K(t,x) = I$ and $|V(t,x)| > 1$ then, one can show that the moving wave equation is not meaningful in general. In fact, assume $N = 2$, $v(t,x)>1$, where $v(t,x)=V(t,x)\cdot n(t,x)$ and $n(t,x)$ is the unit exterior normal vector to Ω_t, and let ϕ be a regular solution of (1.1). By a classical argument, one can check that at a point $P=(t,x)$, $\phi(t,x)$ just depends on of the values of ϕ in the backward cone pointed at P and then ϕ is identically 0 in the exterior of the non cylindrical evolution domain built from Ω_0 with the speed $v(t,x) = +1$.

In the case where $v(t,x) < -1$, from the a-priori estimate in [7] it can be verified that any smooth solution is in fact the restriction to Q_T of the solution on the cylindrical domain $[0,T]\times D$

2 Change of variables.

In this section, we want to tranform problem (1.1) in an evolution equation in the cylindrical domain $Q_0 = [0,T]\times\Omega_0$. To this aim, we set

$$z(t,x) = \phi(t,T(t,x)) \quad , \; p(t,x) = [J(t,x)]^{1/2}u(t,T(t,x)) \tag{2.1}$$

and we use the following notation .If $x\in \mathbf{R}^N$, then $x \circ x$ will represent the linear operator in $\mathcal{L}(\mathbf{R}^N)$ defined by $(x \circ x)\cdot y = \langle x,y\rangle x$ for all $y\in \mathbf{R}^N$.

By using the change of variables (2.1), we find , by straightforward calculations:

(2.2) $$z_{tt}(t,x) = J^{-1}(t,x)\mathrm{div}(A(t,x)\cdot\nabla z(t,x)) - <M(t,x),\nabla[<M(t,x),\nabla z(t,x)]>+ + <M_t(t,x),\nabla z(t,x)>+2<M(t,x),\nabla z_t(t,x)>$$

where

(2.3) $$A(t,x) = J(t,x)\, E(t,x)K(t,T(t,x))E^*(t,x);$$

and

(2.4) $$E(t,x) = [DT(t,x)]^{-1} \;;\quad M(t,x) = E(t,x)W(t,x)\;;\; W(t,x) = V(t,T(t,x))$$

Since

(2.5) $$<D^2z(t,x)\, M(t,x),\, M(t,x)> = \mathrm{Tr}(D^2z(t,x)\cdot M(t,x)oM(t,x))$$

It follows :

(2.6) $$z_{tt}(t,x) = \mathrm{Tr}\left\{[J^{-1}(t,x)A(t,x)- M(t,x)oM(t,x)]D^2z(t,x)\right\} + <M_t(t,x),\nabla z(t,x)>+2<M(t,x),\nabla z_t(t,x)>$$

Since $J^{-1}(t,x)A(t,x) = E(t,x)(K(t,T(t,x))E^*(t,x)$ and

$M(t,x)oM(t,x) = (E(t,x)W(t,x))o(E(t,x)W(t,x))$, we have :

(2.7) $$z_{tt}(t,x) = \mathrm{div}(G(t,x)\cdot\nabla z(t,x)) - \sum_{i,j=1}^{N}\frac{\partial}{\partial x_i}G_{ij}(t,x)\frac{\partial}{\partial x_j}z(t,x) + <M_t(t,x),\nabla z(t,x)>+ +2<M(t,x),\nabla z_t(t,x)>$$

where

(2.8) $G(t,x) = J^{-1}(t,x)K(t,T_t(t,x))E^*(t,x) - (E(t,x)W(t,x))o(E(t,x)W(t,x))$

Finally z is a solution of the following problem

$$(2.9)\quad \begin{cases} z_{tt}(t,x) = J^{-1}(t,x)\mathrm{div}(G(t,x)\cdot\nabla z(t,x)) + <\gamma(t,x),\nabla z(t,x)> + 2<M(t,x),\nabla z_t(t,x)> \\ \qquad + J^{-1/2}(t,x)p(t,x) \\ z(0,x) = z_0(x) \quad \text{in } \Omega_0 \ , \quad z_t(0,x) = z_1(x) \quad \text{in } \Omega_0 \\ z(t,x) = 0 \text{ in } \partial\Omega_0 \end{cases}$$

where :

(2.10) $\gamma(t,x) = M_t(t,x) - \mathrm{div}\, G(t,x)$

Proposition 2.1. Assume that $K(t,x) \geq \alpha_0 I > 0$ and $|V(t,x)| < \alpha_0$ for all $t \in [0,T]$ and all $x \in \overline{D}$. Then there exists $\beta > 0$ such that $G(t,x) \geq \beta I$ for all $t \in [0,T]$ and all $x \in \Omega_0$.

Proof.

Let $\xi \in \mathbf{R}^N$, by the coercivity of K, we have :

(2.11) $<E(t,x)(K(t,T(t,x))E^*(t,x)\xi,\xi> \geq \alpha_0 |E^*(t,x)\xi|^2$

and, by the Cauchy-Schwartz inequality, it follows that

(2.12) $|W(t,x)|\ |E^*(t,x)\xi| \geq <W(t,x),E^*(t,x)\xi>$

Notice that

(2.13) $<(E(t)W(t))o(E(t)W(t))\xi,\xi> = <E(t)W(t),\xi>^2$.

Then, by (2.12) and (2.13), it follows :

(2.14) $<G(t,x)\xi,\xi> \geq |E^*(t,x)\xi|^2(\alpha_0 - |W(t,x)|_{L^\infty(D;R^N)})$

Now, we conclude the proof by observing that

(2.15) $|E^*(t,x\cdot\xi| \geq |E^{-1}(t,x)|^{-1}|\xi|$

and $|W(t,x)|_{L^\infty(D;R^N)} = |V(t,x|_{L^\infty(D;R^N)}$.

We write now problem (2.9) as a first order problem in the space $\mathcal{H} = H_0^1(\Omega_0)\oplus L^2(\Omega_0)$. It is convenient to endowe $\mathcal{H}$ with several equivalent scalar products; for any $t\in[0,T]$ we set

(2.16) $$<\begin{bmatrix} u \\ v \end{bmatrix}, \begin{bmatrix} f \\ g \end{bmatrix}>_t = \int_{\Omega_0} <G(t)\cdot\nabla u,\nabla f>\,dx + \int_{\Omega_0} vg\,dx$$

Moreover we define the linear operator $\mathcal{B}(t)$ in $\mathcal{H}$ by

(2.17) $$D(\mathcal{B}(t)) = [H^2(\Omega_0)\cap H_0^1(\Omega_0)]\oplus H_0^1(\Omega_0)$$

(2.18) $$\mathcal{B}(t)\begin{bmatrix} u \\ v \end{bmatrix} = \begin{bmatrix} 0 & 1 \\ B_{21}(t) & B_{22}(t) \end{bmatrix}\begin{bmatrix} u \\ v \end{bmatrix}$$

where

(2.19) $$B_{21}(t)\zeta = J^{-1}(t,\cdot)\mathrm{div}(G(t,\cdot)\cdot\nabla\zeta + <\gamma(t,\cdot),\nabla\zeta>$$

(2.20) $$B_{22}(t)\zeta = 2<M(t,\cdot),\nabla\zeta>$$

Now problem (2.9) can be written as :

(2.21) $Y'(t) = \mathcal{B}(t)Y(t) + F(t) \quad ; \; Y(0) = Y_0.$

where :

(2.22) $Y(t) = \begin{bmatrix} z(t) \\ z_t(t) \end{bmatrix} \quad Y_0 = \begin{bmatrix} z_0 \\ z_1 \end{bmatrix} \quad F(t) = \begin{bmatrix} 0 \\ J_t^{-1/2}\, p \end{bmatrix}$

We show now that $\mathcal{B}(t)$ is almost dissipative with respect to the inner product $<\cdot>_t$; we have in fact

Proposition 2.2. There exists C>0 such that :

(2.23) $<\mathcal{B}(t)Y,Y>_t \leq C<Y,Y>_t \;, \quad \forall Y \in D(\mathcal{B}(t)) \,, \; \forall t \in [0,T]$

Proof. We have :

(2.24) $$<\mathcal{B}(t)Y,Y>_t = \int_{\Omega_0} <\gamma(t),\nabla y_1>y_2 dx \; + \int_{\Omega_0} \mathrm{div}[G(t,x)\cdot\nabla y_1] y_2 dx \; +$$

$$+ \int_{\Omega_0} <\gamma(t,x),\nabla y_1>y_2 dx \; + 2 \int_{\Omega_0} <M(t,x),\nabla y_2>y_2 dx$$

By integrating by parts, the first two terms in the second hand side of (2.24) cancel. Moreover

(2.25) $$\int_{\Omega_0} <\gamma(t,x),\nabla y_1>y_2 dx \leq |\gamma(t,x)| \frac{1}{\beta} \Big\{ \int_{\Omega_0} <G(t,x)\nabla y_1,\nabla y_1>dx \Big\}^{1/2} \|y_2\|_{L^2(\Omega_0)}.$$

and

(2.26) $$\int_{\Omega_0} <M(t,x),\nabla y_2>y_2 dx = -\frac{1}{2} \int_{\Omega_0} \mathrm{div}(M(t,x))|y_2|^2 dx$$

and the result follows.

Remark 2.3. By Proposition 2.2 it follows that the linear operator $\mathfrak{B}(t)$, is also dissipative in the Hilbert space $D(\mathfrak{B}(t))$, endowed with the inner product :

$$(2.27)\qquad \langle\langle Y,Z\rangle\rangle_t = \langle \mathfrak{B}(t)Y, \mathfrak{B}(t)Z\rangle_t + \langle Y,Z\rangle_t .$$

By (2.17) $D(\mathfrak{B}(t))$ is independent of t. ♠

In order to show that $\mathfrak{B}(t)$ is the infinitesimal generator of a strongly continuous semigroup, it suffices to remark that the equation :

$$(2.28)\qquad \lambda Y - \mathfrak{B}(t)Y = Z$$

has a solution $Y \in \mathfrak{B}(t)$, for any $Z \in \mathcal{H}$ and any λ large enough. We want now to solve problem (2.22) by using the theory of Abstract Evolution Equations(see Appendix below). We shall apply Theorem A.3 with $H = \mathcal{H}$, $B(t) = \mathfrak{B}(t)$ and $K = D(\mathfrak{B}(t)) = [H^2(\Omega_0) \cap H_0^1(\Omega_0)] \oplus H_0^1(\Omega_0)$.

We prove now that the family $\{\mathfrak{B}(t)\}_{t\in[0,T]}$ fulfils hypotheses (A.6).

Proposition 2.4. The family $\{\mathfrak{B}(t)\}_{t\in[0,T]}$ is stable in H.

Proof. For all $Z = \begin{bmatrix} z_1 \\ z_2 \end{bmatrix} \in \mathcal{H}$, we have :

$$|Z|_t^2 = |Z|_s^2 + \int_{\Omega_0} \langle [G(t)-G(s)]\cdot\nabla z_1, \nabla z_2\rangle dx$$

The last term is dominated by $K|t-s| \int_{\Omega_0} \langle|\nabla z_1|^2\rangle dx$; moreover , using the coercivity of G(s), we get

(2.29) $$|Z|_t^2 \le |Z|_s^2\,(1 + \frac{k}{\beta}|t-s|) \le |Z|_s^2\, e^{k(t-s)/\beta}$$

by Proposition A.1 this implies the conclusion, since $\mathcal{B}(t)$ is dissipative with respect to the norm $|\cdot|_t$.

Proposition 2.5. The family $\{\mathcal{B}(t)\}_{t\in[0,T]}$ is stable in

$$D(\mathcal{B}(t)) = [H^2(\Omega_0)\cap H_0^1(\Omega_0)]\oplus H_0^1(\Omega_0).$$

Proof.

Let $Z \in D(\mathcal{B}(t))$, from (2.29) we have :

(2.30) $$\|Z\|_t^2 \le |\mathcal{B}(t)Z|_t^2 + e^{k(t-s)/\beta}\,|\mathcal{B}(t)Z|_s^2\,;$$

We have $|\mathcal{B}(t)Z|_s \le |\mathcal{B}(s)Z|_s + |(\mathcal{B}(t)-\mathcal{B}(s))Z|_s$, and

(2.31) $$|(\mathcal{B}(t)-\mathcal{B}(s))Z|_s^2 = \int_{\Omega_0}[(B_{21}(t)-B_{21}(s))z_1+(B_{22}(t)-B_{22}(s))z_2]^2dx$$

$$\le K|t-s|^2\,\|Z\|_0^2 \le CK|t-s|^2\,\|Z\|_s^2$$

Since this norm is uniformly equivalent to $\|\cdot\|_t$, we get from (2.30) and (2.31) that :

$$\|Z\|_t^2 \le \|Z\|_s^2\,(1 + \frac{CK}{\beta}|t-s|)\,e^{k(t-s)/\beta} \le \|Z\|_s^2\,e^{2CK(t-s)/\beta}$$

As $\mathcal{B}(t)$ is dissipative with respect to the norm $\|\cdot\|_t$, we conclude the proof as before.

Now , it is easy to see that also hypotheses (A.6)iii) holds, thus, by Theorem A.3, we have :

Theorem 2.6. Assume that $K(t,x) \geq \alpha_0 I > 0$ and $|V(t,x)| < \alpha_0$ for all $t \in [0,T]$ and all $x \in \overline{D}$. Let $p \in L^2([0,T] \times \Omega_0)$, $z_0 \in H_0^1(\Omega_0)$ and $z_1 \in L^2(\Omega_0)$. Then, problem (2.22) has a unique strong solution $Y \in C([0,T];\mathcal{H})$ such that $Y(0) = Y_0$. Moreover, if $z_0 \in H^2(\Omega_0) \cap H_0^1(\Omega_0)$, $z_1 \in H_0^1(\Omega_0)$ and $p \in L^2(0,T;H_0^1(\Omega_0))$ then Y is a strict solution.

Remark 2.7. We go back now to problem (2.9). We shall define a strong solution of (2.9) as a function z such that $Y(t) = \begin{bmatrix} z(t) \\ z_t(t) \end{bmatrix}$, is a strong solution of (2.21). This means that $z \in C([0,T];H_0^1(\Omega_0)) \cap C^1([0,T];L^2(\Omega_0))$ and there exist two sequences $\{u_k\}$ and $\{v_k\}$, with $u_k \in C([0,T];H^2(\Omega_0) \cap H_0^1(\Omega_0)]$ and $v_k \in C([0,T];H_0^1(\Omega_0))$ such that

(2.32) $$u_k \to z, \quad \frac{d}{dt} u_k - v_k \to 0 \ \text{ in } C([0,T];H_0^1(\Omega_0))$$

(2.33) $$v_k \to z_t, \quad \frac{d}{dt} v_k - B_{21} u_k - B_{22} v_k \to 0 \ \text{ in } C([0,T];L^2(\Omega_0)).)$$

We remark finally that, if Y is a strict solution of (2.22), then we have

(2.34) $$z \in C([0,T];H^2(\Omega_0) \cap H_0^1(\Omega_0)) \cap C^1([0,T];H_0^1(\Omega_0)) \cap W^{2,\infty}([0,T];L^2(\Omega_0))$$ ♠

We go back now to problem (1.1). For any smooth domain Ω included in D let $E(\Omega)$ be any Sobolev space (For example $E(\Omega)$ will be one of the following spaces : $L^2(\Omega), H^1(\Omega), H_0^1(\Omega), \ldots$). For any one parameter family of domain Ω_t, included in D, built by the transformation T_t associated to the field V, we define the *moving domain* Banach spaces

$$C^i([0,T];E(\Omega_t)) = \{\psi \in L^2(Q_T) \,;\, \exists\, \Psi \in C^i([0,T];E(D)) \text{ with } \Psi_{|Q_T} = \psi\}$$

equipped with the norm

$$\|\psi\|_{C^i([0,T];E(\Omega_t))} = \inf\{\|\Psi\|_{C^i([0,T];E(D))},\ \text{such that } \Psi_{|Q_T} = \psi\}$$

Proposition 2.8. Let $V \in C([0,T];C^k(D;R^N))$ with $k \geq 1$ (so that the mapping $\psi \to \psi o T_t$, is one to one from $E(\Omega_t)$ onto $E(\Omega_0)$). Then, for i=0,1, we have

$$\psi \in C^i([0,T];E(\Omega_t)) \Leftrightarrow \psi o T_t \in C^i([0,T];E(\Omega_0))$$

We give now the definition of a weak solution of problem (1.1). Let us first recall the Green formula for the noncylindrical domain Q_T; let ϕ and ψ be two smooth functions, then

(2.35)
$$\int_0^T \int_{\Omega_t} \{\phi_{tt} - \mathrm{div}(K\cdot\nabla\phi)\}\psi\, dx\, dt =$$

$$= \int_0^T \int_{\Omega_t} \{-\phi_t\psi_t + \langle K\cdot\nabla\phi, \nabla\psi\rangle\}\psi\, dx\, dt + \int_{\Omega_T} \phi_t(T)\psi(T)dx - \int_{\Omega_0} \phi_t(0)\psi(0)dx$$

$$- \int_0^T dt \int_{\Gamma_t} \{\langle K\cdot n, \nabla\phi\rangle\psi + \phi\psi v\}\, d\Gamma_t$$

Then, if ϕ is a smooth solution of the problem (1.1) then , for any test function ψ,such that

(2.36) $\psi \in C^\infty(\overline{Q_T})$ which $\psi(T) = 0$ and $\psi = 0$ on Σ_T

we have

(2.37) $$\int_0^T\int_{\Omega_t}\{-\phi_t\psi_t+<K\cdot\nabla\phi,\nabla\psi>\}\psi dxdt = \int_{\Omega_0} z_1\psi(0)dx + \int_0^T\int_{\Omega_t} u\psi dxdt$$

Thus we give the following definition. A function ϕ is a weak solution of the problem (1.1) if

$\phi\in C^1([0,T];L^2(\Omega_t))\cap C([0,T];H_0^1(\Omega_t))$, $\phi(0) = z$ and ϕ verifies (2.37).

We prove now a change of variables formula.

Proposition 2.9. ϕ is a weak solution of (1.1) if and only if $z = \phi(t,T(t,\cdot))$belongs to $C([0,T];H_0^1(\Omega_0))\cap C^1([0,T];L^2(\Omega_0))$ and fulfils :

(2.38) $$\int_0^T\int_{\Omega_0}\{-[z_t-z_x\cdot M][\zeta_t-\zeta_x\cdot M]J+<JK(T(\cdot,))E^*\cdot\nabla z,E^*\cdot\nabla\zeta>\}dxdt =$$

$$= \int_{\Omega_0} z_1\zeta(0)dx + \int_0^T\int_{\Omega_t} zu(T(\cdot,\cdot))Jdxdt$$

for all $\zeta\in C^\infty([0,T]\times\Omega_0)$ such that $\zeta(T,x) = 0$ in Ω_0 and $\zeta(t,x) = 0$ in $[0,T]\times\partial\Omega_0$

Proof. The result follows by integration by parts.

We have now the result :

Theorem 2.10. Assume that $K(t,x)\geq\alpha_0 I>0$ and $|V(t,x)| <\alpha_0$ for all $t\in[0,T]$ and all $x\in\overline{D}$. Let $u\in L^2(Q_T)$, $z_0\in H_0^1(\Omega_0)$ and $z_1\in L^2(\Omega_0)$. Then, problem (1.1) has a unique weak solution $\phi\in C^1([0,T];L^2(\Omega_t))\cap C([0,T];H_0^1(\Omega_t))$. Moreover, if $z_0\in H^2(\Omega_0)\cap H_0^1(\Omega_0)$, $z_1\in H_0^1(\Omega_0)$ and $u\in L^2(0,T;H_0^1(\Omega_t))$ then $\phi\in W^{2,\infty}([0,T];L^2(\Omega_t))\cap C^1([0,T];H_0^1(\Omega_t))\cap C([0,T];H^2(\Omega_t)\cap H_0^1(\Omega_t))$.

Proof. Let z be the strong solution of problem (2.9). As easily checked , z fulfils (2.38) for all $\zeta \in C^{\infty}([0,T]\times\Omega_0)$ such that $\zeta(T,x) = 0$ in Ω_0 and $\zeta(t,x) = 0$ in $[0,T]\times\partial\Omega_0$. Thus the conclusion follows from Proposition 2.9.

Remark 2.11 (Dissipativity results for increasing domains)

We assume here smooth data $z_0 \in H^2(\Omega_0)\cap H_0^1(\Omega_0)$, $z_1 \in H_0^1(\Omega_0)$ and set u = 0. We assume also K independent of t. By Theorem (2.9) the problem :

$$\text{(2.39)} \quad \begin{cases} \phi_{tt}(t,x) = \operatorname{div}(K(t,x)\nabla\phi(t,x)) & t\in\,]0,T],\ x\in\Omega_t \\ \phi(0,x) = \phi_0(x) \quad \phi_t(0,x) = \phi_1(x) & x\in\Omega_0 \\ \phi(t,x) = 0 & t\in\,]0,T],\ x\in\Gamma_t \end{cases}$$

has a solution $\phi\in W^{2,\infty}([0,T];L^2(\Omega_t))\cap C^1([0,T];H_0^1(\Omega_t))\cap C([0,T];H^2(\Omega_t)\cap H_0^1(\Omega_t))$. We will show now that the energy estimate in [7] and [8] can be stated. Consider the energy :

$$\text{(2.40)} \qquad E(t) = \frac{1}{2}\int_{\Omega_t} <K(x)\nabla\phi(t,x),\nabla\phi(t,x)>dx + \frac{1}{2}\int_{\Omega_t} |\phi_t(t,x)|^2>dx$$

By (2.39) we get :

$$\text{(2.41)} \qquad E'(t) = \int_{\Gamma_t} <K(x)n_t(t,x),\nabla\phi(t,x)>\phi_t(t,x)d\Gamma_t +$$

$$+ \frac{1}{2}\int_{\Gamma_t} [<K(x)\nabla\phi(t,x),\nabla\phi(t,x)>+|\phi_t(t,x)|^2]v(t,x)d\Gamma_t$$

where n_t is the outward normal field on Γ_t and $v(t) = \langle V(t),n_t \rangle$ is the normal component of the speed on Γ_t. But, for any $x \in \Gamma_0$ we have $\phi(t,T_t(x)) = 0$; then, taking the derivative with respect to t we get

(2.42) $$\phi_t(t,T_t(x)) + \frac{\partial}{\partial n_t}\phi(t,T_t(x))\, V(t,T_t(x)) = 0$$

Using also

(2.43) $$\langle K(x)n_t(t,x),\nabla\phi(t,x)\rangle = \langle K(x)n_t(t,x),n_t(t,x)\rangle\frac{\partial}{\partial n_t}\phi(t,x)$$

we get

(2.44) $$E'(t) = \frac{1}{2}\int_{\Gamma_t} \left|\frac{\partial}{\partial n_t}\phi(t,x)\right|^2\{v^3(t,x)-\langle K(x)n_t(t,x),n_t(t,x)\rangle v(t,x)\}d\Gamma_t$$

Obviously with the condition $|V(t,x)| < \alpha_0$ of Proposition 2.1 we get $v(t,x) \leq \langle K(x)n_t(t,x),n_t(t,x)\rangle$; it follows that $E'(t) \leq$ if and only if $v(t,x) \geq 0$ which corresponds to the expansion domain.

3 Dynamic Programming.

We consider here the state space $\mathcal{H} = H_0^1(\Omega_0)\oplus L^2(\Omega_0)$, and the control space $\mathcal{U} = L^2(\Omega_0)$. The state equation is :

(3.1) $$Y'(t) = \mathcal{B}(t)Y(t) + \mathcal{C}p(t) \quad ; \quad Y(0) = Y_0.$$

where Y is defined by (2.23) and $\mathcal{C}$ is the linear operator in $\mathcal{L}(\mathcal{U};\mathcal{H})$ defined by

(3.2) $$\mathcal{C}v = \begin{bmatrix} 0 \\ J_t^{-1/2}\, v \end{bmatrix}, \quad \forall v \in \mathcal{U}$$

Also we transform the cost function (1.4) by the change of variables (2.2). We find

(3.3) $$J_1(p)=:J(u)=\int_0^T\{<S(t)Y(t),Y(t)>_{\mathcal{H}} + \|p(t)\|^2_{\mathcal{U}}\}dt+<S(T)Y(T),Y(T)>_{\mathcal{H}}$$

where the linear operator $S(t)\in \mathcal{L}(\mathcal{H})$ is defined by :

(3.4) $$<S(t)Z,V> =$$

$$= \int_{\Omega_0}\{<A(t)\nabla z_1,\nabla v_1>+J_t[z_2-<E^*(t)\cdot\nabla z_1,W(t)>][v_2-<E^*(t)\cdot\nabla v_1,W(t)>]dx$$

for any $Z = \begin{bmatrix} z_1 \\ z_2 \end{bmatrix}$, $V = \begin{bmatrix} v_1 \\ v_2 \end{bmatrix}$ in $\mathcal{H}$.

Now the problem (1.4) is equivalent to the following :

(3.5) Minimize $J_1(p)$ over all controls $p\in L^2(0,T;\mathcal{U})$, subject to (3.1).

We consider now the Riccati equation :

(3.6) $$Q' + \mathbb{B}^*Q + Q\mathbb{B} - Q\mathbb{C}\mathbb{C}^*Q +S = 0 \ ; \ Q(T) = S(T)$$

which we write in the integral form :

(3.7) $$Q(t)x = U^*(T,t)S(T)U(T,t)x + \int_t^T U^*(s,t)\{S(s)-Q(s)\mathbb{C}\mathbb{C}^*Q(s)\}U(s,t)xds \ ,$$

$x\in\mathcal{H}$

We shall also consider the approximating equation :

$$(3.8)\qquad Q_n(t)x = U_n^*(T,t)S(T)U_n(T,t)x + \int_t^T U_n^*(s,t)\{S(s)-Q_n(s)\mathcal{C}\mathcal{C}^*Q_n(s)\}U_n(s,t)x\,ds$$

$, x\in\mathcal{H}$

where $U_n(s,t)$ is the evolution operator associated to the family $\{\mathcal{B}_n(t)\}$ and $\mathcal{B}_n(t) = n\mathcal{B}(t)(n-\mathcal{B}(t))^{-1}$. In order to solve (3.7) we will generalize the method introduced in [1]. We need the following notations.

a) $\Sigma^+(\mathcal{H}) = \{ T\in \mathcal{L}(\mathcal{H})\ ;\ T = T^*\ ,\ T \geq 0\}$

b) $\Sigma(\mathcal{H}) = \{ T\in \mathcal{L}(\mathcal{H})\ ;\ T = T^*\ \}$

c) $C_s([0,T];\ \Sigma(\mathcal{H})) = \{ Z : [0,T]\to \Sigma(\mathcal{H})\ ;\ Z(.)x\in C([0,T];\ \mathcal{H})\}$

d) $C_s([0,T];\ \Sigma^+(\mathcal{H})) = \{Z\in C_s([0,T];\ \Sigma(\mathcal{H}));\ Z(t)\geq 0\ \forall t\in [0,T]\}$

Proposition 3.1. Under the hypotheses of Theorem 2.6, equations (3.7) and (3.8) have unique global solutions, Q and Q_n in $C_s([0,T];\ \Sigma^+(\mathcal{H}))$. Moreover

$$(3.9)\qquad \lim_{n\to\infty} Q_n(\cdot)x = Q(\cdot)x \quad \text{in } C([0,T]:\mathcal{H}), \text{ for all } x\in\mathcal{H}$$

Proof.

We only sketch the proof.

Step 1 (Local existence).

Equations (3.6) and (3.7) can be written as :

$$(3.10)\qquad Q = F - \gamma(Q) \qquad Q_n = F - \gamma_n(Q_n)$$

where :

(3.11) $$F(t)x = \int_t^T U^*(s,t)S(s)U(s,t)x\,ds$$

(3.12) $$F_n(t)x = \int_t^T U_n^*(s,t)S(s)U_n(s,t)x\,ds$$

and γ, γ_n are defined analogously.

As easily seen γ (resp.γ_n) maps $C_s([T-d,T]; \Sigma^+(\mathcal{H}))$ into itself for any $d\in]0,T]$ and it is locally Lipschitz continuous. By the Contractions Principle, there exists $d\in]0,T]$ (resp.$d_n \in]0,T]$) such that equation (3.7) (resp. (3.8)) has a unique solution in $C_s([T-d,T]; \Sigma^+(\mathcal{H}))$. Moreover, recalling Theorems A.2 and A.3, one can show that

(3.13) $$\lim_{n\to\infty} Q_n(\cdot)x = Q(\cdot)x \quad \text{in } C([T-d,T]:\mathcal{H}), \text{ for all } x\in\mathcal{H}$$

Step 2.(Positivity of the maximal solution)

By a standard argument we can define the solutions Q and Q_n of (3.7) and (3.8) in maximal intervals $]T-\delta,T]$ and $]T-\delta_n,T]$ respectively. We have

(3.14) $$Q_n' + (\mathcal{B} - \tfrac{1}{2}\mathcal{C}\mathcal{C}^*Q_n)^*Q_n + Q_n(\mathcal{B} - \tfrac{1}{2}\mathcal{C}\mathcal{C}^*Q_n) + S = 0 \quad \text{in }]T-\delta_n,T]$$

Let us denote by $Z_n(t,s)$ the evolution operator associated to $\mathcal{B} - 1/2\mathcal{C}\mathcal{C}^*Q_n$; by (3.14) it follows

(3.15) $$Q_n(t)x = Z_n^*(T,t)S(T)Z_n(T,t)x + \int_t^T Z_n^*(s,t)S(s)Z_n(s,t)x\,ds \quad , \ x\in\mathcal{H} \ , t\in]T-\delta_n,T]$$

Thus $Q_n(t)\geq 0$, $\forall t\in]T-\delta_n,T]$ which implies $Q(t)\geq 0$, $\forall t\in]T-\delta,T]$.

Step 3. (a priori estimate and conclusion).

Let Q be the maximal solution of (3.7). For all $x\in\mathcal{H}$ and $t\in]T-\delta,T]$, we have

$$<Q(t)x,x> \leq < S(T)U(T,t)x,U(T,t)x> + \int_t^T <S(s)U(s,t)x,U(s,t)x>ds$$

Since Q is positive, we find $\|Q(t)\| \leq$ Const. $\forall t\in]T-\delta,T]$, thus the global existence follows. The last statement follows by (3.13).

Now we go back to the control problem (3.5).

Theorem 3.2. Assume that $K(t,x)\geq\alpha_0 I>0$ and $|V(t,x)| <\alpha_0$ for all $t\in[0,T]$ and all $x\in\overline{D}$. Let $z_0\in H_0^1(\Omega_0)$ and $z_1\in L^2(\Omega_0)$. Then there exists a unique optimal control p* for problem (3.5), given by :

(3.16) $\quad p^*(t) = -\mathcal{C}^*(t)Q(t)Y^*(t)$

where Y* is the solution to the closed loop equation

(3.17) $\quad Y'(t) = \mathcal{B}(t)Y(t) - \mathcal{C}(t)\mathcal{C}^*(t)Q(t)Y(t) \;\; ; \;\; Y(0) = Y_0$

and Q is the solution to (3.7) Moreover the optimal cost is given by :

(3.18) $\quad J(p^*) = <Q(0)Y_0,Y_0>_H$

Proof. Let $p\in L^2(0,T]\times\Omega_0)$, $z_0\in H_0^1(\Omega_0)$ and $z_1\in L^2(\Omega_0)$. Let Q_n be the solution to (3.8) and Y_n the solution to the problem

(3.19) $\quad Y_n'(t) = B_n(t)Y_n(t) + Cp(t) \;\; ; \;\; Y_n0) = Y_0.$

Then, by computing $\frac{d}{dt}<Q_n(t)Y_n(t),Y_n(t)>$, completing squares and integrating in [0,T] we find the identity

$$(3.20)\qquad \langle Q_n(0)Y_0,Y_0\rangle + \int_0^T \|p(s)+C^*Q_n(s)Y_n(s)\|^2 ds =$$

$$\int_0^T \{\langle S(t)Y_n(t),Y_n(t)\rangle + \|p(t)\|^2\}dt+\langle S(T)Y_n(T),Y_n(T)\rangle$$

By (3.9) and Theorem A.2, we can let n tend to infinity in (3.20). We obtain

$$(3.21)\qquad \langle Q(0)Y_0,Y_0\rangle + \int_0^T \|p(s)+C^*Q(s)Y(s)\|^2 ds = J_1(p)$$

for all $p \in L^2(0,T]\times\Omega_0)$. By (3.21) the conclusion follows by standard arguments.

Appendix. Abstract hyperbolic evolution equations.

The results of this section are due essentially to T.Kato (see [4] , [5] and [6]; see also [2] for some generalization).

Let H be a Hilbert space (norm $|\cdot|$) and let $\{B(t)\}_{t\in[0,T]}$ be a family of infinitesimal generators of strongly continuous semi-groups in H. We assume that the domain of B(t) is constant and coincides with a Hilbert space K (norm $\|\cdot\|$) continuously and densely embedded in H.We consider the evolution equation :

$$(A.1)\qquad \frac{d}{dt}u(t) = B(t)u(t) + f(t)\ ;\ \ u(0) = x$$

where $x\in H$ and $f\in L^2(0,T;H)$. A function $u\in W^{1,2}(0,T;H)\cap L^2(0,T;D(A))$ is called a *strict solution* if fulfils (A.1). A function $u\in L^2(0,T;H)$ is called a *strong solution* if there exists a sequence $\{u_k\}$ in $W^{1,2}(0,T;H)\cap L^2(0,T;D(A))$ such that $u_k \to u$, $\frac{d}{dt}u_k - B(t)u_k \to f$ in $L^2(0,T;H)$ and $u_k(0)\to x$ in H. We also consider the approximating problem :

(A.2) $$\frac{d}{dt}u_n(t) = B_n(t)u_n(t) + f(t) \ ; \ u(0) = x$$

where $B_n(t) = nB_n(t)R(n,B_n(t))$.

The family $\{B(t)\}_{t\in[0,T]}$ is called *stable* if there are constants M>1 and ω (called the stability constants) such that

(A.3) $$\begin{cases} \rho(B(t)) \supset \,]\omega,+\infty[\\ \left|\prod_{i=1}^{n} R(\lambda,B(t_i))\right| \le C(\lambda-\omega)^{-n} \\ \text{for any sequence } 0\le t_1\le\ldots\le t_n\le T \text{ and } n\in\mathbb{N}. \end{cases}$$

A sufficient condition for stability is the following ([4], Prop.3.4).

Proposition A.1 For each $t\in[0,T]$, let $|\cdot|_t$ be a new norm in H equivalent to the original one, such that , for some C>0

(A.4) $$|x|_t \le e^{C|t-s|}\,|x|_s \quad \forall t,s\in[0,T], \ \ \forall x\in H$$

Assume moreover that there exists ω>0 such that $B(t)-\omega I$ is maximal dissipative in H, endowed with the norm $|\cdot|_t$. Then the family $\{B(t)\}_{t\in[0,T]}$ is stable with constants e^{CT} and ω .

We shall denote by $\tilde{B}(t)$ the part of B(t) in K, that is :

(A.5) $$D(\tilde{B}(t)) = \{x \in D(B(t))\ ;\ B(t) \in K\}\ ,\ \tilde{B}(t)x = B(t)x\ \ \forall x \in D(\tilde{B}(t))$$

The following result is proved in [2] (Theorem 2)

Theorem A.2. Assume

(A.6)
$$\begin{cases} \text{i) } \{B(t)\}_{t\in[0,T]} \text{ is stable with constants } C \text{ and } \omega \\ \text{ii) } \{\tilde{B}(t)\}_{t\in[0,T]} \text{ is stable with constants } C_1 \text{ and } \omega_1 \\ \text{iii) For any } y \in K,\ B(\cdot)y \text{ is continuous in } H \end{cases}$$

Then, for any $x \in H$ and $f \in L^2(0,T;H)$, problem (A.1) has a unique strong solution $u \in C([0,T];H)$ such that $u(0) = x$. Moreover, if $x \in K$ and $f \in L^2(0,T;K)$ then u is a strict solution and $u \in W^{1,2}(0,T;H) \cap L^\infty(0,T;H)$. Finally, if u_n is the solution to (A.2), we have $u_n \to u$ in $C([0,T];H)$.

Consider now the evolution operator U(t,s) defined by U(t,s)x = u(t) , where u is the strong solution of the problem :

(A.7) $$\frac{d}{dt}u(t) = B(t)u(t)\ ;\ \ u(s) = x$$

The following result is proved in [6] (Theorem 5.3.1).

Theorem A.3. Assume (A.6), then there exists a unique evolution operator U(t,s), $0 \le s \le t \le T$ in H, such that U(t,s)x is continuous for any x in H, and

(A.8) $\frac{\partial^+}{\partial t}U(t,s)y|_{t=s} = B(s)y$ for all $y \in K$, $0 \le s \le t \le T$

(A.9) $\frac{\partial}{\partial s}U(t,s)y = -U(t,s)B(s)y,$ for all $y \in K$, $0 \le s \le t \le T$

References.

[1] G.Da Prato, Quelques résultat d'existence, unicité et régularité pour un problème de la théorie du contrôle, J. Math. Pures Appl. **52**, (1973), 353-375.

[2], G.Da Prato and M.Iannelli, On a method for studying abstract evolution equations in the hyperbolic case, Comm.in Partial Differential Equations, 1(6), (1976),585-608

[3] J.L.Lions and E.Magenes, Problèmes aux limites non homogènes, Dunod, Paris (1968).

[4] T.Kato, Linear evolution equations of "hyperbolic" type, J.Fac.Sci.Univ.Tokyo, **17**,(1970),241-258

[5] T.Kato, Linear evolution equations of "hyperbolic" type, II, J.Math.Soc.Japan,**25**,(1973),648-666.

[6], A.Pazy, Semigroups of Linear Operators and Applications to Partial Differential Equations, Springer-Verlag (1983)

[7] C.Truchi , Stabilisation par variation de domain, Thèse de Doc. d'état, Nice (1987).

[8] C.Truchi and J.P.Zolésio, Shape stabilization of wave equation, Proceedings of the IFIP WG 7.2 Conference on "Boundary Control and Boundary Variations" (J.P. Zolésio Editor). Lecture Notes in Control and Information Sciences n° 100, Springer-Verlag, (1987), 372-398.

[9] J.P.Zolésio, Identification de domaines par déformation, Thèse de Doc. d'état, Nice (1979).

[10] J.P.Zolésio, The material derivative. In : Céa J.and Haug E.J. Editors. "Optimization of distributed parameter structures: Vol. II. Sijthoff and Nordhoff, Alpen aan den Rijn, 5, (1981), 1089-1151.

[11] Zolésio, Galerkin approximation forwave equation in moving domains. These Proceedings.

GALERKINE APPROXIMATION FOR WAVE EQUATION IN MOVING DOMAIN

J.P. ZOLESIO
U.S.T.L., Place Eugène Bataillon
34060 MONTPELLIER Cedex - FRANCE

INTRODUCTION.

We consider the wave equation in a moving domain Ω_t. The Dalembert equation $\Box y = \ddot{y} - \Delta y = f$ will be solved in $Q = \bigcup_{0<t<T} \{t\} \times \Omega_t$ for various Boundary conditions on the lateral Boundary $\Sigma = \bigcup_{0<t<Y} \{t\} \times \Gamma_t$. The main point is that we use here a constructive Galerkine-penalization approximation using a dense family $\omega_1, \dots, \omega_m$ in $H_0^1(D)$ where D is a smooth bounded domain which contains all the domains Ω_t, $0 \leqslant t \leqslant T$. An other method would be to consider for each t, $\omega_1(t), \dots \omega_m(t), \dots$ a dense family in the moving domain $H_0^1(\Omega_t)$ (if we are concerned by the Dirichlet Boundary condition).
But the difficulty arises when we have the time derivative $\dot{\omega}_i(t)$ that we are not in general able to estimate.
This work is a first part of a study devoted to non cylindrical problem. We also refer to G. Da Prato-J.P. Zolesio [4] where using the change of variable technic we prove (by a non constructive method) the existence and smoothness of solution, in particular we obtain the evolution operator. A completely different technic is used in P. Cannarsa-G. Da Prato-J.P. Zolesio [1],[2],[3].

1. Dirichlet Boundary condition and monotone increasing domain.
2. Dirichlet Boundary condition. The General case.
3. Contracting domains.
4. Newman Boundary condition and monotone decaying domain
5. A dissipative boundary condition for a moving domain.

I - WAVE EQUATION WITH DIRICHLET BOUNDARY CONDITION IN MOVING DOMAIN

Let Ω_t be a moving domain such that for any time t, $0 \leqslant t \leqslant t$, $\Omega_t \subset D$ where D is a bounded domain in $\mathbb{R}^N$ with smooth boundary. We denote by Γ_t the boundary of Ω_t and by $\Omega_t^c = D \setminus \bar{\Omega}_t$ the complementary open domain in D; $I = [0,T]$ and Q is the non cylindrical subdomain of $I \times D$ defined as

$$Q = \bigcup_{0<t<T} \{t\} \times \Omega_t$$

with lateral boundary Σ given by

$$\Sigma = \bigcup_{0<t<T} \{t\} \times \Gamma_t$$

Being given two elements y_0, y_1 in $H_0^1(\Omega_0) \times L^2(\Omega_0)$ we consider the following wave equation

$$\square\, y = \ddot{y} - \Delta y = 0 \quad \text{in } Q \tag{1.1}$$

$$y = 0 \quad \text{on } \Sigma \tag{1.2}$$

$$y(0) = y_0 \quad \text{in } \Omega_0 \tag{1.3}$$

$$\dot{y}(0) = y_1 \quad \text{in } \Omega_0 \tag{1.4}$$

where $\dot{y}$ stands for the time derivative while Δy is the Laplace operator.

n_t denotes the unit normal vector field on Γ_t outgoing to Ω_t while ν_t denotes the unit normal vector field to Σ outgoing to Q ; $V_t(x)$ is an element of $\mathbb{R}^{n+1}$, its horizontal componant is proportional to $n_t(x)$ but its time componant is not zero for Q is not cylindrical. In fact $\nu_t(x)$ can be written as

$$\nu_t(x) = (1 + v^2(t,x))^{-1/2}\,(v(t,x),\, n_t(x)) \tag{1.5}$$

for each $t \in I$, $n \in \Gamma_t$ (i.e. $(t,x) \in \Sigma$).

It can be easily verified that $v(t,x)$ is the normal componant of any speed vector field $V(t,x)$ whose flow T_t builts the non cylindrical evolution domain Q (i.e. $\Omega_t = T_t(\Omega_0)$, $\forall\, t \in I$).

To built solution to the problem (1.1)-(1.4) we shall use a Galerkine technic and for this purpose we introduce a family $\omega_1, \omega_2, \ldots, \omega_n, \ldots$ of element of $H^1_0(D)$. The ω_i being choosen such that the family is dense in $H^1_0(D)$ (see for example J.L. Lions and E. Magenes [5]). To obtain the Dirichlet condition (1.2) on the lateral boundary Σ we shall also use a penalty technic and at this point it is convenient to treat first the specific situation corresponding to monotone increasing domain Ω_t, we shall consider the general situation after.

1.1 Increasing domain and Dirichlet condition

We consider the specific situation of an increasing domain, that is when $t_2 > t_1$ implies $\Omega_{t_1} \subset \Omega_{t_2}$. When the boundaries Γ_t are smooth enough this assumption is equivalent to assume that $v(t,x) \geqslant 0$ for all $t \in I$ and n in Γ_t (i.e. all (t,x) in Σ). But in this section the smoothness assumption of the boundary Γ_t, could be avoid . We consider, for each integer m and positive number ϵ, the expansion

$$y_{m,\epsilon}(t,x) = \sum_{i=1}^{m} \varphi_{m,i}(t)\ \omega_i(x) \tag{1.5}$$

where the function $\varphi_{m,i}(t)$ are defined as the solution of the ordinary differential system $0 < t < T$, $i=1,\ldots,m$;

$$\int_D \left(\nabla y_{m,\epsilon} \nabla \omega_i + \ddot{y}_{m,\epsilon} \omega_i + \frac{1}{\epsilon} \chi^c_t \ y_{m,\epsilon} \ \omega_i \right) dx = 0 \tag{1.6}$$

$$\varphi_{m,i}(0) = a_{m,i} \quad , \quad \dot{\varphi}_{m,i}(0) = b_{m,i} \tag{1.7}$$

where χ_t^c is the characteristic function of $D\backslash\Omega_t$ i.e. $\chi_t^c\ (x) = \begin{cases} 0 & \text{if } x\in\Omega_t \\ 1 & \text{otherwise,} \end{cases}$

$a_{m,i}$ and $b_{m,i}$ are such that we have the following convergences :

$$\sum_{i=1}^{m} a_{m,i}\omega_i \longrightarrow y_0^o \quad \text{in} \quad H_0^1(D) \quad , \quad m \longrightarrow \infty \tag{1.8}$$

$$\sum_{i=1}^{m} b_{m,i}\ \omega_i \longrightarrow y_1^o \quad \text{in} \quad L^2(D) \quad , \quad m \longrightarrow \infty, \tag{1.9}$$

where y_0^o and y_1^o denotes the extension by zero of y_0 and y_1 to the domain D. Considering the energy term

$$E_{m,\epsilon}(t) = \frac{1}{2}\int_D\left[|\nabla y_{m,\epsilon}|^2 + \left(\dot{y}_{m,\epsilon}\right)^2\right]dx + \frac{1}{2\epsilon}\int_{\Omega_t^c}(y_{m,\epsilon})^2\ dx \tag{1.10}$$

we get

$$E_{m,\epsilon}(t') = \int_D\left(\nabla y_{m,\epsilon}\nabla\dot{y}_{m,\epsilon} + \ddot{y}_{m,\epsilon}\dot{y}_{m,\epsilon} + \frac{1}{\epsilon}\chi_t^c\ y_{m,\epsilon}\ \dot{y}_{m,\epsilon}\right)dx$$

$$- \frac{1}{2\epsilon}\int_{\Gamma_t}(y_{m,\epsilon})^2\ v(t)\ d\Gamma_t \tag{1.11}$$

From (1.11) and (1.6) we get

$$E_{m,\epsilon}(t') = -\frac{1}{2\epsilon}\int_{\Gamma_t}(y_{m,\epsilon})^2\ v(t)\ d\Gamma_t \tag{1.12}$$

and then

$$E_{m,\epsilon}(t) + \frac{1}{2\epsilon}\int_0^t \int_{\Gamma_s} (y_{m,\epsilon})^2 \; v(s) \; d\Gamma_s \; ds = E_{m,\epsilon}(0) \tag{1.13}$$

where we have

$$E_{m,\epsilon}(0) = \frac{1}{2}\int_D \left[\left(\nabla \left(\sum_{i=1}^{m} a_{m,i}\omega_i \right) \right)^2 + \left(\sum_{i=1}^{m} b_{m,i}\omega_i \right)^2 \right] dx$$

$$+ \frac{1}{2\epsilon}\int_{\Omega_0^c} \left(\sum_{i=1}^{m} a_{m,i}\omega_i \right)^2 dx \tag{1.14}$$

the boundness of the family $y_{m,\epsilon}$ will derives from the behavior of the term $E_{m,\epsilon}(0)$. From (1.8), (1.9) and (1.14) we get, for each ϵ,

$$E_{m,\epsilon}(0) \longrightarrow \frac{1}{2}\int_E \left(\left| \nabla y_0^o \right|^2 + \left(y_1^o \right)^2 \right) dx + \frac{1}{\epsilon}\int_{\Omega_0^c} \left(y_0^o \right)^2 dx$$

when $m \longrightarrow \infty$.

But y_0^0 is identiquely zero on Ω_0^c, then the term affected by ϵ vanishes in the last expression. We introduce

$$E(0) = \frac{1}{2}\int_D \left(\left| \nabla \, y_0^o \right|^2 + \left(y_1^o \right)^2 \right) dx. \tag{1.15}$$

We have obtain the :

Lemma 1.1

$\forall \; \epsilon > 0$, $E_{m,\epsilon}(0)$ converges to $E(0)$ when $m \longrightarrow \infty$ and the obvious corollary.

Corollary 1.2

For any ϵ positive there exists an integer $M(\epsilon)$ such that for any integer m, m ⩾ M(ε) we have :

$$\max_{0\leqslant t\leqslant T} E_{m,\epsilon}(t) \leqslant 2E(0) \tag{1.16}$$

$$\frac{1}{\epsilon}\int_0^T\int_{\Gamma_t}(y_{m,\epsilon})^2\, v(t)\, d\Gamma_t\, dt \leqslant 2E(0) \quad \Box \tag{1.17}$$

Notice that from (1.16) we get

$$\max_{0\leqslant t\leqslant T} |\nabla y_{m,\epsilon}(t)|_{L^2(D)^n} \leqslant 2E(0) \tag{1.18}$$

and

$$\max_{0\leqslant t\leqslant T} \left|\dot{y}_{m,\epsilon}(t)\right|_{L^2(D)} \leqslant 2E(0) \quad . \tag{1.19}$$

Then one can find subsequences $\epsilon_n \to 0$, $M_n = M(\epsilon_n)$ and $y_n = y_{m_n,\epsilon_n}$ and the following convergences :

$$y_n \rightharpoonup y \quad \text{for the weak-* topology of } L^\infty(I,L^2(D)) \tag{1.20}$$

$$\nabla_n \rightharpoonup \nabla y \quad \text{for the weak-* topology of } L^\infty(I,L^2(D)^n\ ) \tag{1.21}$$

$$\dot{y}_n \rightharpoonup \dot{y} \quad \text{for the weak-* topology of } L^\infty(I,L^2(D)) \tag{1.22}$$

and the extra conditions (from the penalty term) :

$$\frac{1}{\sqrt{\epsilon_n}}\chi_{\Omega_t^c}\, y_n \rightharpoonup \mu \quad \text{for the weak-* topology of } L^\infty(I,L^2(D)) \tag{1.23}$$

$$\frac{1}{\sqrt{\epsilon_n}}\, y_n\Big|_\Sigma \rightharpoonup \lambda \quad \text{for the weak topology of } L^2(\Sigma) \tag{1.24}$$

where y, μ and λ are respectively elements of L^∞, $(I,\ H^1(D)) \cap W^{1,\infty}(I,L^2(D))$, $L^\infty(I,L^2(D))$ and $L^2(\Sigma)$.

In the limit we get the estimate

$$\|\nabla y\|^2_{L^\infty(I,L^2(D)^n)} + \left\|\dot{y}\right\|^2_{L^\infty(I,L^2(D))}$$

$$+ \|\mu\|^2_{L^\infty(I,L^2(D))} + \|\lambda\|^2_{L^2(\Sigma)} \leqslant E(0) \quad . \tag{1.25}$$

A problem is now to characterize the limiting elements y, μ and λ. Obviously from the boundness (1.16), (1.17) we obtain

$$\int_\Sigma y_n^2 \, d\Sigma \leqslant \epsilon_n \, E(0) \tag{1.26}$$

and

$$\max_{0\leqslant t\leqslant T} \int_{\Omega_t^c} y_n^2 \, dx \leqslant \epsilon_n \, E(0) \tag{1.27}$$

and than

$$y_n\Big|_\Sigma \longrightarrow 0 \quad \text{in} \quad L^2(\Sigma) \tag{1.28}$$

while

$$y_n\Big|_{Q^c} \longrightarrow 0 \quad \text{in} \quad L^\infty\left(I,\ L^2\left(\Omega_t^c\right)\right)$$

Then the limiting element y verifies the Dirichlet boundary condition (1.2) as

$$y \in L^\infty(I,\ H^1(D)) \cap W^{1,\infty}(I,\ L^2(D))$$

then in particular $y \in H^1(I \times D)$ and $y|_{Q^c}$ is zero in $H^1(Q^c)$ so its trace on Σ is zero, $y|_\Sigma$ being a priori an element of $H^{1/2}(\Sigma)$, where we recall here that $Q^c = \bigcup_{0<t<T} \{t\} \times \Omega_t^c$ is an open (non cylindrical) subdomain of the cylindrical domain $I \times D$.

Characterization of the problem whose y is solution.

We introduce the linear space

$$S = \left\{ \psi = \sum_{i=1}^{p} \varphi_i(t)\, \omega_i(x) \;\middle|\; p \in \mathbb{N},\ \varphi_i \in C^\infty([0,T]),\ \varphi_i(0) = \varphi_i(T) = 0 \right\}, \qquad (1.29)$$

S is dense in $H_0^1(I \times D)$. For any ψ in S we consider the Green's theorem as follows :

$$\int\!\!\int_{I\times D} \left(\nabla y_n \cdot \nabla\psi - \dot{y}_n \dot{\psi}\right) dt\, dx + \frac{1}{\epsilon_n} \int_0^T\!\!\int_{\Omega_t^c} y_n \psi\, dt\, dx$$

$$= \int\!\!\int_{I\times D} \left(\Box\, y_n + \frac{1}{\epsilon_n} y_n \chi_{\Omega_t^c}\right) \psi\, dt\, dx \quad . \qquad (1.30)$$

From (1.6) and (1.29) the right hand side of (1.30) is reduced to zero. As a first result we obtain that for any ψ in $H'_o(I\times D)$

$$\int\!\!\int_{I\times D} \frac{1}{\epsilon_n} y_n \chi_{\Omega_t^c} \psi\, dt\, dx$$

is bounded when n goes to infinity. By the uniform bound theorem we get that $\frac{1}{\epsilon_n} \chi_{\Omega_t^c} y_n$ is bounded in $H^{-1}(I\times D) = H_0^1(I\times D)'$. Then there exists a sub-sequence $\epsilon_n \to 0$ and an element θ of $H^{-1}(I\times D)$ such that

$$\frac{1}{\epsilon_n} \chi_{\Omega_t^c} y_n \rightharpoonup \theta \quad \text{weakly in} \quad H^{-1}(I\times D) \qquad (1.31)$$

In the limit we get

$$\int\int_{I\times D} \left(\nabla y\ \nabla\psi - \dot{y}\ \dot{\psi}\right) dt\ dx + \langle \theta,\psi \rangle = 0, \quad \forall\ \psi \in H'_0(I\times D) \qquad (1.32)$$

then as an element of $H^{-1}(I\times D)$, $\square\ y$ solves the waves problem

$$\square\ y + \theta = 0 \quad \text{in } I \times D \qquad (1.33)$$

$$y = 0 \quad \text{in} \quad Q^c \ . \qquad (1.34)$$

Obviously θ is a distribution over $I\times D$ which is supported by Q^c in the following sens : $\forall\ \varphi \in H^1_0(I\times D)$ with support of $\varphi \subset Q$ we have $\langle \theta,\varphi \rangle = 0$. We introduce the restriction of y to the open domain Q :

$$Y = y|_Q \ . \qquad (1.35)$$

By the Green's theorem we obtain

$$\dot{Y}(0) = y_1\Big|_{\Omega_0} \quad \text{in} \quad H^{1/2}_{00}(\Omega_0)'$$

and also we have

$Y(0) = y_0$ in Ω_0 by construction of the y_n.
Finaly we have the

Theorem 1.3

Let the non cylindrical domain $Q = \bigcup_{0<t<T} \{t\} \times \Omega_t$ be smooth enough so that the horizontal normal componant v of the speed is positive (which is then equivalent to the fact that the domain Ω_t is monotoniquely increasing with t). For any $(y_0, y_1) \in H'_0(\Omega_0) \times L^2(\Omega_0)$ the element y obtained and (1.35) verifies

$$Y \in W^{1,\infty}\left(I,\ L^2(\Omega_t)\right) \cap L^\infty\left(I,\ H^1_0(\Omega_t)\right) \qquad (1.36)$$

$$\square Y = 0 \quad \text{in} \quad Q \tag{1.37}$$

$$Y = 0 \quad \text{in} \quad \Sigma \tag{1.38}$$

$$Y(0) = y_o \tag{1.39}$$

$$\dot{Y}(0) = y_1 \tag{1.40}$$

Notice that from (1.36) and (1.37) by the Green's theorem the traces $\dot{Y}(0)$ is a priori defined in $H^{1/2}_{00}(\Omega_o)'$.

Characterization of the limiting element μ from (1.23) and (1.31) we get $\mu = 0$ so the estimate (1.25) is reduced to

$$E(t) + \int_0^t \int_{\Gamma_2} \lambda^2 v \, d\Gamma_s \, ds \leqslant E(0) \tag{1.41}$$

with

$$E(t) = \|\nabla Y\|_{L^\infty\left(I, L^2(\Omega_t)^n\right)} + \left\|\dot{Y}\right\|_{L^\infty(I, L^2(D))} \tag{1.42}$$

where we recall that

$$\frac{1}{\sqrt{\epsilon_n}} \; y_n \rightharpoonup \lambda \quad \text{weakly in} \quad L^2(\Sigma)$$

so λ is *the boundary dissipation.*

We have also an other characterization for λ as follows :

Proposition 1.4

There exists subsequence ϵ_n going to zero and an element $\eta \in H^{-2}(I\times D)$ such that

$$\frac{1}{\sqrt{\epsilon_n}} \chi_{\Omega_t^c} \dot{y}_n \rightharpoonup \eta \quad \text{weakly in } H^{-2}(I\times D) \tag{1.43}$$

and

$$t\gamma_\Sigma . \lambda = \frac{1}{T-t} \eta \tag{1.44}$$

where γ_Σ is the trace (or restriction) operator on the lateral boundary Σ while $t\gamma_\Sigma$ is its transposed operator ; (1.44) is equivalent to :

$$\forall \psi \in H_0^2(I\times D) \qquad \int_\Sigma (T-t)\lambda\, \psi \, d\Sigma = \langle \eta, \psi \rangle \tag{1.45}$$

Proof : We introduce the bilinear form associated to the energy term.

$$B_n(y,z)(t) = \frac{1}{2} \int_D \left(\nabla y \, \nabla z + \dot{y} \, \dot{z} \right) dx + \frac{1}{2\epsilon_n} \int_{\Omega_t^c} yz \, dx \tag{1.46}$$

we get for $\psi \in H_0^2\ (I\times D)$,

$$B(y_n, \psi)'(t) = \frac{1}{2} \int_D (\nabla y_n \, \nabla\dot{\psi} + \ddot{y}_n \, \dot{\psi}) \, dx + \frac{1}{2\epsilon_n} \int_{\Omega_t^c} y_n \, \dot{\psi} \, dx$$

$$+ \frac{1}{2} \int_D \left(\nabla\dot{y}_n \, \nabla\psi + \dot{y}_n \, \ddot{\psi} \right) dx + \frac{1}{2\epsilon_n} \int_{\Omega_t^c} \dot{y}_n \, \psi \, dx - \frac{1}{2\epsilon_n} \int_{\Gamma_t} y_n \, \psi \, v(t) \, d\Gamma_t \tag{1.47}$$

And from (1.6) we get in (1.47) that the two first terms in the right hand side of (1.47) are equal to zero and then

$$\sqrt{\epsilon_n} \int_0^T B(y_n, \psi)\,(t)\, dt = \frac{\sqrt{\epsilon_n}}{2} \int_0^T (T-t) \int_D \left(\nabla\dot{y}_n \, \nabla\psi + \dot{y}_n \, \dot{\psi} \right) dt \, dx$$

$$+\frac{1}{2\sqrt{\epsilon_n}}\int_0^T(T-t)\int_D \chi_{\Omega_t^c}\,\dot{y}_n\,\psi\,dx-\frac{1}{2\sqrt{\epsilon_n}}\int_0^T(T-t)\int_{\Gamma_t} y_n\,\psi\,v\,d\Gamma_t \tag{1.48}$$

Obviously the left hand side of (1.48) goes to zero as n goes to infinity, and also the first term in the right hand side (to verify this one has to perform a by part integration in time in view to get $y_{m,\epsilon}$ but $\dot{\psi}$ in this term. Now as we know from (1.25) that the second term in the right hand side of (1.48) is bounded when n goes to infinity and we obtain (145). □

Remark 1.6

We have considered the homogeneous Dirichlet condition on the moving boundaries Γ_t. The fundamental assumption in this section was that the domains Ω_t are increasing, that is (under smoothness assumption on the boundary Σ) that the normal speed v(t) on each Γ_t (at each time t) is positive. But the Dirichlet condition could have been generalised to

$$y = G \quad \text{on} \quad \Sigma \tag{1.49}$$

where G is given in $H^1(I\times D)$. Then we just have to modify the energy term as follows

$$E_{m,\epsilon}(t)=\frac{1}{2}\int_D\left(|\nabla y_{m,\epsilon}|^2+\left(\dot{y}_{m,\epsilon}\right)^2\right)dx+\frac{1}{2\epsilon}\int_{\Omega_t^c}(y_{m,\epsilon}-G)^2\,dx \tag{1.50}$$

where $y_{m,\epsilon}$ is now the solution of the modified linear system :

$$\int_D\left(\nabla y_{m,\epsilon}\,\nabla\omega_i+y_{m,\epsilon}\,\omega_i\right)dx+\frac{1}{\epsilon}\int_{\Omega_t^c}(y_{m,\epsilon}-G)\,\omega_i\,dx$$

$$=\int_D f\,\omega_i\,dx \qquad i=1,\ldots,n \tag{1.51}$$

for some given f in $L^2(I\times D)$ and all the previous results and analogous results concerning the terms λ, μ, θ and η are preserved and we solved in the same spaces the wave problem.

$$\begin{aligned} \Box\, y &= f \quad \text{in} \quad Q, \\ y &= G \quad \text{on} \quad \Sigma \qquad\qquad (1.52) \\ y(0) = y_0, \quad \dot{y}(0) &= y_1 \quad \text{in} \quad \Omega_0 \end{aligned}$$

An other penalization for the Dirichlet problem.
In this section we consider the situation when the domain Ω_t is expending, that is that the normal speed $v(t)$ is positive on the boundary Γ_t.
The family $\omega_i(x)$ being choosen as previously we consider the penalized Galerkine expansion

$$z_{m,\epsilon} = \sum_{i=1}^{m} \varphi_{m,i}(t)\ \omega_i(x) \tag{1.53}$$

where the $\varphi_{m,i}$ are solution of the following ordinary differential system

$$\begin{aligned} &\begin{matrix} i = 1,\dots,m \\ 0<t<T \end{matrix} \quad \int_D \nabla z_{m,\epsilon}\ \nabla\omega_i\ dx + \frac{1}{\epsilon}\int_{\Omega_t^c} \nabla z_m\ \nabla\omega_i\ dx \\ &\qquad + \int_D \ddot{z}_{m,\epsilon}\ \omega_i\ dx = 0 \end{aligned} \tag{1.54}$$

and the associated energy term

$$W_{m,\epsilon}(t) + \frac{1}{2}\int_D \left\{ |\nabla z_{m,\epsilon}|^2 + \left(\dot{z}_{m,\epsilon}\right)^2 \right\} dx + \frac{1}{2\epsilon}\int_{\Omega_t^c} |\nabla z_{m,\epsilon}|^2\ dx \tag{1.55}$$

As previously we obtain the derivative of the energy :

Proposition 1.7

$$W_{m,\epsilon}(t)' = -\frac{1}{2\epsilon}\int_{\Gamma_t} |\nabla z_{m,\epsilon}|^2\, v(t)\, d\Gamma_t \qquad (1.56)$$

and we get the following estimate.

Corollary 1.8

$$W_{m,\epsilon}(t) + \frac{1}{2\epsilon}\int_0^t\int_{\Gamma_t} |\nabla z_{m,\epsilon}|^2\, v(t)\, d\Gamma_t = W_{m,\epsilon}(0) \qquad (1.57)$$

As previously from (1.57) we obtain the boundness of $\nabla z_{m,\epsilon}$, $\dot{z}_{m,\epsilon}$, $\nabla z_{m,\epsilon}|_\Sigma$ and $z_{m,\epsilon}$ as ϵ and m goes to zero and infinity, with $m \geq M(\epsilon)$ for as previously it can be verified that $W_{m,\epsilon}(0) \longrightarrow W(0) = E(0)$ when $m \longrightarrow \infty$, for each $\epsilon > 0$.

We exactly use the same argument as in the previous situation we introduce $z_n = z_{m_n,\epsilon_n}$ and we get the following weak-* convergence in $L^\infty(0,T,L^2(D))$

$$z_n \rightharpoonup z,\ \nabla z_n \rightharpoonup \nabla z,\ \dot{z}_n \rightharpoonup \dot{z} \qquad (1.58)$$

$$(\epsilon_n)^{-1/2}\, \nabla z_n\Big|_\Sigma \rightharpoonup \tilde{\lambda} \quad \text{weakly in } L^2(\Sigma) \qquad (1.59)$$

$$(\epsilon_n)^{-1/2}\, \chi_{\Omega_t^c}\, \nabla z_n \rightharpoonup \tilde{\mu}$$

It can also be verified that z, as an element of $H_0^1\,(I\times D)$ is reduced to zero in Q^c, so that ∇z and $\dot{z}$ are almost everywhere equal to zero in Q^c ;

$$\frac{1}{\epsilon_n}\operatorname{div}\left[\chi_{\Omega_t^c}\, \nabla z_n\right] \rightharpoonup \tilde{\theta} \quad \text{weakly in} \quad H^{-1}(I\times D) \qquad (1.60)$$

and then z solves the problem

$$\square z + \tilde{\theta} = 0 \quad \text{in} \quad I{\times}D \quad \text{and} \quad z = 0 \quad \text{in} \quad Q^c ,$$

finely introducing the restriction of z to Q,

$$Z = z|_Q \tag{1.61}$$

we get the :

Theorem 1.9

Being given y_0, y_1 in $H_0^1(\Omega_0) \times L^2(\Omega_0)$ the element z obtained at (1.61), $Z \in L^\infty\left(I, H_0^1(\Omega_t)\right) \cap W^{1,\infty}\left(I, L^2(\Omega_t)\right)$, verifies

$$\square z = 0 \quad \text{in} \quad Q \tag{1.62}$$

$$z|_\Sigma = 0 \quad \text{in} \quad Q \tag{1.63}$$

$$z(0) = y_0 \quad , \quad \dot{y}(0) = y_1 \quad . \qquad 1.64)$$

Notice that from the regularity of Z and (1.62), (1.63) by the Green's theorem we define $\dot{z}(0)$ being an element of $H_{00}^{1/2}(\Omega_0)'$.

Remark 1.10

We used the linear spaces $W^{1,\infty}\left(I, L^2(\Omega_t)\right)$ whose the domain Ω_t moves with time. As Z is the restriction of z to Q we define this linear space as being composed of all the restrictions to Q (the non cylindrical domain) of element of $W^{1,\infty}(I, L^2(D))$.

Remark 1.11

Again here $\tilde{\lambda}^2$ is the dissipative term, for in the limit the estimate (1.57) becomes

$$W(t) + \int_0^t \int_{\Gamma_s} (\tilde{\lambda})^2 \; v(s) \; d\Gamma_s \leqslant W(0) = E(0)$$

where

$$W(t) = \frac{1}{2} \int_{\Omega_t} \left(|\nabla z|^2 + \left(\dot{z}\right)^2 \right) dx$$

and we recall that $\tilde{\lambda}$ is the $L^2(\Sigma)$ weak limit of $(\epsilon_n)^{-1/2} \nabla z_n$.

II - DIRICHLET CONDITION, THE GENERAL CASE

The first section was devoted to increasing domains. By Galerkine penalty technics we obtain approximate solutions y_n and z_n for weak the energies $E_n(t)$ and $W_n(t)$ are decaying with time and we also had some informations on the dissipative terms λ and $\tilde{\lambda}$ which turn to be supported by the lateral boundary Σ.

In the general case, i.e. when the domain Ω_t is not increasing we have not such estimate nor dissipation, nevertheless by modification of the penalization technic using some weighting function $\theta(t,x)$ and the Gronwall lemma we obtain some boundness for the approximating solution which unable us to pass to the limit and obtain solution to the Dirichlet problem. We consider the moving domain Ω_t included in D, and $\omega_1,\ldots,\omega_m,\ldots$ a base of $H^1_0(D)$ and the expansion

$$y_{m,\epsilon}(t,x) = \sum_{i=1}^{m} y^{\epsilon}_{m,i}(t)\ \omega_i(x) \tag{2.1}$$

where the $y^{\epsilon}_{m,i}(t)$ are solution of the ordinary differential system

$0 < t < T$,

$i = 1,\ldots,m$

$$\int_D \left(\nabla y_{m,\epsilon} \nabla \omega_i + \ddot{y}_{m,\epsilon}\ \omega_i \right) dx + \frac{1}{\epsilon} \int_{\Omega^c_t} \theta(t)\ y_{m,\epsilon}\ \omega_i\ dx = 0 \tag{2.2}$$

with the usual initial conditions

$$y_{m,i}(0) = a_i \quad , \quad \dot{y}_{m,i}(0) = b_i \tag{2.3}$$

with

$$\sum_{i=1}^{m} a_i \omega_i \longrightarrow y_0^0 \quad , \quad \sum_{i=1}^{m} b_i \omega_i \longrightarrow y_1^0 \text{ respectively in } H_0^1(D) \text{ and } L^2(D)$$

$$\text{when } m \longrightarrow \infty \tag{2.4}$$

$\theta(t,x)$ is a weight function having the following properties :

$$\theta(t)\ (x) = \theta(t,x) \geqslant 0 \quad \text{on} \quad I \times D \tag{2.5}$$

$$\theta(t)\ (.) = 0 \quad \text{on} \quad \Gamma_t \tag{2.6}$$

$$\left| \frac{\partial}{\partial t} \theta(t,.) \right| \leqslant M\ \theta(t,.) \tag{2.7}$$

of some constant M positive. The properties (2.5) and (2.7) are required over I×D while (2.6) is only required on the lateral boundary Σ in fact (2.6) is essential to avoid the boundary integral term which was the key of the dissipation at the previous section. Now as the normal componant v(t) of the speed field on Γ_t has no prescribed sign the boundary term could not be controlable in any estimate, so from (2.6) no such term will appears in the derivative of the energy term.

In fact the weight θ needs to force the solution to zero (with ϵ) in Q^c for this we require more that (2.5) and precisely :

$$x \mapsto \theta(t,x) \text{ is continuous and } \theta(t,s) > 0 \quad \text{for} \quad x \in Q^{\,c} \tag{2.8}$$

And in fact no attention is made for $\theta(t,.)$ on Q, as $\theta(t,.)$ is zero on the lateral boundary it can be choosen zero on Q but it is not necessary for the restriction of θ on Q never appears in the sequel.

To obtain a constructive method for θ we can use a convective approach : Let $V(t,x)$ be a smooth speed vector field which builts the non cylindrical domain Q, i.e. that

$$\Omega_t = T_t(V)(\Omega_0) \qquad (2.9)$$

where $T_t(V)$ is the flow associated to V, equivalentely if a smooth transformation T_t of $\bar{D}$ (onto itself) maps Ω_0 onto Ω_t (and Γ_0 onto Γ_t) then T_t is the flow of the speed vector field defined by

$$V(t,x) = \left(\frac{\partial}{\partial t} T_t\right) \circ T_t^{-1}(x) \quad . \qquad (2.10)$$

Now let $\theta_0(x)$ be a smooth function defined on $\bar{D}$ and having the properties (2.5), (2.6) and (2.8) relatively to $t = 0$ and Ω_0, T_0. We define $\theta(t)(.)$ as follows :

$$\theta(t,x) = \theta_0\left(T_t(V)^{-1}(x)\right) \qquad (2.11)$$

this definition means that θ is obtained by convection of θ_0. The only property to be verified is now (2.7) (2.5, 2.6 and 2.8 being verified by construction).

$$\frac{\partial}{\partial t}\theta(t,x) = (\nabla\theta_0)\left(T_t(V)^{-1}(x)\right) \cdot \frac{\partial}{\partial t} T_t(V)^{-1}(x) \quad , \qquad (2.12)$$

We assume that θ_0 verifies the following property

$$|\nabla\theta_0(x)|_{\mathbb{R}^n} \leq C_1 \; \theta_0(x) \qquad (2.13)$$

obviously there exist a constant C_2 such that

$$\left\| \frac{\partial}{\partial t} T_t(V)^{-1} \right\|_{L^{\infty}(I \times D)^n} \leqslant C_2 \tag{2.14}$$

with $M = C_1 \; C_2$ we get

$$\left| \frac{\partial}{\partial t} \theta(t,x) \right| \leqslant M \; \theta_0 \left(T_t(V)^{-1}(x) \right) \tag{2.15}$$

but (2.15) and (2.11) imply (2.7).

So the convection technic reduces the problem to built a θ_0 verifying (2.6), (2.8) and (2.13) at time $t = 0$.

We assume that there exists such function θ_0 on $\bar{D}$ and $\theta(t,.)$ is defined by convection. We then introduce the energy term as being

$$E_{m,\epsilon}(t) = \frac{1}{2} \int_D \left(|\nabla y_{m,\epsilon}|^2 + \left(\dot{y}_{m,\epsilon}\right)^2 \right) dx + \frac{1}{2\epsilon} \int_{\Omega_t^c} \theta(t) \; (y_{m,\epsilon})^2 \; dx \tag{2.16}$$

and the time derivative, using (2.2) is given by

$$E_{m,\epsilon}(t)' = - \frac{1}{2\epsilon} \int_{\Gamma_t} \theta(t) \; (y_{m,\epsilon})^2 \; v(t) \; d\Gamma_t + \frac{1}{2\epsilon} \int_{\Omega_t^c} \theta'(t) \; (y_{m,\epsilon})^2 \; dx \tag{2.17}$$

As we already said, by (2.6) the first term in the right hand side of (2.17) vanishes and from (2.7) we get

$$E_{m,\epsilon}(t)' \leqslant \frac{M}{2\epsilon} \int_{\Omega_t^c} \theta(t) \; (y_{m,\epsilon})^2 \; dx \tag{2.18}$$

Considering (2.18) and (2.16) we get

$$E_{m,\epsilon}(t)' \leqslant M \; E_{m,\epsilon}(t) \tag{2.19}$$

and then

$$E_{m,\epsilon}(t) \leqslant E_{m,\epsilon}(0) \quad e^{Mt} \tag{2.20}$$

As in the first section (i.e. the monotone dissipative situation) we fix ϵ and let n goes to infinity. From the initial conditions (2.4) we have

$$E_{m,\epsilon}(0) \longrightarrow E(0) + \frac{1}{2\epsilon} A, \ m \longrightarrow \infty \tag{2.21}$$

where

$$E(0) = \frac{1}{2} \int_D \left(\left| \nabla y_0^0 \right|^2 + \left(y_1^0 \right)^2 \right) dx \tag{2.22}$$

$$A = \int_{\Omega_0^c} \theta_c \left(y_0^0 \right)^2 dx = 0 \tag{2.23}$$

for any $\rho \geqslant 1$, there exists $M(\rho,\epsilon)$ such that $m \geqslant M(\rho,\epsilon)$ implies

$$E_{m,\epsilon}(0) \leqslant \rho \ E(0) \quad . \tag{2.24}$$

Let ϵ_n be a sequence going to zero and $m_n = M(\rho,\epsilon_n)$, from (2.24) and (2.20) we get

$$0 \leqslant t \leqslant T, \quad E_n(t) = E_{m_n,\epsilon_n}(t) \leqslant \rho \ e^{MT} \ E(0) \tag{2.25}$$

with $y_n = y_{m_n,\epsilon_n}$, (2.25) shows that the functions ∇y_n, $\dot{y}_n$ (and then y_n itself) remain bounded in $L^\infty(I, L^2(D))$ while $\frac{1}{\sqrt{\epsilon_n}} \chi_{\Omega_t^c} \ \theta(t)^{1/2} y_n$ remains bounded in $L^2(I\times D)$ we still denote by ϵ_n, y_n a subsequence and an element $y \in W^{1,\infty}(I, L^2(D)) \cap L^\infty\left(I, H_0^1(D)\right)$ such that the following convergences hold :

$$\nabla y_n \rightharpoonup \nabla y, \ \dot{y}_n \rightharpoonup \dot{y}, \ y \rightharpoonup y \ \text{ weakly-* in } \ L^\infty(I, L^2(D)) \tag{2.26}$$

and

$$(\epsilon_n)^{-1/2} \chi_{\Omega_t^c} \theta^{1/2} y_n \rightharpoonup \lambda \text{ weakly-* in } L^\infty(I, L^2(D)) \qquad (2.27)$$

In the limit the estimate (2.25) which involves lower semi continuous terms with respect to the convergences described at (2.26) and (2.27) leads to the following estimate

$$\forall \rho > 1, \quad E(t) = \frac{1}{2}\int_D \left(|\nabla y|^2 + \left(\dot{y}\right)^2\right) dx + \frac{1}{2}\int_D \lambda^2 \, dx \leqslant \rho \, e^{MT} E(0) \qquad (2.28)$$

as (2.28) holds for any $\rho > 1$ then we get

$$E(t) \leqslant e^{MT} E(0), \; \forall \, t \in [0,T] \quad , \qquad (2.29)$$

We consider the linear space S defined at (1.29), $S \subset H_0^1(I\times D)$. For any element ψ in S, by (2.2) we get

$$\int\int_{I\times D} \left(\nabla y_n \nabla\psi - \dot{y}_n \dot{\psi}\right) dt \, dx = -\frac{1}{\epsilon}\int_0^T \int_{\Omega_t^c} \theta(t) \, y_n \psi \, dt \, dx \qquad (2.30)$$

Using (2.26) and (2.27) in the limit (2.30) leads to

$$\int\int_{I\times D} \left(\nabla y \nabla\psi - \dot{y} \, \dot{\psi}\right) dt \, dx = -\int\int_{I\times D} \lambda \, \psi \, dt \, dx \, , \qquad \forall \, \psi \in H_0^1(I\times D) \qquad (2.31)$$

with the Green's theorem we obtain

$$\Box \, y = \ddot{y} - \Delta y = -\lambda \; . \qquad (2.32)$$

But $\lambda \in L^\infty(I, L^2(D))$ has its support in the complementary Q^c of the non cylindrical domain Q.

Lemma 2.1

Support of $\lambda \subset Q^c$, i.e. $\forall \, \psi \in \mathcal{D}(Q)$ we have

$$\int_0^T \int_{\Omega_t} \lambda \, \psi \, dt \, dx = 0 \quad . \tag{2.33}$$

Proof : It is sufficient to verify that

$$\frac{1}{\sqrt{\epsilon_n}} \int_0^T \int_{\Omega_t} \chi_{\Omega_t^c} \, \theta^{1/2} \, y_n \, \psi \, dt \, dx = 0 \tag{2.34}$$

for all n, in the limit in (2.34), by (2.27) we obtain (2.33) □

On the other hand we obtain that the restriction to Q^C of y, element of $H_0^1(I\times D)$, is zero.

Lemma 2.2

$$\int_0^T \int_{\Omega_t^c} \theta \, y^2 \, dt \, dx = 0 \quad . \tag{2.35}$$

Proof : From the estimate (2.25) we get

$$\frac{1}{\epsilon_n} \int_0^T \int_{\Omega_t^c} \theta(t) \, (y_n)^2 \, dx \, dt \leqslant C$$

the compact injection of $H_0^1(I\times D)$ in $L^2(I\times D)$ implies that y_n converges strongly to y in $L^2(I\times D)$, then in the limit we get (2.35) □

We introduce, as at the previous section, the restriction of y to Q:

$$Y = y|_Q \quad . \tag{2.36}$$

Corollary 2.3

$$Y \in H^1(Q) \quad \text{with} \quad Y|_\Sigma = 0 \qquad \square$$

and Y solves the wave equation.

Proposition 2.4

Let $(y_o, y_1) \in H_0^1(\Omega_o) \times L^2(\Omega_o)$ and θ_o verifying (2.13), then Y is solution of

$$Y \in L^\infty\left(I,\ H_0^1(\Omega_t)\right) \cap W^{1,\infty}\left(I,\ L^2(\Omega_t)\right)$$
$$\square\, Y = 0 \quad \text{in} \quad Q,\ Y|_\Sigma = 0\ , \tag{2.37}$$
$$Y(0) = y_0,\ \dot{Y}(0) = y_1 \quad \text{in} \quad \Omega_0$$

The general situation can be considered with the same technic : if $f \in L^1\left(0,T,\ L^2(\Omega_t)\right)$ and $g \in H^{1/2}(\Sigma)$ are given (i.e. $\|f(t)\|_{L^2(\Omega_t)}$ belongs to $L^1(0,T)$) then we consider F and G, extensions of f and G to IxD such that $F \in L^2(I\times D)$ and $G \in H^1(I\times D)$ and we modify (2.2) and (2.16) as follows

$$\int_D \left(\nabla y_{m,\epsilon}\ \nabla\omega_1 + \ddot{y}_{m,\epsilon}\ \omega_1\right)dx + \frac{1}{\epsilon}\int_{\Omega_t^c} \theta(t)\ (y_{m,\epsilon} - G)\omega_1\ dx = \int_D F\ \omega_1\ dx \tag{2.38}$$

and

$$E_{m,\epsilon}(t) = \frac{1}{2}\int_D |\nabla y_{m,\epsilon}|^2 + \left(\dot{y}_{m,\epsilon}\right)^2\ dx + \frac{1}{2\epsilon}\int_{\Omega_t^c} \theta(t)\ (y_{m,\epsilon} - G)^2\ dx \tag{2.39}$$

The proof of Proposition 2.4 can be repeated and we obtain the :

Theorem 2.5

Let D be a bounded domain in $\mathbb{R}^N$ with Lipchitzian boundary, I = [0,T] and Ω_t be a moving domain in D defined by $\Omega_t = T_t\ (\Omega_0)$ for some smooth one to one mapping t_t of $\bar{D}$ into itself. $\bar{\Omega}_0$ being included in D let θ_0 a smooth function verifying (2.) and $\theta(t)$ be defined by (.). Then the restriction Y to Q of any limiting element y (obtained by the extraction process y_n) is solution of $Y \in W^{1,\infty}\left(0,T, L^2(\Omega_t)\right) \cap L^\infty\left(0,T, H^1(\Omega_t)\right)$,

$$\square Y = f \quad \text{in} \quad Q \tag{2.40}$$

$$Y = g \quad \text{on} \quad \Sigma \tag{2.41}$$

$$Y(0) = y_0, \quad \dot{Y}(0) = y_1 \quad \text{in} \quad \Omega_0 \tag{2.42}$$

where f, y, y_0, y_1 are respectively given elements in

$$L^1\left(0,T\ L^2(\Omega_t)\right),\ H^{1/2}(\Sigma),\ H^1(\Omega_0),\ L^2(\Omega_0)\ \square$$

III - CONTRACTING DOMAINS. (TIME LIKE BOUNDARY)

We consider now the case where the domain Ω_t is monotone and more precisely decaying. As we assume the lateral boundary Σ to be smooth this assumption is equivalent to

$$v < -1 \quad \text{on} \quad \Sigma \tag{3.1}$$

We have $\Omega_t \subset \Omega_0 \subset D$, $\omega_1, \ldots, \omega_m$, is a base of $H^1_0(D)$ and

$$y_m^{(t,x)} = \sum_{i=1}^{m} \varphi_{m,i}(t)\ \omega_i(x) \tag{3.2}$$

is define by the non penalized system

$$i = 1, \ldots, m \quad 0<t<T \qquad \int_{\Omega_t} \square\, y_m\ \omega_i\ dx = 0 \tag{3.3}$$

with the initial condition as usual

$$\varphi_{m,i}(0) = a_i \quad , \quad \dot{\varphi}_{m,i}(0) = b_i \tag{3.4}$$

with

$$\sum_{i=1}^{m} a_i \omega_i \underset{m\to\infty}{\longrightarrow} y_o^0 \quad \text{in} \quad H_o^1(D)$$

$$\sum_{i=1}^{m} b_i \omega_i \longrightarrow y_1^0 \quad \text{in} \quad L^2(D)$$

where (y_o, y_1) is given in $H_o^1(\Omega_o) \times L^2(\Omega_o)$ the energy term is

$$E_m(t) = \frac{1}{2} \int_{\Omega_t} \left(|\nabla y_m|^2 + \dot{y}_m^2\right) dx \tag{3.5}$$

we get, with (3.3)

$$E_m(t)' = \int_{\Gamma_t} \left[\frac{\partial y_m}{\partial ,} \dot{y}_m + \frac{1}{2}\left(|\nabla y_m|^2 + \left(\dot{y}_m\right)^2\right) v\right] d\Gamma_t \quad . \tag{3.6}$$

We assume that Ω_t is rapidely contracting and more precisely

$$\forall t, \quad \forall\, x \in \Gamma_t, \; v(t,x) = -1 - \alpha(t,x) \quad \text{with} \quad \alpha(t,x) > 0 \tag{3.7}$$

and then we get

$$E_m(t)' = \frac{1}{2} \int_{\Gamma_t} \left(\left(\frac{\partial}{\partial n} y_m\right)^2 + \left(\dot{y}_m\right)^2\right) \left((-1-\alpha) + 2 \frac{\partial y_m}{\partial n} \dot{y}_m\right) dt$$

$$+ \frac{1}{2} \int_{\Gamma_t} |\nabla_\Gamma\, y_m|^2 \;(-1 - \alpha)\; d\Gamma \tag{3.8}$$

that is

$$E_m(t)' = -\frac{1}{2}\int_{\Gamma_t}\left(\frac{\partial}{\partial n}y_m - \dot{y}_m\right)^2 d\Gamma_t - \frac{1}{2}\int_{\Gamma_t}|\nabla_\Gamma y_m|^2\,(1+\alpha)\,d\Gamma_t$$

$$-\frac{1}{2}\int_{\Gamma_t}\left(\dot{y}_m\right)^2 \alpha\, d\Gamma_t - \frac{1}{2}\int_{\Gamma_t}\left(\frac{\partial y_m}{\partial n}\right)^2 \alpha\, d\Gamma_t \qquad (3.9)$$

and it is obvious in (3.9) that $v < -1$ implies a dissipation as $E_m'(t)$ is negative.

Let

$$2\beta = \min_{t,x\in\Sigma} \alpha(t,x) > 0 \qquad (3.10)$$

we get

$$E_n(t) + \beta\int_0^t\int_{\Gamma_t}\left[\left(\frac{\partial}{\partial n}y_m - \dot{y}_m\right)^2 + |\nabla_\Gamma y_m|^2 + \dot{y}_m^2 + \left(\frac{\partial y}{\partial n}\right)^2\right]dt\,d\Gamma_t \leqslant \rho E(0) \qquad (3.11)$$

where ρ is any number, $\rho > 1$ and m larger than $M(\rho)$. As in the two previous section we have the boundness of the sequences ∇y_m and $\dot{y}_m$ in $L^\infty\left(I, L^2(\Omega_t)\right)$, then by Poincaré inequality of y_m itself in $L^\infty\left(I, L^2(\Omega_t)\right)$, ($y_m$ being given on Ω_0 at $t = 0$). Then we consider subsequences y_m, ∇y_m, $\dot{y}_m$ and an element $y \in W^{1,\infty}\left(I, L^2(\Omega_t)\right) \cap L^\infty\left(I, H^1(\Omega_t)\right)$ such that the weak-* convergences in $L^\infty(I, L^2(D))$ hold :

$$y_m \rightharpoonup y,\ \nabla y_m \rightharpoonup \nabla y,\ \dot{y}_m \rightharpoonup y \qquad (3.12)$$

(in fact y_m converges strongly to y in $L^2(Q)$.

For any ψ in Q we have

$$\int_0^T\int_{\Omega_t}\left(\nabla y_m \nabla\psi - \dot{y}_m\dot{\psi}\right)dt\,dX = \int_0^T\int_{\Gamma_t}\frac{\partial}{\partial n}y_m\,\psi\,dt\,d\Gamma_t \quad . \qquad (3.13)$$

By (3.12) the left hand side of (3.13) is bounded then by the uniform bound theorem $\frac{\partial}{\partial n} y_m$ remains bounded in $H^{-1/2}(\Sigma)$ (more precisely in $H^{1/2}_{00}(\Sigma)'$) then there exist $\lambda \in H^{1/2}_{00}(\Sigma)'$ such that $\frac{\partial}{\partial n} y_m \rightharpoonup \lambda$ weak convergence in this Hilbert space. In the limit in (3.13) and with Green's theorem we get

$$\int_0^T\int_{\Omega_t} (\nabla y\, \nabla\psi - \dot{y}\dot{\psi})\, dt\, dX = \langle \Box\, y, \psi\rangle + \int_\Sigma \frac{\partial y}{\partial n}\, \psi\, d\Sigma = \int_\Sigma \lambda\, \psi\, d\Sigma \qquad (3.14)$$

then

$$\Box y = 0 \quad \text{in} \quad Q \quad \text{and} \quad \frac{\partial y}{\partial n} = \lambda \quad \text{on} \quad \Sigma \qquad (3.15)$$

but λ being not prescribed this does not mean that y solves a Newman problem. Also we have no indication on any Dirichlet Boundary condition on Σ ; simply we clame the

Proposition 3.1

The sequence y_m is bounded in $E = W^{1,\infty}(I, L(\Omega_t)) \cap L^\infty\left(I, H^1(\Omega_t)\right)$ and each weak-* limiting point y in E solves the wave equation

$$\Box y = 0 \quad \text{in} \quad Q \qquad (3.16)$$

$$y(0) = y_0 \quad \text{in} \quad \Omega_0 \qquad (3.17)$$

$$\dot{y}(0) = y_1 \quad \text{in} \quad \Omega_0 \qquad (3.18)$$

Notice that as $\frac{\partial y}{\partial n} = \lambda \in H^{1/2}_{00}(\Sigma)'$ and $\Box y \in L^2(Q)$, $y \in E$, the Green's theorem give sens to $\dot{y}(0)$ in $H^{1/2}_{00}(\Omega_0)'$. □

Remark 3.2

We shall see at Section that in fact the problem (3.16), (3.17), (3.18) is well posed in E (i.e. that the solution y is unique). One can think that as the boundary Γ_t is moving at speed v, |v| larger than 1, speed of propagation of the information, no boundary condition can be imposed. So in In fact one can solves the wave problem in I×D, with any boundary condition on I×∂D, compatible with the initial data (y_0, y_1) given on Ω_0. Let z be the solution and set $y = z|_Q$. Then y is the unique solution of problem (3.16), (3.17), (3.18).

IV - NEWMAN CONDITION AND CONTRACTING DOMAIN

We consider now that the normal componant v(t) of the speed vector is negative

$$0 \leqslant t \leqslant T, \quad v(t) \leqslant 0 \quad \text{on} \quad \Gamma_t \tag{4.1}$$

that is Ω_t is decaying with time and $\Omega_t \subset \Omega_0 \subset D$; $\omega_1, \ldots, \omega_m, \ldots$ being a dense family in $H_0^1(D)$ we consider

$$y_m = \sum_{i=1}^{m} \varphi_{m,i}(t)\ \omega_i(x) \tag{4.2}$$

where the $\varphi_{m,i}$ are determined by the ordinary differential system :

$$\int_{\Omega_t} \left(\nabla y_m \nabla \omega_i + \ddot{y}_m \omega_i\right) dx = \int_{\Omega_t} f\ \omega_i\ dx + \int_{\Gamma_t} g\ \omega_i\ d\Gamma_t \tag{4.3}$$

and the energy term :

$$E_m(t) = \frac{1}{2} \int_{\Omega_t} \left(|\nabla y_m|^2 + \left(\dot{y}_n\right)^2\right) dx \tag{4.4}$$

we have

$$E_m(t') = \int_{\Omega_t} \left(\nabla y_m \nabla \dot{y}_m + \ddot{y}_m \dot{y}_m\right) dx + \frac{1}{2} \int_{\Gamma_t} \left\{ |\nabla y_m|^2 + \left(\dot{y}_m\right)^2 \right\} v \, d\Gamma_t \qquad (4.5)$$

with (4.3) and (4.5) we get

$$E_m(t)' + \frac{1}{2} \int_{\Gamma_t} \left(|\nabla y_m|^2 + \left(\dot{y}_m\right)^2 \right) |v| \, d\Gamma_t = \int_{\Omega_t} f \, \dot{y}_m \, dx + \int_{\Gamma_t} g \, \dot{y}_m \, d\Gamma_t \qquad (4.6)$$

The initial conditions associated to (4.3) are similar to (2.3) and (2.4). Then we have

$$E_m(0) = \frac{1}{2} \int_{\Omega_0} \left(|\nabla y_m(0)|^2 + y_m(0)^2 \right) dx$$

$$E_m(0) \xrightarrow[m \to \infty]{} E(0) = \int_{\Omega_0} \frac{1}{2} \left(|\nabla y_0|^2 + y_1^2 \right) dx \quad . \qquad (4.7)$$

Then ∇y_n, $\dot{y}_n$ remain bounded in $L^{\infty}\left(I, L^2(\Omega_t)\right)$ while $\frac{\partial}{\partial n} y_m$ remains bounded in $L^2(\Sigma)$.

So we consider a subsequence y_m and an element y such that ∇y_m, $\dot{y}_m$ (resp. $\frac{\partial}{\partial n} y_m$) converge weakly-* (resp. weakly) to ∇y, $\dot{y}$ $\left(\text{resp. } \frac{\partial}{\partial n} y\right)$ and we shall now characterise the problem solved by the limiting element y. We use the

Lemma 4.1

Let y and z be two smooth functions defined on $\bar{I} \times \bar{D}$ then

$$\int_0^T \int_{\Omega_t} f \, \dot{g} \, dt \, dx = \int_{\Omega_T} f \, g \, dx - \int_{\Omega_0} fg \, dx$$

$$- \int_0^T \int_{\Omega_t} \dot{f} \, g \, dt \, dx - \int_0^T \int_{\Gamma_t} fg \, v(t) \, d\Gamma_t \, dt \qquad (4.8)$$

Proof : We have

$$\frac{d}{dt}\int_{\Omega_t} fg\ dx = \int_{\Omega_t} \frac{d}{dt}(fg)\ dx + \int_{\Gamma_t} fg\ v(t)\ d\Gamma_t \qquad \square$$

Let ψ be an element of S defined at (.) then

$$\int_0^T\int_{\Omega_t}\left(\nabla y_m\ \nabla\psi - \dot{y}_m\ \dot{\psi}\right) dt\ dx =$$

$$= \int_0^T\int_{\Gamma_t} \dot{y}_m\ \psi\ v(t)\ d\Gamma_t\ dt + \int_0^T\int_{\Omega_t}\left(\nabla y_m \nabla\psi + \ddot{y}_m \psi\right) dt\ dx \qquad (4.9)$$

with (4.9) and (4.3) we get

$$\int_0^T\int_{\Omega_t}\left(\nabla y_m\ \nabla\psi - \dot{y}_m \dot{\psi}\right) dt\ dx$$

$$= \int_0^T\int_{\Omega_t} f\ \psi\ dt\ dx + \int_0^T\int_{\Gamma_t}\left(\dot{y}_m\ v + g\right)\psi\ dt\ d\Gamma_t \quad . \qquad (4.10)$$

The left hand side of (4.10) converges to $\int_0^T\int_{\Omega_t}(\nabla y \nabla\psi - \dot{y}\dot{\psi}) dt\ dx$, then using the Green's theorem we have

$$\int_0^T\int_{\Omega_t}(\nabla y\ \nabla\psi - \dot{y}\dot{\psi}) dt\ dx$$

$$= \int\int_Q (-\Delta y + \ddot{y})\psi\ dQ + \int_0^T\int_{\Gamma_t}\left(\frac{\partial y}{\partial n}\psi + \dot{y}\psi\ v(t)\right) d\Gamma_t\ dt \qquad (4.11)$$

then with (4.10) and (4.11) we get

$$\int\int_Q \square\ y\ \psi\ dQ + \int_0^T\int_{\Gamma_t}\frac{\partial y}{\partial n}\psi\ dt\ d\Gamma_t$$

$$= \int\int_Q f \psi \, dQ + \int_0^T \int_{\Gamma_t} y \psi \, dt \, d\Gamma_t \qquad (4.12)$$

finely we have the solution for the Newman problem in the non cylindrical domain Q.

Proposition 4.2

Let f, g, y_0, y_1 be given elements respectively in $L^1\left(0,T, L^2(\Omega_t)\right), L^1\left(0,T,L^2(\Gamma_t)\right)$, $H^1(\Omega_0)$, $L^2(\Omega_0)$ and $v(t) \leqslant 0$ on Γ_t, then the limiting element y verifies

$$y \in W^{1,\infty}\left(0,T,L^2(\Omega_t)\right) \cap L^\infty\left(0,T,H^1(\Omega_t)\right) \quad - \Delta y + \ddot{y} = f \quad \text{in} \quad Q \qquad (4.13)$$

$$\frac{\partial y}{\partial n_t} = g \quad \text{on} \quad \Sigma \qquad (4.14)$$

$$y(0) = y_0 \; , \; \dot{y}(0) = y_1 \quad \text{on} \quad \Omega_0 \; . \qquad (4.15)$$

Remark 4.3

From (4.6) we also have the term $|\nabla y_m|^2 + \left(\dot{y}_m\right)^2 \dfrac{|v|}{(1+v^2)^{1/2}}$ which remains bounded as m goes to infinity in $L^2(\Sigma)$. One must pay attention to the fact that $d\Sigma = (1+v^2)^{1/2} \, d\Gamma_t \, dt$.

V - A DISSIPATIVE BOUNDARY CONDITION FOR A MOVING DOMAIN

We briefly discuss now the so called "moving actuator" problem for wave equation [6]. D is still a fixed domain in which we consider moving domain (or moving object) B(t). The moving domain Ω_t is now the complementary in D of $\bar{B}(t)$. Γ_t is the boundary of B(t) and n_t is the unitary normal field on Γ_t out going to B(t). Let V(t,.) be any speed vector field which moves B(t), i.e. $B(t) = T_t(V)(B_0)$, $t \in [0,T]$ where $T_t(V)$ is the flow of V. Then $v(t,.) = V(t,.).n_t(.)$ is uniquely defined on Γ_t (Γ_t is smooth enough then v is not depending on the choice of the speed V). The boundary Γ_t is decomposed in three parts

$$\Gamma_t = \Gamma_t^+ \cup \Gamma_t^- \cup \Gamma_t^0$$

with

$$\Gamma_t^{+,-,0} = \{x \in \Gamma_t \,|\, v(t,x) > 0, < 0, = 0\} \tag{5.1}$$

And we consider the mixed non cylindrical wave problem :

$$\ddot{y} - \Delta y = 0 \quad \text{in} \quad \Omega_t \tag{5.2}$$

$$y = 0 \quad \text{on} \quad \Gamma_t^0 \cup \Gamma_t^- \cup \partial D \tag{5.3}$$

$$\frac{\partial y}{\partial n_t} = 0 \quad \text{on} \quad \Gamma_t^+ \tag{5.4}$$

$$y(0) = y_0 \quad \text{in} \quad H^1(\Omega_0 \ ) \tag{5.5}$$

$$\dot{y}(0) = y_1 \quad \text{in} \quad L^2(\Omega_0) \tag{5.6}$$

to use the Galerkine-penalty technics which have been developed at the previous sections we introduce to functions $R_d(t)$ and $R_n(t)$ (respectively for Dirichlet condition and Newman condition) defined and smooth on $\bar{D}$ and verifying

$$0 \leqslant R(t)\ (.) \leqslant 1 \quad . \tag{5.7}$$

Let as usual $\omega_1,\ldots,\omega_m\ldots$ be a dense family in $H_0^1(D)$ and consider

$$y_{m,\varepsilon}(t) = \sum_{i=1}^{m} \varphi_i(t)\ \omega_i(x) \tag{5.8}$$

where the φ_i are defined as being the solution of the ordinary differential system

$$\int_D (1-\chi_{B(t)} R_n(t)) \left(\nabla y_{m,\epsilon} \nabla\omega_i + \ddot{y}_{m,\epsilon}\omega_i\right)dx + \frac{1}{\epsilon}\int_{B(t)} R_d(t)\ (y_{m,\epsilon})^2\ dx \qquad (5.9)$$

We assume that

$$R_d(t)(.) = 0 \quad \text{on} \quad \Gamma_+(t) \qquad (5.10)$$

$$R_d(t)(.) > 0 \quad \text{on} \quad B_-(t) \qquad (5.11)$$

where $B_-(t)$ is a neighborhood in $\overline{B(t)}$ of Γ_t^-. We consider the energy term

$$E_{m,\epsilon}(t) = \frac{1}{2}\int_D (1 - \chi_{B(t)}\ R_n(t)) \left(|\nabla y_{m,\epsilon}|^2 + \left(\dot{y}_{m,\epsilon}\right)^2\right)dx$$
$$+ \frac{1}{2\epsilon}\int_{B(t)} R_d(t)\ (y_{m,\epsilon})^2\ dx \qquad (5.12)$$

we get

$$E_{m,\epsilon}(t)' = -\frac{1}{2}\int_{\Gamma_t} R_n(t)\ \left(|\nabla y_{m,\epsilon}|^2 + \left(\dot{y}_{m,\epsilon}\right)^2\right)\ v(t)\ d\Gamma_t$$
$$+ \frac{1}{2\epsilon}\int_{\Gamma_t} R_d(t)\ (y_{m,\epsilon})^2\ v(t)\ d\Gamma_t$$
$$- \int_{B(t)} R_n'(t)\ \left(|\nabla y_{m,\epsilon}|^2 + \left(\dot{y}_{m,\epsilon}\right)^2\right)\ dx$$
$$+ \frac{1}{2\epsilon}\int_{B(t)} R_d'(t)\ \left(|\nabla y_{m,\epsilon}|^2 + \left(\dot{y}_{m,\epsilon}\right)^2\right)\ dx \qquad (5.13)$$

We formulate the convective assumption we had at Section 3 for the Dirichlet situation $\exists\, M > 0$ such that

$$|R_n'(t)\ (.)| \leqslant M\ (1 - R_n(t)\ t.)) \tag{5.14}$$

$$|R_d'(t)\ (.)| \leqslant M\ R_d(t)\ (.) \quad . \tag{5.15}$$

Using the Gronwall Lemma, from (5.13)-(5.15) we get

$$E_{m,\epsilon}(t) \leqslant C\ M$$

by the same boundness arguments used in the previous section we can construct a "diagonal" subsequence y_1 which remains bounded so that any of its limit point y is solution of the problem (5.2)-(5.6) :

Proposition 5.1

Assuming that the interior of Γ_t^0 in Γ_t is non empty. With initial data (y_0, y_1) in $H^1(\Omega_0) \times L^2(\Omega_0)$, $y_0\Big|_{\Gamma_0^+ \cup \Gamma_0^0} = 0$, there exists an element $y \in L^\infty\left(0,T,\ H^1(D\backslash\bar{B}(t))\right)$ such that $\dot{y} = y_t \in L^\infty(0,T,\ L^2(D\backslash B(t)))$ which solves the problem (5.2)-(5.6) □

Remark 5.2

The fundamental assumption that the interior of Γ_t^0 is non empty is imposed to make possible the existence of the function $R_d(t)(.)$ verifying (5.11) with $B_-(t)$ strictly containing Γ_t^-.

Remark 5.3

In fact on the part Γ_t^0 of the boundary any well posed condition can be reached by this technic.

Remark 5.4

The fundamental assumption that interior of Γ_t^0 is non empty means that at each time a part of the boundary must be tangent to the speed, which can be achieved on examples by specific rotating geometry or by deformation motion of B(t). In that it seams that in a further deeper analysis this assumption could be avoid.

Remark 5.5. Dissipative situation

Assuming smoothness of the solution y we obtain the energy term

$$W(t) = \frac{1}{2} \int_{\Omega_t} \left(|\nabla y|^2 + (\dot{y})^2 \right) dx \tag{5.16}$$

which is differentiable and

$$W'(t) = -\frac{1}{2} \int_{\Gamma_t^-} \left(\frac{\partial y}{\partial n_t} \right)^2 \left[(v(t)^-)^3 = (v(t)^-) \right] d\Gamma_t$$

$$- \frac{1}{2} \int_{\Gamma_t^+} \left(|\nabla_\Gamma \ y(t)|^2 + (\dot{y}(t))^2 \right) v(t)^+ \ d\Gamma_t$$

that is that W'(t) is strictly negative.

REFERENCES

P. Cannarsa - G. Da Prato - J.P. Zolesio [1]
in these Proceedings.

P. Cannarsa - G. Da Prato - J.P. Zolesio [2]
Evolution Equation in non cylindrical domains, to appear.

P. Cannarsa - G. Da Prato - J.P. Zolesio [3]
The damped wave equation in a moving domain. Journal of Differential Equation, to appear.

G. Da Prato - J.P. Zolesio [4]
Existence and Optimal Control for wave equation in moving domain. (Preprint Scuola Normale Sup. Pisa, n°23, 1988).

J.L. Lions - E. Magenes [5]
Problèmes aux limites non homogènes, Dunod, Paris, 1968.

J.P. Zolesio [6]
Moving actuors in wave equation, Proc. IFAC Symposium, Perpignan, June 1989, EL JAI ed.

Further results on exact controllability of the Euler-Bernoulli equation with controls on the Dirichlet and Neumann boundary conditions⁺

I. Lasiecka and R. Triggiani

Department of Applied Mathematics, Thornton Hall

University of Virginia, Charlottesville, VA 22903

1. Introduction and main result

The present note is a successor to (part of) our recent paper [L-T.1] which studies, by a direct method, the exact controllability of the Euler-Bernoulli problem

$$\begin{cases} w_{tt} + \Delta^2 w = 0 & \text{in } (0,T] \times \Omega = Q & \text{(1.1a)} \\ w(0,x) = w_0(x);\ w_t(0,x) = w_1(x) & \text{in } \Omega & \text{(1.1b)} \\ w|_\Sigma = g_1 & \text{in } (0,T] \times \Gamma = \Sigma & \text{(1.1c)} \\ \frac{\partial w}{\partial \nu}\Big|_\Sigma = g_2 & \text{in } (0,T] \times \Gamma = \Sigma & \text{(1.1d)} \end{cases}$$

Here, Ω is an open bounded domain in R^n, n typically ≥ 2, with sufficiently smooth boundary Γ, and ν is the outward unit vector to Γ. We decompose Γ as follows: let $x^0 \in R^n$ be a fixed but otherwise arbitrary point and let

$$\Gamma = \Gamma_+(x^0) \cup \Gamma_-(x^0) \tag{1.2}$$

$$\Gamma_+(x^0) = \{x \in \Gamma : (x-x^0) \cdot \nu(x) > 0\}; \quad \Gamma_-(x^0) = \{x \in \Gamma : (x-x^0) \cdot \nu(x) \leq 0\} \tag{1.3}$$

Throughout the paper, we let A be the positive, self-adjoint operator on $L_2(\Omega)$ defined by $Af = \Delta^2 f$, $\mathcal{D}(A) = H^4(\Omega) \cap H_0^2(\Omega)$. Several exact controllability results were obtained in [L-T.1] for problem (1.1), but not however the following which, among other things, is needed in our semilinear study in [L-T.2].

⁺Research partially supported by the Air Force Office of Scientific Research under Grant AFOSR-87-0321.

Theorem 1.1 a) For any $T > 0$, given any pair of initial data $\{w_0, w_1\} \in Y$, with

$$Y = H_0^1(\Omega) \times H^{-1}(\Omega) \equiv \mathcal{D}(A^{1/4}) \times [\mathcal{D}(A^{1/4})]' \text{ (equivalent norms)} \tag{1.4}$$

there exist boundary control functions

$$g_1 \in H_0^1(0,T; L_2(\Gamma)) \tag{1.5}$$

$$g_2 = \begin{cases} 0 & \text{on } (0,T) \times \Gamma_+(x^0) = \Sigma_+(x^0) \quad (1.6a) \\ \in H_0^1(0,T; [H^1(\Gamma_-(x^0))]'), & \text{on } (0,T] \times \Gamma_-(x^0) = \Sigma_-(x^0) \quad (1.6b) \end{cases}$$

such that the corresponding solution of problem (1.1) satisfies $w(T,\cdot) = w_t(T,\cdot) = 0$

b) The exact controllability statement in part a) is equivalent to the following: there is a constant $c_T > 0$ such that

$$\int_\Sigma \left[\frac{\partial(\Delta\phi)}{\partial\nu}\right]^2 d\Sigma + \int_{\Sigma_-(x^0)} (|\nabla_\sigma(\Delta\phi)|^2 + |\Delta\phi|^2)\, d\Sigma_-(x^0) \ge c_T \| \{\phi_0, \phi_1\} \|^2_{\mathcal{D}(A^{3/4}) \times \mathcal{D}(A^{1/4})} \tag{1.7}$$

where ∇_σ denotes the tangential gradient and ϕ is the solution of the corresponding homogeneous problem, backward in time

$$\begin{cases} \phi_{tt} + \Delta^2 w = 0 & \text{in } Q \quad (1.8a) \\ \phi|_{t=T} = \phi_0, \quad \phi_t|_{t=T} = \phi_1 & \text{in } \Omega \quad (1.8b) \\ \phi|_\Sigma \equiv 0 & \text{in } \Sigma \quad (1.8c) \\ \dfrac{\partial\phi}{\partial\nu}|_\Sigma \equiv 0 & \text{in } \Sigma \quad (1.8d) \end{cases}$$

c) Problem (1.1) is exactly controllable over $[0, T]$ with respect to the state space (1.4) and control spaces (1.5), (1.6) as described in part a), if and only if problem (1.1) is exactly controllable over $[0, T]$ on the state space

$$X = [\mathcal{D}(A^{1/4})]' \times [\mathcal{D}(A^{3/4})]' \tag{1.9}$$

with respect to the control functions

$$g_1 \in L_2(0,T; L_2(\Gamma)) \tag{1.10}$$

$$g_2 = \begin{cases} 0 & \text{on } \Sigma_+(x^0) \\ \in L_2(0,T; [H^1(\Gamma_-(x^0))]'), & \text{on } \Sigma_-(x^0) \quad \square \end{cases} \tag{1.11a}$$

Remarks 1.1 (i) We must prove only parts a) and b) of Theorem 1.1, for part c) follows then from [L-T.1, Theorem 4.5]. We note explicitly that no geometrical conditions are assumed on Ω (except smoothness of Γ). Thus, Theorem 1.1 complements and completes a previous result [L-T.1, Theorem 1.2] which provided exact controllability on Y (see (1.4)) with g_1 as in (1.5), however with T "sufficiently large" and with $g_2 \equiv 0$ under the additional geometrical condition on Ω: that there exists x^0 such that $\Gamma_-(x^0)$ is empty and so $\Gamma = \Gamma_+(x^0)$. The proof of Lemma 2.2 of our present note permits to improve the time of exact controllability from "T sufficiently large" to " T arbitrary" also in Theorem 1.2 of [L-T.1] when $g_2 = 0$.

ii) When $g_1 \equiv 0$, [L.1, Section 3, plus Appendix] gives two exact controllability results for problem (1.1) both for $T > 0$ arbitrary and without geometrical conditions on Ω: one on

$$\text{the state space } L_2(\Omega) \times H^{-2}(\Omega), \quad \text{with controls } g_2 \in L_2(\Sigma_+(x^0)) \tag{1.12}$$

and the other on

$$\text{the state space } H_0^2(\Omega) \times L_2(\Omega) \quad \text{with controls } g_2 \in H_0^1(0,T; L_2(\Gamma_+(x^0)) \tag{1.13}$$

Applying, as in [L-T.3], the interpolation theorem to the continuous inverse of the surjective map: $\{g_1 \equiv 0, g_2\} \to \{w(T,\cdot), w_t(T,\cdot)\}$ of problem (1.1) with $w_0 = w_1 = 0$ in both cases (1.12) and (1.13), one obtains exact controllability for any $T > 0$ and without geometrical conditions on Ω on the same space $Y = H_0^1(\Omega) \times H^{-1}(\Omega)$ as in (1.4), this time with controls

$$g_1 = 0 \quad \text{and } g_2 \in H_{00}^{1/2}(0,T; L_2(\Gamma_+(x^0))) \tag{1.14}$$

a result which neither contains, nor is contained by, our Theorem 1.1 above. See [L-M.1 p.66] for the definition of the space in (1.14). Thus, two non-comparable classes of controls, the one in (1.5), (1.6), and the one in (1.14), produce the same exact controllability result on Y both for any $T > 0$ and without geometrical conditions in Ω. $\square$

2. Proof of Theorem 1.1

As remarked before, we only need to show parts a) and b), for then part c) follows from [L-T.1]. For sake of brevity, we shall merely complement the arguments in [L-T.1], and assume that the reader has [L-T.1] at his hand. Thus, it will suffice to show parts a) and b) of Theorem 1.1 with $\Gamma_+(x^0) = $ empty, and thus with g_1 as in (1.5) and $g_2 \in H_0^1(0,T; H^{-1}(\Gamma))$, for then for the minor variations needed to handle the case with g_2 as in (1.6) we may refer to the proof of [L-T.1, Theorem 4.5].

Step 1. The solution at time T to problem (1.1) with $w_0 = w_1 = 0$ can be written as [L-T.1]

$$\begin{vmatrix} w(T;\ t=0;\ w_0 = w_1 = 0) \\ w_t(T;\ t=0;\ w_0 = w_1 = 0) \end{vmatrix} = \mathcal{L}_T \begin{vmatrix} g_1 \\ g_2 \end{vmatrix} = \mathcal{L}_{1T} g_1 + \mathcal{L}_{2T} g_2 \tag{2.1}$$

$$\mathcal{L}_{iT} g_i = \begin{vmatrix} A\int_0^T S(T-t) G_i g_i(t)dt \\ A\int_0^T C(T-t) G_i g_i(t)dt \end{vmatrix} \tag{2.2}$$

The sought after exact controllability result, i.e. the statement that the operator $\mathcal{L}_T = [\mathcal{L}_{1T}, \mathcal{L}_{2T}]$ is surjective from the space $U = H_0^1(0,T;\ L_2(\Gamma)) \times H_0^1(0,T;\ H^{-1}(\Gamma))$ onto the space $Y = H_0^1(\Omega) \times H^{-1}(\Omega)$, is equivalent to the property that the Hilbert space adjoint $\mathcal{L}_T^*$ has a continuous inverse: i.e. there is $C_T > 0$ such that

$$\left\| \mathcal{L}_T^* \begin{vmatrix} z_1 \\ z_2 \end{vmatrix} \right\|_U^2 = \left\| \frac{d}{dt} \mathcal{L}_{1T}^* \begin{vmatrix} z_1 \\ z_2 \end{vmatrix} \right\|_{L_2(0,T;L_2(\Gamma))}^2 + \left\| \frac{d}{dt} \mathcal{L}_{2T}^* \begin{vmatrix} z_1 \\ z_2 \end{vmatrix} \right\|_{L_2(0,T;H^{-1}(\Gamma))}^2$$

$$\geq C_T \| \{z_1, z_2\} \|_{\mathcal{D}(A^{1/4}) \times [\mathcal{D}(A^{1/4})]'}^2 \tag{2.3}$$

where on the left hand side of (2.3) we have used the equivalence between the H_0^1–norm and the gradient norm in the time variable.

Step 2. An equivalent partial differential equation characterization of inequality (2.3) is given in the next Lemma

Lemma 2.1 a) For

$$z = \{z_1, z_2\} \in \mathcal{D}(A^{1/4}) \times [\mathcal{D}(A^{1/4})]' \tag{2.4}$$

we have

$$\left[\frac{d\mathcal{L}_{1T}^{*}z}{dt}\right](t) = -\frac{\partial(\Delta\phi(t))}{\partial\nu} + K_{1T} \tag{2.5}$$

where $\phi(t) = \phi(t, \phi_0, \phi_1)$ is the solution of the homogeneous problem (1.8) with initial data

$$\phi_0 = A^{-1/2} z_1 \in \mathcal{D}(A^{3/4});\ \phi_1 = A^{-1/2} z_2 \in \mathcal{D}(A^{1/4}) \tag{2.6}$$

explicitly given by

$$\phi(t) = C(t-T)\phi_0 + S(t-T)\phi_1 \tag{2.7}$$

while K_{1T} is the vector

$$K_{1T} = \frac{G_1^{*}}{T}\{[C(T) - I]\,\phi_1 + AS(T)\phi_0\}. \tag{2.8}$$

b) For z as in (2.4) we have

$$\Lambda^{-1}\left[\frac{d}{dt}\,\mathcal{L}_{2T}^{*}z\right](t) = \Lambda\,\Delta\phi(t)\,|_{\Gamma} + \Lambda K_{2T} \tag{2.9}$$

where $\phi(t)$ is the solution of (1.8); Λ is an isomorphism $H^{s}(\Gamma)$ onto $H^{s-1}(\Gamma)$, selfadjoint on $L_2(\Gamma)$ so that

$$(f_1, f_2)_{H^{-1}(\Gamma)} = (\Lambda^{-1} f_1, \Lambda^{-1} f_2)_{L_2(\Gamma)} = (f_2, \Lambda^{-2} f_2)_{L_2(\Gamma)} \tag{2.10}$$

while K_{2T} is the vector

$$K_{2T} = \frac{G_2^{*}}{T}\{[C(T) - I]\,\phi_1 + AS(T)\phi_0\}. \tag{2.11}$$

c) Thus, by (2.5) and (2.9), inequality (2.3) - which characterizes exact controllability of problem (1.1) with controls $\{g_1, g_2\} \in H_0^1(0,T; L_2(\Gamma) \times H^{-1}(\Gamma))$ on the state space $Y = H_0^1(\Omega) \times H^{-1}(\Omega)$ over the interval $[0,T]$ - is equivalent to saying: there is a constant $C_T' > 0$ such that

$$\int_{\Sigma}\left[-\frac{\partial(\Delta\phi)}{\partial\nu} + K_{1T}\right]^2 d\Sigma + \int_{\Sigma} |\Lambda\Delta\phi + \Lambda K_{2T}|^2 d\Sigma \geq C_T' \|\{\phi_0, \phi_1\}\|^2_{\mathcal{D}(A^{3/4}) \times \mathcal{D}(A^{1/4})} \tag{2.12}$$

<u>Proof</u> Part a) was already shown in [L-T.1, Lemma 3.1]. Part b) is proved in a similar way. By definition of $\mathcal{L}_{2T}^{*}$, via (2.10)

$$(\mathcal{L}_{2T}g_2, z)_{\mathcal{D}(A^{1/4})\times[\mathcal{D}(A^{1/4})]'} = (g_2, \mathcal{L}_{2T}^{*}z)_{H_0^1(0,T;\, H^{-1}(\Gamma))} = (\Lambda^{-1}\frac{d}{dt}g_2, \Lambda^{-1}\frac{d}{dt}\mathcal{L}_{2T}^{*}z)_{L_2(0,T;L_2(\Gamma))}$$

$$= -\int_0^T (g_2(t), \Lambda^{-2}\left[\frac{d^2}{dt^2}\mathcal{L}_{2T}^{*}z\right](t))_{L_2(\Gamma)}dt \tag{2.13}$$

after integration by parts in t using the assumption $g_2(0) = g_2(T) = 0$. On the other hand, starting from (2.2) and proceeding as usual [L-T.1], one obtains

$$(\mathcal{L}_{2T}g_2, z)_{\mathcal{D}(A^{1/4})\times[\mathcal{D}(A^{1/4})]'} = \int_0^T (g_2(t), G_2^{*}A[S(T-t)A^{1/2}z_1 + C(T-t)A^{-1/2}z_2])_{L_2(\Gamma)}dt \tag{2.14}$$

Comparing (2.13) with (2.14) yields (compare also with [L-T.1, Eq (3.13)])

$$-\Lambda^{-2}\left[\frac{d^2}{dt^2}\mathcal{L}_{2T}^{*}z\right](t) = G_2^{*}[C(t-T)A^{1/2}z_2 + S(t-T)(-A^{\frac{3}{2}}z_1)]$$

Integrating in t and requiring that $\mathcal{L}_{2T}^{*}z \in H_0^1(0,T;\, H^{-1}(\Gamma))$ vanishes at t=0 and t=T yields (2.9), after recalling that $G_2^{*}Af = -\Delta f|_{\Gamma}$ [L-T.1, Eq (4.5)], and also (2.11).

Part c): By (2.5) and (2.9) via (2.10), the left hand side of (2.3) becomes precisely the left hand side of (2.12), while the right hand side of (2.3) is equivalent to the right hand side of (2.12) via (2.4), (2.6) (see also [L-T.1 Eq (3.17)]) □

Step 3. Here we "absorb" the "lower order terms" K_{1T} and K_{2T} via an argument of the same type as e.g. in [L.1], [L.2], [L-T.1], [L-T.3], adapted to present circumstances.

Lemma 2.2 Inequality (2.12) is equivalent to the following inequality: there is $C_T'' > 0$ such that

$$\int_{\Sigma}\left[\frac{\partial(\Delta\phi)}{\partial\nu}\right]^2 d\Sigma + \int_{\Sigma} |\Lambda\Delta\phi|^2 d\Sigma \geq C_T''\| \{\phi_0, \phi_1\} \|^2_{\mathcal{D}(A^{3/4})\times\mathcal{D}(A^{1/4})} \tag{2.15}$$

Proof. We first assume (2.12) and prove, as a consequence, (2.15). By contradiction, suppose that there is a sequence $\{\phi_{0n}, \phi_{1n}\} \in \mathcal{D}(A^{3/4})\times\mathcal{D}(A^{1/4})$ with

$$\phi_n(t) = C(t-T)\phi_{0n} + S(t-T)\phi_{1n} = \text{solution of problem (1.8)} \tag{2.16}$$

for which we have

$$\left\| -\frac{\partial(\Delta\phi_n)}{\partial\nu} + K_{1Tn} \right\|^2_{L_2(\Sigma)} + \| \Lambda\Delta\phi_n + \Lambda K_{2Tn} \|^2_{L_2(\Sigma)} \equiv 1 \tag{2.17}$$

$$\left\| \frac{\partial(\Delta\phi_n)}{\partial\nu} \right\|^2_{L_2(\Sigma)} + \| \Lambda\Delta\phi_n \|^2_{L_2(\Sigma)} \to 0 \text{ as } n \to \infty \tag{2.18}$$

But $\phi_n(t)$ satisfies (2.12) by assumption, and so by (2.17) there is a subsequence $\{\phi_{0n}, \phi_{1n}\}$ converging to some $\{\tilde{\phi}_0, \tilde{\phi}_1\}$ weakly in $\mathcal{D}(A^{3/4}) \times \mathcal{D}(A^{1/4})$ and hence, by compactness of A^{-1}, strongly in $\mathcal{D}(A^{3/4-\delta}) \times \mathcal{D}(A^{1/4-\delta})$ for $\delta > 0$. As a consequence, we return to (2.8) and (2.11) and see that:

i)

$$\begin{cases} K_{1Tn} = \dfrac{G_1^*}{T} A^{-(1/4-\delta)} \{[C(T) - I]A^{1/4-\delta}\phi_{1n} + A^{1/2}S(T)A^{3/4-\delta}\phi_{0n}\} & (2.19a) \\ \text{converges strongly in } L_2(\Gamma) \text{ to} \\ \tilde{K}_{1T} = \dfrac{G_1^*}{T} \{[C(T) - I]\,\tilde{\phi}_1 + AS(T)\tilde{\phi}_0\}, & (2.19b) \end{cases}$$

G_2^* being a bounded operator $L_2(\Omega) \to L_2(\Gamma)$;

ii)

$$\begin{cases} K_{2Tn} = \dfrac{G_2^* A^{-(1/4-\delta)}}{T} \{[C(T) - I]A^{1/4-\delta}\phi_{1n} + A^{1/2}S(T)A^{3/4-\delta}\phi_{0n}\} & (2.20a) \\ \text{converges strongly in at least } H^1(\Gamma) \text{ (in fact, in } H^{\frac{5}{2}-\varepsilon}(\Gamma)) \text{ to} \\ \tilde{K}_{2T} = \dfrac{G_2^*}{T} \{[C(T) - I]\,\tilde{\phi}_1 + AS(T)\tilde{\phi}_0] & (2.20b) \end{cases}$$

(Since $\mathcal{D}(A^{1/4-\delta}) = H_0^{1-4\delta}(\Omega)$ [G.1] and G_2^*: continuous $H_0^{1-4\delta}(\Omega) \to H^{5/2-4\delta}(\Gamma)$ by duality of G_2 : continuous $H^{-5/2+4\delta}(\Gamma) \to H^{-1+4\delta}(\Omega)$ [L-M p. 189].) As a consequence

$$\Lambda K_{2Tn} \to \Lambda\tilde{K}_{2T} \text{ strongly in (at least) } L_2(\Gamma) \tag{2.21}$$

Using (2.19), (2.21) and (2.18), we obtain that

$$-\frac{\partial(\Delta\phi_n)}{\partial\nu} + K_{1Tn} \to \tilde{K}_{1T} \text{ strongly in } L_2(0,T;\ L_2(\Gamma)) \tag{2.22}$$

$$\Lambda\Delta\phi_n + \Lambda K_{2Tn} \to \Lambda\tilde{K}_{2T} \text{ strongly in } L_2(0,T;\ L_2(\Gamma)) \tag{2.23}$$

and (2.22) and (2.23) along with (2.17) yield

$$\|\tilde{K}_{1T}\|_{L_2(\Sigma)} + \|\Lambda\tilde{K}_{2T}\|_{L_2(\Sigma)} \equiv 1 \tag{2.24}$$

On the other hand $\tilde{\phi}(t) = C(t-T)\tilde{\phi}_0 + S(t-T)\tilde{\phi}_0$ satisfies

$$\begin{cases} \tilde{\phi}_{tt} + \Delta^2\tilde{\phi} = 0 \\ \tilde{\phi}|_\Sigma = \frac{\partial\tilde{\phi}}{\partial\nu}|_\Sigma \equiv 0 \\ \frac{\partial(\Delta\tilde{\phi})}{\partial\nu}|_\Sigma = \Delta\tilde{\phi}|_\Sigma \equiv 0 \quad \text{(from (2.18))} \end{cases} \tag{2.25}$$

for $0 \le t \le T$. The Holmgren Uniqueness Theorem [H.1, p.129] then yields $\tilde{\phi}_0 = \tilde{\phi}_1 = 0$, or $\tilde{\phi} \equiv 0$ in Q, and hence $\tilde{K}_{1T} = \tilde{K}_{2T} = 0$, by (2.19b) and (2.20b). But this contradicts (2.24).

The proof that inequality (2.15) implies inequality (2.12) is identical. □

Corollary 2.3. Exact controllability over [0, T] of problem (1.1) on the state space $Y = H_0^1(\Omega) \times H^{-1}(\Omega)$ within the class of controls $\{g_1, g_2\} \in H_0^1(0,T; L_2(\Gamma) \times H^{-1}(\Gamma))$ is equivalent to (2.3), hence to (2.12), hence to (2.15), i.e. to

$$\int_\Sigma \left[\frac{\partial(\Delta\phi)}{\partial\nu}\right]^2 d\Sigma + \int_\Sigma (|\nabla_\sigma(\Delta\phi)|^2 + |\Delta\phi|^2) d\Sigma \ge C_T \|\{\phi_0, \phi_1\}\|^2_{\mathcal{D}(A^{3/4}) \times \mathcal{D}(A^{1/4})} \tag{2.26}$$

for some $C_T > 0$, where ϕ solves problem (1.8) □

Step 4. That inequality (2.26) holds true is proved in [L-T.1; Proposition 4.2]. Thus, Theorem 1.1 is proved at least when $\Gamma(x^0) = \Gamma$. In the general case as in (1.3), modifications as in the proof of Theorem 4.5 of [L-T.1] apply. □

References

[G.1] P. Grisvard, Characterization des qualques especes d'interpolation, Arch. Rational Mech. Anal. 25 (1967), pp. 40-63.

[L-T.1] I. Lasiecka and R. Triggiani, "Exact controllability of the Euler-Bernoulli equation with controls in the Dirichlet and Neumann boundary conditions: a non-conservative case," SIAM J. Control and Optimization, vol. 27, (1989), pp. 339-373.

[L-T.2] I. Lasiecka and R. Triggiani, "Exact controllability for semilinear abstract systems with application to wave and plate equations," Applied Mathem. & Optimiz., to appear.

[L-T.3] I. Lasiecka and R. Triggiani, "Exact controllability for the wave equation with Neumann boundary control," Applied Mathem. and Optimiz., 19 (1989), pp. 243-290..

[L.1] J. L. Lions, "Controlabilitë exacte des systems distribues," vol. 1, Masson, to appear.

[L.2] W. Littman, "Near Optimal time boundary controllability for a class of hyperbolic equations," pp. 307-312, Springer-Verlag Lecture Notes LNCIS #97, 1987.

[L-M.1] J. L. Lions and E. Magenes, Non-homogeneous boundary value problems, vol. I, Springer-Verlag 1972.

[H.1] L. Hormander, "Linear partial differential operators," Springer-Verlag 1969.

Some Properties of the Value Function of a Nonlinear Control Problem in Infinite Dimensions

PIERMARCO CANNARSA*

Dipartimento di Matematica
Università di Pisa
Via F. Buonarroti, 2, 56127 Pisa, Italy

GIUSEPPE DA PRATO**

Scuola Normale Superiore
56126 Pisa, Italy

1. Introduction

Let X be a separable reflexive Banach space with norm $|\ |$, which is assumed to be continuously differentiable in $X\backslash\{0\}$. We denote by X* the dual of X, the duality pairing being represented by $\langle , \rangle$. For any $R>0$ and $x \in X$ we set $B_R(x) = \{ y \in X : |y-x|<R \}$ and abbreviate B_R for $B_R(0)$.

Let Y be another Banach space and k a non-negative integer.

We denote by $\mathcal{L}(X,Y)$ the Banach space of all bounded linear operators $L:X \rightarrow Y$, equipped with the usual norm $\|\cdot\|$.

We denote by $C^k(X,Y)$ the set of all the mappings $f:X \rightarrow Y$ which are continuous and bounded on all bounded subsets of X, together with their derivatives up to the order k.

* Work (partially) supported by the Italian National Project "Equazioni Differenziali e Calcolo delle Variazioni".

** Work (partially) supported by the Italian National Project "Equazioni di Evoluzione e Applicazioni Fisico-Matematiche"

We denote by $C^{0,1}(D;Y)$ the space of all functions $f:D\to Y$ that are Lipschitz continuous in D, i. e.

$$\|f\|_L = \sup_{x,z\in D, x\neq z} \frac{|f(x)-f(z)|_Y}{|x-z|_X} < +\infty$$

Moreover, we define $C^{0,1}_{loc}(X;Y)$ as the space of all mappings $f\in C(X;Y)$ such that $f\in C^{0,1}(B_R;Y)$ for all $R>0$.

Let $T>0$, $(t,x)\in[0,T]\times X$ and $u\in L^2(t,T;U)$, where U is another Banach space (the *Control Space*). We are concerned with a system governed by the *State Equation*

$$\begin{cases} y'(s)=Ay(s)+F(y(s))+Bu(s), & t\leq s\leq T \\ y(t)=x \end{cases} \tag{1.1}$$

under the following assumptions:

i) $A:D(A)\subset X\to X$ *is the generator of an analytic semigroup*, e^{tA}, *and there exists* $\omega\in\mathbf{R}$ *such that* $\|e^{tA}\|\leq e^{\omega t}$ *for all* $t\geq 0$; (1.2)

ii) $F\in C^{0,1}_{loc}(X;X)$ *and there exist* $a,b\in\mathbf{R}$ *such that*
$\langle F(x),x^*\rangle \leq a|x|+b,\ \forall x\in X,\ \forall x^*\in\partial|x|$

iii) $B\in\mathcal{L}(U;X)$.

Then, (1.1) has a unique mild solution $y(\cdot;t,x,u)\in C([t,T];X)$, which is called the *trajectory associated to* u. Moreover, by using the Gronwall Lemma, assumptions (1.2) yield the following estimate

$$|y(s;t,x,u))|\leq e^{(\omega+a)T}\{\ |x| + bT + \sqrt{T}\,\|B\|\,\|u\|_{L^2(0,T;U)}\} \tag{1.3}$$

for all $s\in[t,T]$.

We are interested in the *Optimal Control* problem:

$$\textit{Minimize}\quad J(t,x;u) := \int_t^T [g(y(s;t,x,u))+h(u(s))]ds + \phi(y(T;t,x,u)) \tag{1.4}$$

over all controls $u\in L^2(t,T;U)$. A control u^* is *optimal* if the above minimum is attained at u^*.

The *Cost Functional* J will be assumed to satisy the following:

i) $g, \phi \in C^{0,1}_{loc}(X;R)$ *and there exists* $c_0>0$ *such that* (1.5)

$$g(x) \geq - c_0 (|x| + 1), \phi(x) \geq - c_0 (|x| + 1), \forall x \in X;$$

ii) $h:U \rightarrow \mathbf{R}$ *is continuous, convex and there exist constants* $\lambda_1, \lambda_0>0$ *such that*

$$h(u) \geq \lambda_0|u|^2 - \lambda_1, \quad \forall u \in U,$$

The *value function* V of problem (1.4) is defined for all $(t,x) \in [0,T] \times X$ as

$$V(t,x)= \inf \{ J(t,x;u) : u \in L^2(t,T;U) \} \tag{1.6}$$

This function V has many remarkable properties. For example, the following regularity results are well known (see e.g. [2], [1]):

(i) V maps bounded sets into bounded sets; (1.7)

(ii) V is continuous on $[0,T] \times X$;

(iii) V is locally Lipschitz continuous in x, uniformly for $t \in [0,T]$;

(iv) V is Lipschitz continuos in t for all $x \in D(A)$.

As a matter of fact, the above results hold true even if the semigroup generated by A is just strongly continuous. If X is a Hilbert space, however, it can be proved that, under our assumptions, V is locally Lipschitz in (t,x) on $[0,T) \times X$ (see [3]).

Our interest in this function derives from the fact that it relates problem (1.4) to the Hamilton-Jacobi equation

$$\begin{cases} -\frac{\partial V}{\partial t}(t,x) + H(B^*\nabla V(t,x)) - \langle Ax+F(x),\nabla V(t,x)\rangle - g(x) = 0 \\ \\ V(T,x) = \phi(x) \end{cases} \tag{1.8}$$

where B^* denotes the adjoint of B, $\nabla V(t,x)$ denotes the x-gradient of V and the hamiltonian H is defined by

$$H(v) = \sup\{-\langle u,v\rangle-h(u) \; ; u \in U\}, \; \forall v \in U^* \tag{1.9}$$

Obviously, equation (1.8) has to be properly interpreted for at least two reasons. In the first place, V is not differentiable, in general, but just locally Lipschitz. In addition, the term Ax in (1.8) is not everywhere defined, but only on D(A).

The lack of differentiability of V can be treated by intruducing the concept of viscosity solution ([6], [7]), which we recall in the next section.

The second difficulty, that is the unboundedness of A, has led to a generalization of the concept of viscosity solution (see [7], [8]). Here, we prefer to follow the procedure developed in [4], which consists of solving a regularized Hamilton-Jacobi equation, in which A is replaced by its Yosida approximation and then showing the convergence of the solutions of these regularized problems to the value function V. In fact, in this paper we improve the result of [4], by proving that the above convergence is uniform on all bounded subsets of $[0,T]\times X$.

For linear state equations we also obtain a regularity results for the gradient of V. More specifically, we show that the generalized gradient $\partial_x V(t,x)$ is contained in $D((-A)^{\alpha})$ for all $\alpha\in(0,1)$, where $(-A)^{\alpha}$ denotes the α-fractional power of A.

The interest of the above results derives from the use of the value function to obtain the synthesis of optimal controls. Indeed, under suitable assumptions, it can be proved that the trajectory $y^*(s)$ associated to an optimal control $u^*(s)$ satisfies the feedback law

$$u^*(s) \in - DH(B^*\partial_x V(s,y^*(s))).$$

2. Viscosity solutions of Hamilton-Jacobi equations

In this section we recall, first, the notion of viscosity solution for equation (1.8) in the case of $A\in \mathcal{L}(X)$. Then, we prove an approximation result for the general case.

Let Λ be an open subset of a Banach space For any function $f:\Lambda\to \mathbf{R}$ and $x\in\Lambda$ we denote by

$$D^+f(x) = \{ \zeta\in X^* : \limsup_{y\to x} \frac{f(y)-f(x)-\langle y-x,\zeta\rangle}{|y-x|} \le 0 \}$$

$$D^-f(x) = \{ \zeta\in X^* : \liminf_{y\to x} \frac{f(y)-f(x)-\langle y-x,\zeta\rangle}{|y-x|} \ge 0 \}$$

the *superdifferential* and the *subdifferential* of f at x, respectively. Clearly, either one of these convex sets may be empty. However, if they are both non-empty, then f is Frechet differentiable at x and $D^{\pm}f(x)=\{Df(x)\}$.

The semi-differentials above are used to define viscosity solutions of the Hamilton-Jacobi equation

$$-\frac{\partial v}{\partial t}(t,x) + H(B^*\nabla v(t,x)) - \langle Ax+F(x),\nabla v(t,x)\rangle - g(x) = 0 \tag{2.1}$$

Definition 2.1 - *Assume (1.2), (1.5) and suppose that* $A\in \mathcal{L}(X)$. *A function* $v\in C((0,T)\times X;\mathbf{R})$ *is said to be a viscosity solution of (2.1) if, for all* $(t,x)\in (0,T)\times X$,

$$\begin{cases} -p_t + H(B^*p_x) - \langle Ax+F(x),p_x\rangle \leq g(x) \ \textit{for all} \ (p_t,p_x)\in D^+v(t,x) \\ \\ -p_t + H(B^*p_x) - \langle Ax+F(x),p_x\rangle \geq g(x) \ \textit{for all} \ (p_t,p_x)\in D^-v(t,x) \end{cases} \tag{2.2}$$

An equivalent definition can be given by using test functions instead of semi-differentials (see [6]).

The following result is derived in [4] as a consequence of the existence and uniqueness theorems of [7].

Proposition 2.2 - *Assume (1.2), (1.5) and suppose that* $A\in \mathcal{L}(X)$. *Then, there exists a unique viscosity solution* $v\in C([0,T]\times X;\mathbf{R})$ *of (2.1) satisfying the terminal condition* $v(T,x) = \phi(x)$, *for all* $x\in X$. *Moreover,* v *is given by the value function* V *defined in (1.6).*

Let us now turn to the case of an unbounded operator A. Following the approach of [4], we will approximate the problem

$$\begin{cases} -\frac{\partial v}{\partial t}(t,x) + H(B^*\nabla v(t,x)) - \langle Ax+F(x),\nabla v(t,x)\rangle - g(x) = 0 \\ \\ v(T,x) = \phi(x) \end{cases} \tag{2.3}$$

by the "regularized problems"

$$\begin{cases} -\frac{\partial v}{\partial t}(t,x) + H(B^*\nabla v(t,x)) - \langle A_nx+F(x),\nabla v(t,x)\rangle - g(x) = 0 \\ \\ v(T,x) = \phi(x) \end{cases} \tag{2.3$_n$}$$

where $A_n = nA(n-A)^{-1}\in \mathcal{L}(X)$ is the Yosida approximation of A. Clearly, problem $(2.3)_n$ fulfills the assumptions of Proposition 2.2 and so it possesses a unique viscosity solution.

In [4] V_n was shown to converge pointwise to V on [0,T]×D(A). The main purpose of this paper is to improve the above convergence result as follows.

Theorem 2.3 - *Assume (1.2), (1.5) and let* V_n *be the viscosity solution of problem* $(2.3)_n$. *Then,* V_n *converges to the value function* V *of problem (1.4), uniformly on all bounded subsets of* [0,T]×X.

The main technical tool of the proof of Theorem 2.3 is the approximation result for solution of evolution equation that we give below. Let us denote by $y_n(\cdot;t,x,u)$ the solution of the Cauchy problem

$$\begin{cases} y'(s) = A_n y(s) + F(y(s)) + Bu(s), & t \le s \le T \\ y(t) = x \end{cases} \tag{2.4}$$

where $A_n = nA(n-A)^{-1}$ and $u \in L^2(t,T;U)$. As in (1.3), we can estimate $y_n(\cdot;t,x,u)$ as follows: for all $s \in [t,T]$

$$|y_n(s;t,x,u))| \le e^{(\omega_n+a)T}\{ |x| + bT + \sqrt{T}\,\|B\|\,\|u\|_{L^2(0,T;U)}\} \tag{2.5}$$

where $\omega_n = \frac{n\omega}{n-\omega}$.

Lemma 2.4 - *Assume (1.2), (1.5) and let* $y(\cdot;t,x,u)$ *(resp.* $y_n(\cdot;t,x,u)$*) be the solution of problem (1.1)) (resp. (2.4)). Then,*

$$\lim_{n\to\infty} \sup\{ |y_n(s;t,x,u)-y(s;t,x,u)| \mid \tau \le s \le T,\ \|u\|_{L^2(t,T;U)} \le R,\ |x| \le R\} = 0 \tag{2.6}$$

for all $\tau \in (t,T)$ *and* $R>0$.

Proof - As the initial time has no influence in any computation, we may assume t=0 without loss of generality. Accordingly, we set $y(s;0,x,u) = y(s;x,u)$ and $y_n(s;0,x,u) = y_n(s;x,u)$. For all $s \in [0,T]$ we have

$$| y_n(s;x;u) - y(s;x;u) | \le | (e^{sA_n} - e^{sA})\, x | + \tag{2.7}$$

$$+ \left| \int_0^s \{ e^{(s-r)A_n} - e^{(s-r)A} \} Bu(r)dr \right| +$$

$$+ \left| \int_{\sigma}^{s} \left\{ e^{(s-r)A_n} - e^{(s-r)A} \right\} f(y_n(r,x;u))dr \right| +$$

$$+ \left| \int_{\sigma}^{s} e^{(s-r)A} \left\{ f(y_n(r,x;u)) - f(y(r,x;u)) \right\} dr \right|$$

$$=: I_1(s) + I_2(s) + I_3(s) + I_4(s)$$

Now, set $\sigma_n(s) = \| e^{sA_n} - e^{sA} \|$. Then,

$$I_1(s) \leq R\, \sigma_n(s) \tag{2.8}$$

$$I_2(s) \leq \left\{ \int_{\sigma}^{s} \| e^{(s-r)A_n} - e^{(s-r)A} \|^2 dr \right\}^{1/2} \|B\| \, \|u\|_{L^2(0,T;U)} \tag{2.9}$$

$$\leq R \, \|B\| \, \|\sigma_n\|_{L^2(0,T)}$$

$$I_3(s) \leq \int_{\sigma}^{s} \| e^{(s-r)A_n} - e^{(s-r)A} \| \, |f(y_n(r,x;u))| dr \leq \tag{2.10}$$

$$\leq R\,(T + T^{1/2}) \, \|f\|_L \, \|\sigma_n\|_{L^2(0,T)}.$$

$$I_4(s) \leq \|f\|_L \int_{\sigma}^{s} |y_n(r,x;u) - y(r,x;u)| dr \tag{2.11}$$

From (2.7),..., (2.11) we obtain, by the Gronwall Lemma,

$$| y_n(s,x;u) - y(s,x;u) | \leq R \left\{ \sigma_n(s) + C_T \|\sigma_n\|_{L^2(0,T)} \right\} + \tag{2.12}$$

$$+ R \int_{\sigma}^{s} e^{(s-r)\|f\|_L} \left\{ \sigma_n(r) + C_T \|\sigma_n\|_{L^2(0,T)} \right\} dr$$

where $C_T = (T+T^{1/2}) \, \|f\|_L$.

Now, by known properties of the Yosida approximation , we have that

$$0 \leq \sigma_n(s) \leq 2 \text{ and } \lim_{n\to\infty} \sigma_n(s) = 0 \tag{2.13}$$

for all $s\geq 0$. Moreover, the limit is uniform on each interval of the form $[\tau,T]$, $\tau>0$. Then, the conclusion follows from (2.12) and (2.13) in view of Lebesgue's Dominated Convergence Theorem.

Proof of Theorem 2.3 - Let us fix $R>0$ and let $x\in B_R$. By the definition of V there is a sequence $\{u_n\}\subset L^2(t,T;U)$ such that

$$J(t,x;u_n) < V(t,x) + \frac{1}{n} \leq M_R + \frac{1}{n} \tag{2.14}$$

where $M_R = \sup_{t\in[0,T],\,|x|\leq R} |V(t,x)|$ (recall (1.7)(i)). Now, (2.14) and assumptions (1.5) yield

$$\lambda_0 \int_t^T |u_n(s)|^2 ds \leq \tag{2.15}$$

$$\leq M_R + \frac{1}{n} + \lambda_1 T + c_0 \left(|y(T;t,x,u_n)| + 1 \right) + c_0 \int_t^T (|y(r;t,x,u_n)|+1)dr$$

Hence, by using (1.3), it is easy to show that

$$\int_t^T |u_n(s)|^2 ds \leq C_R \tag{2.16}$$

for some constant $C_R>0$. Consequently, (2.16) and (1.3) imply that

$$|y(s;t,x,u_n)| \leq K_R \tag{2.17}$$

for all $s\in[t,T]$ and some constant K_R. Similarly, by (2.5),

$$|y_n(s;t,x,u_n)| \leq K'_R \tag{2.18}$$

Now, by Proposition 2.2,

$$V_n(t,x) = \inf \{ J_n(t,x;u) : u\in L^2(t,T;U) \}$$

where

$$J_n(t,x;u) := \int_t^T [g(y_n(s;t,x,u)) + h(u(s))]ds + \phi(y_n(T;t,x,u))$$

Then, we can use u_n to estimate $V_n(t,x)$. From (2.14) we obtain

$$V_n(t,x) \le J_n(t,x;u_n) < J_n(t,x;u_n) - J(t,x;u_n) + V(t,x) + \frac{1}{n}$$

So, for any fixed $0<\varepsilon<T-t$,

$$V_n(t,x) - V(t,x) < \tag{2.19}$$

$$< \frac{1}{n} + \int_t^T [g(y_n(s;t,x,u_n)) - g(y(s;t,x,u_n))]ds + \phi(y_n(T;t,x,u_n)) - \phi(y(T;t,x,u_n))$$

$$< \frac{1}{n} + 2\varepsilon \|g\|_{\infty,R} + \|g\|_{L,R} \int_{t+\varepsilon}^T |y_n(s;t,x,u_n) - y(s;t,x,u_n)|ds +$$

$$+ \|\phi\|_{L,R} |y_n(T;t,x,u_n)) - y(T;t,x,u_n)|$$

where $\|g\|_{\infty,R}$ (resp. $\|g\|_{L,R}$) denotes the supremum of $|g(x)|$ (resp. a Lipschitz constant for g) over the set $\{ |x| \le \max[K_R,K'_R] \}$.

Similarly, there exists a sequence $\{\hat{u}_n\} \subset L^2(0,T_\varepsilon;H)$ such that

$$J_n(t,x;\hat{u}_n) < V_n(t,x) + \frac{1}{n} \le M_R + \frac{1}{n}$$

whence

$$V(t,x) \le J(t,x;\hat{u}_n) < J(t,x;\hat{u}_n) - J_n(t,x;\hat{u}_n) + V_n(t,x) + \frac{1}{n}$$

Thus, arguing as we did above, it follows that

$$V(t,x) - V_n(t,x) < \tag{2.20}$$

$$< \frac{1}{n} + 2\varepsilon \|g\|_{\infty,R} + \|g\|_{L,R} \int_{t+\varepsilon}^T |y_n(s;t,x,\hat{u}_n) - y(s;t,x,\hat{u}_n)|ds +$$

$$+ \|\phi\|_{L,R} |y_n(T;t,x,\hat{u}_n) - y(T;t,x,\hat{u}_n)|$$

Finally, since ε is arbitrary, (2.19), (2.20) and Lemma 2.4 yield the conclusion.

3. Space regularity of the value function

In this section we assume that X is a Hilbert space and

i) $A:D(A)\subset X\to X$ *is the generator of an analytic semigroup,* e^{tA}, *and there exists* $\omega<0$ *such that* $\|e^{tA}\| \le e^{\omega t}$ *for all* $t\ge 0$; (3.1)

ii) $B\in \mathcal{L}(U;X)$;

iii) $g, \phi\in C^{0,1}(X;R)$ *and*

$$g(x) \ge 0,\ \phi(x) \ge 0,\ \forall x\in X;\ g(0) = \phi(0) = 0;$$

iv) $h:U\to R$ *is continuous, convex,* $h(0) = 0$ *and there exists* $\lambda_0>0$ *such that*

$$h(u) \ge \lambda_0|u|^2\ ,\ \forall u\in U.$$

For any $\alpha\in]0,1[$, we denote by $(-A)^\alpha$ the fractional powers of A. We recall that , for any $\alpha\in (0,1)$ there exists a constant M_α such that

$$\|(-A)^\alpha e^{tA}\| \le M_\alpha t^{-\alpha}\ ,\ t>0 \tag{3.2}$$

We consider the linear state equation

$$\begin{cases} y'(s)=Ay(s)+Bu(s), & t\le s\le T \\ y(t)=x \end{cases} \tag{3.3}$$

The main result of this section is the following.

Theorem 3.1. *Assume* (3.1) *and let* V *be the value function of problem (1.3)-(3.3). Then*

$$\partial_x V(t,x)\subset D((-A)^\alpha) \tag{3.4}$$

for any $\alpha\in (0,1)$ *and any* $(t,x)\in [0,T]\times X$.

Here, the generalized gradient ([5]) $\partial_x v(t,x)$ is defined as follows

$$\partial V(t,x) = \{p\in X :\ \limsup_{x'\to x,\lambda\downarrow 0} \frac{V(t,x'+\lambda\theta)-V(t,x')}{\lambda} \ge \langle p,\theta\rangle,\ \forall\theta\in X\}$$

In order to prove Theorem 3.1, we introduce an auxiliary control problem . Let $t \in [0,T]$ and $x \in X$. Minimize :

$$J_\alpha(t,x;u) = \int_t^T \{ g((-A)^\alpha z(s,t,x;u)) + h(u(s)) \} ds$$

with respect to all $u \in L^2(t,T;X)$ where $z(s) = z(s,t,x;u)$ is the solution of the problem

$$\begin{cases} z'(s) = Az(s) + (-A)^{-\alpha} Bu(s) \\ z(t) = x \end{cases}$$

We denote by $V_\alpha(t,x)$ the value function

$$V_\alpha(t,x) = \inf\{ J_\alpha(t,x;u) \; ; \; u \in L^2(t,T;U) \}$$

The following propositions establish some basic properties of V_α.

Proposition 3.2 -*The value function* $V_\alpha(t,x)$ *is bounded on bounded sets of* $[0,T] \times X$.

Proof - From (3.1) we have

$$V_\alpha(t,x) \le J_\alpha(t,x;0) \le \|g\|_L \int_t^T |(-A)^\alpha e^{(s-t)A} x| ds \le M_\alpha \|g\|_L |x| \int_t^T |s-t|^{-\alpha} ds$$

and the conclusion follows.

Proposition 3.3 - *Let* $t \in [0,T]$, $\alpha \in]0,1[$.*Then*

$$|V_\alpha(t,x_0) - V_\alpha(t,x_1)| \le M_\alpha \|g\|_L |x_1 - x_0| \tag{3.5}$$

for all $x_0, x_1 \in X$.

Proof - First we assume that $x_0, x_1 \in D((-A)^\alpha)$. Fix $\varepsilon > 0$ and let $u_\varepsilon \in L^2(t,T;U)$be such that

$$V_\alpha(t,x_0) > J_\alpha(t,x_0,u_\varepsilon) - \varepsilon$$

Then,

$$V_\alpha(t,x_1) - V_\alpha(t,x_0) \le \int_t^T \{ g((-A)^\alpha y(s,t,x_1;u_\varepsilon)) - g((-A)^\alpha y(s,t,x_0;u_\varepsilon)) \} ds + \varepsilon$$

$$\leq \|g\|_L \int_t^T |(-A)^\alpha y(s,t,x_1;u_\varepsilon)-(-A)^\alpha y(s,t,x_0;u_\varepsilon)| ds + \varepsilon$$

$$= \|g\|_L \int_t^T |(-A)^\alpha e^{(s-t)A}(x_1-x_0)| ds + \varepsilon$$

$$\leq M_\alpha \|g\|_L |x_1-x_0| \int_t^T |s-t|^{-\alpha} ds + \varepsilon$$

and (3.5) holds since ε is arbitrary in the case of x_0, $x_1 \in D((-A)^\alpha)$.The general case follows from the fact that $D((-A)^\alpha)$ is dense in X.

We can finally prove the theorem.

Proof of Theorem 3.1 - As easily checked, for all $x \in D((-A)^\alpha)$, we have

$$z(s,t,(-A)^\alpha x;u) = (-A)^\alpha y(s,t,x;u)$$

which implies in turn

$$V_\alpha(t,x) = V(t,(-A)^\alpha x) \ , \ \forall x \in D((-A)^\alpha)$$

or

$$V(t,x) = V_\alpha(t,(-A)^{-\alpha} x) \ , \ \forall x \in X \tag{3.6}$$

Let now $x \in X$, $p \in \partial_x V(t,x)$; in order to prove (3.4) it suffices to show that there exists a constant C_α such that

$$\langle p,(-A)^\alpha y\rangle \leq C_\alpha |y| \ , \quad \text{for all } y \in D((-A)^\alpha)$$

We have in fact, recalling (3.6),

$$\langle p,(-A)^\alpha y\rangle \leq \limsup_{x' \to x, \lambda \downarrow 0} \frac{V(t,x'+\lambda(-A)^\alpha y)-V(t,x')}{\lambda} =$$

$$= \limsup_{x' \to x, \lambda \downarrow 0} \frac{V_\alpha(t,(-A)^{-\alpha}x'+\lambda y)-V_\alpha(t,(-A)^{-\alpha}x')}{\lambda}$$

This concludes the proof.

References

[1] V. BARBU, *Hamilton-Jacobi equations and non linear control problems,* J. Math. Anal. Appl. **120** (1986), 494-509.

[2] V. BARBU AND G. DA PRATO, Hamilton-Jacobi equations in Hilbert spaces, Pitman, Boston, 1983.

[3] P. CANNARSA, *Lipschitz Regularity of the Value Function of Nonlinear Control Problems in Infinte Dimensions,* in preparation.

[4] P. CANNARSA AND G. DA PRATO, *Some results on nonlinear optimal control problems and Hamilton-Jacobi equations in infinite dimensions,* J. Funct. Anal. (to appear).

[5] F. CLARKE, *Optimization and non-smooth analysis,* Wiley, New York, 1983

[6] M. G. CRANDALL AND P. L. LIONS, *Hamilton-Jacobi equations in infinite dimensions I. Uniqueness of Viscosity Solutions.,* J. Funct. Anal. **62** (1985), 379-396.

[7] M. G. CRANDALL AND P. L. LIONS, *Hamilton-Jacobi equations in infinite dimensions.III.,* J. Funct. Anal. **68** (1986), 368-405.

[8] M. G. CRANDALL AND P. L. LIONS, *Hamilton-Jacobi equations in infinite dimensions. IV,* pre-print.

IDENTIFICATION OF COEFFICIENTS WITH BOUNDED VARIATION IN THE WAVE EQUATION

J.P. ZOLESIO
Laboratoire de Physique Mathématique
U.S.T.L., Place Eugène Bataillon 34095 MONTPELLIER Cedex 02
F R A N C E

Key words : Transcient wave equation, unbounded domain D, coefficients in BV(D), Existence of Minimum for quadratic cost.

INTRODUCTION.

The identification of coefficient in Partial Differential Equations is a well known problem which has been intensively studied these last ten years. A robust method is to introduce the minimization of a cost function J(K), K being the parameter to identify, $J(K) = \|y(K) - y_{measurements}\|$ where y(K) is the solution of the PDE associated to the coefficient K. It turns out that numerically it is much more efficient to solve the equation $\nabla J(K) = 0$ rather then to minimize J(K). In general for Laplace equation, $\operatorname{div}(K\nabla y) = f$, to obtain existence results one have to introduce some constraints on K. These constraints are of two kinds : smoothness, for example $K \in H^1(D)$, boundness, for example $0 \leqslant \alpha \leqslant K \leqslant \beta$. In this paper we shall consider the transcient wave equation in a non necessary bounded domain D.

Using the BV(D) regularity and boundness for the coefficients K we shall prove at Proposition 10 that the functional J possesses a minimum. To reach this result we shall give a priori estimate to obtain boundness of y(K)in $W^{1,\infty}([0,T], L^2(D) \cap L^\infty(0,T, H^1(D)/\mathbb{R})$ and establish the continuity of $K \mapsto y(K)$ from $L^1(D)$ to weak-x topologies.

In fact the identification of K is limitated to the fact that only a part of K, $K|D_1$ can be identified (D_1 is a bounded subset of D). For example if K_0 is given then K should be equal to K_0 in $D\backslash D_1$ and the cost J will be regularized in $J(K) + \sigma\|K_0-K\|_{BV(D_1)}$. If D is bounded then D_1 can be taken as D.

We are concerned with an unbounded domain $D \subset \mathbb{R}^N$ with boundary ∂D which is smooth. $I = [0,T]$ is the time interval, the space variable $x \in D$ and $Q = I{\times}D$ is the evolution cylinder in $\mathbb{R}^{N+1}$.

Let be given a matrix function K(x), $K_{ij}(x) \in L^\infty(D)$, $i\leqslant i,j\leqslant N$. We consider for any $S \in L^1(0,T, L^2(D))$ the wave equation

$$\frac{\partial^2}{\partial t^2} y - \operatorname{div}(K.\nabla y) = S \quad \text{in} \quad Q \tag{1}$$

with Newmann boundary condition on the lateral boundary

$$\langle K.\nabla y,\ n \rangle = g \quad \text{on} \quad I \times \partial D = \Sigma \quad . \tag{2}$$

The initial condition being takes as

$$y(0,n) = \frac{\partial}{\partial t} y(0,n) = 0 \quad \text{in} \quad D \quad . \tag{3}$$

We give an a priori estimate for the problem (1)-(3). We denote by $H = H(K)$ the wave operator $H.\varphi = \frac{\partial^2}{\partial t^2} \varphi - \operatorname{div}\ (K\nabla\varphi)$ for any matrix function $K \in L^2(D)^{N^2}$. Consider the vector space

$$\mathcal{H} = \{\varphi \in H^1(Q) \quad \text{such that} \quad H\varphi \in L^2(Q)\}$$

it turns out that $\mathcal{H}$ is a closed subspace of $H^1(Q)$ depending on K.

Using the Green formula we have the

LEMMA 1.

$$\forall\ (y;\varphi)\ \mathcal{H} \times H^1(Q)^2 \quad \text{such that} \quad y(0) = \varphi(T) = 0 \quad \text{on} \quad D$$

we have

$$\int_\Sigma \langle K\ \nabla y.n,\varphi \rangle\, d\Sigma = \int_Q \left(K\ \nabla y\ \nabla\varphi - \frac{\partial}{\partial t} y \frac{\partial}{\partial t} \varphi\right) dQ - \int_Q Hy\ \varphi\, dQ \tag{4}$$

In view of (4) we have the

COROLLARY 2.

For any $y \in \mathcal{H}$, $\langle K\ \nabla y,n \rangle$ is defined as an element of $H^{1/2}_{00}(\Sigma)'$ ■

In view of Corollary 2 it makes sens to look, with S and g given in $L^2(D)$ and $H^{1/2}_{00}(\Sigma)'$, for y solution of the wave problem (1)-(3), $y \in \mathcal{H}(Q)$.

In fact for smooth data, $S \in L^2(D)$ and $g = 0$, we give now an a priori estimate.

For any t, $0 \leqslant t \leqslant T$, consider the energy term

$$E(t) = \frac{1}{2}\int_D \langle K.\nabla y, \nabla y\rangle\, dx + \frac{1}{2}\int_D \left(\frac{\partial}{\partial t} y\right)^2 dx$$

for given y in $C^\infty(\bar{Q})$, its derivative is

$$E'(t) = \int_D Hy \frac{\partial}{\partial t} y\, dx + \int_{\partial D} \langle K.\nabla y, n\rangle \frac{\partial}{\partial t} y\, ds$$

by Cauchy Schwartz inequality

$$E'(t) \leqslant \|Hy\|_{L^2(D)} \left\|\frac{\partial}{\partial t} y\right\|_{L^2(\partial)} + \|\gamma_K . y\|_{L^2(\partial D)} \left\|\frac{\partial}{\partial t} y\right\|_{L^2(\partial D)}$$

and when $\gamma_K y = \langle K.\nabla y, n\rangle = 0$ on ∂D we obtain, in view of $E(0) = 0$, for any t, $0 \leqslant t \leqslant T$

$$E(t) \leqslant \|Hy\|_{L^2(0,T,L^2(D))} \left\|\frac{\partial}{\partial t} y\right\|_{L^\infty(0,T,\ L^2(D))}$$

$$\leqslant \|Hy\|_{L^1(0,T,L^2(D))} \max_{0\leqslant t\leqslant T} E(t)^{1/2} \tag{5}$$

and then

LEMMA 3

For any function $y \in C^\infty(Q)$ such that $\langle K.\nabla y, n\rangle = 0$ on ∂D we have, assuming $K >> \alpha\, Id$

$$\alpha\, \|y\|_{L^\infty(0,T,H^1(D)/\mathbb{R})} \leqslant \|Hy\|_{L^1(0,T,L^2(D))} \tag{6}$$

$$\alpha \left\|\frac{\partial}{\partial t} y\right\|_{L^\infty(0,T,\ L^2(D)/\mathbb{R})} \leqslant \|Hy\|_{L^1(0,T,L^2(D))} \tag{7}$$

∎

COROLLARY 4.

The solution y of problem (1)-(3), with $S \in L^1(0,T,\ L^2(D))$ and $g = 0$, in the space $L^\infty(0,T,\ H^1(D)) \cap W^{1,\infty}(0,T,\ L^2(D))$ is unique.

REMARK 5.

From (5) we could have, instead of (6)

$$\|K.\nabla y\|_{L^\infty(0,T,\ L^2(D))} \leqslant \|Hy\|_{L^1(0,T,\ L^2(D))}$$

We consider now the continuity property on the mapping $K \to y = y(K)$, the solution of problem (1)-(3) for given S in $L^1(0,T, L^2(D))$. Let $K_n \to K$ in $L^1(D)^N$ with the following conditions

$$\alpha\, I_d \ll K_n \ll M\, I_d \tag{8}$$

and denote y_n the solution of problem (1)-(3) associated to K_n, for given $S \in L^1(0,T,L^2(D))$, $g = 0$. In view of the estimations (5),(6)(7) we have y_n and $\frac{\partial}{\partial t} y_n$ which remain bounded. Then one can consider a subsequence (still denoted y_n for simplicity) such that,

$$\forall \varphi \in H^1(D)/\mathbb{R}, \int_D K_n\, \nabla y_n\, \nabla\varphi\, dx \longrightarrow \int_D K\, \nabla y.\nabla\varphi\, dx \tag{9}$$

weakly -* in $L^\infty(0,T)$

$$\forall \psi \in L^2(D), \int_D \frac{\partial}{\partial t} y_n\, \psi\, dx \longrightarrow \int_D \frac{\partial}{\partial t} y\, \psi\, dx, \tag{10}$$

weakly * in $L^\infty(0,T)$

multiplying the problem (1)-(3) associated to K_n by a function $\varphi \in \mathcal{D}(Q)$ we obtain the weak form of the wave problem as :

$$\forall \varphi \in \mathcal{D}(Q), \int_Q K_n\, \nabla y_n.\nabla\varphi\, dQ - \int_Q \frac{\partial}{\partial t} y_n \frac{\partial}{\partial t}\varphi\, dQ = \int_Q S\varphi\, dQ \tag{11}$$

in view of (9), (10), (11) in the limit we get, $n \to \infty$:

$$\int_Q K\, \nabla y\, \nabla\varphi\, dQ - \int_Q \frac{\partial}{\partial t} y \frac{\partial}{\partial t}\varphi\, dQ = \int_Q S\varphi\, dQ \quad . \tag{12}$$

Then $y = \text{limite } y_n$, $n \to \infty$, is the solution of problem (1)-(3) associated to the matrix K,

$$y \in L^\infty(0,T, H^1(D)/\mathbb{R}) \cap W^{1,\infty}(0,T, L^2(\Omega))$$

with $H(K).y = S$ and $\langle K.\nabla y, n\rangle = 0$ on Σ. (here again the term $\langle K.\nabla y, n\rangle$ is considered in the weak sens, see Corollary 2, for which K can be given in $L^\infty(D)^{N^2}$).

Finely we obtain the

PROPOSITION 6.

Let K_n verifies the condition (8) and $K_n \to K$ in $L^1(D)^{N^2}$, then $y_n \rightharpoonup y(K)$

weakly-* in $L^\infty(0,T, H^1(D)/\mathbb{R})$ while $\frac{\partial}{\partial t} y_n \rightharpoonup \frac{\partial}{\partial t} y$ weakly-* in $L^\infty(0,T, L^2(D))$ ∎

We consider the identification problem associated to the wave equation in the spatial domain D. The general approach is to consider a cost functional $J = y \mapsto J(y)$ which is continuous on $L^2(Q)$. For example being given some measurements y_m of the solution we have the quadratic cost

$$J(K) = \frac{1}{2} \int_0^T \int_D (y(t,x) - y_m(t,x))^2 \, dt \, dx$$

and the minimization of J relatively to the matrix K is the identification problem of K.

In general the measurement y_m is not done for all x in D, but on a (may be) little subdomain D_0 on D and the cost takes the form

$$J(K) = \frac{1}{2} \int_0^T \int_{D_0} (y - y_m)^2 \, dt \, dx \quad . \tag{11}$$

To obtain on existence for the minimization of the cost J over the set of matrices K we have to restrict the minimization to a compact family of matrices. The compacity being compatible with the continuity properties of the cost J and of the state y(K) which is described at Proposition 6. Using the compact embedding of $H^1(0,T)$ in $L^2(0,T)$ the Proposition 6 leads to

PROPOSITION 7.

Let K_n the conditions (8) and $K_n \longrightarrow K$ in $L^1(D)^{N^2}$, then $y_n \longrightarrow y$ strongly in $L^2(Q)$.

Proof. We have $\frac{\partial}{\partial t} y_n \rightharpoonup \frac{\partial}{\partial t} y$ weakly.* in $L^\infty(0,T, L^2(D))$ then also weakly in $L^2(0,T, L^2(D))$ but $y_n(0)$ is given by the initial data then the result is classical. ∎

PROPOSITION 8.

Let J be any cost functional continuously defined on $L^2(Q)$. Let D_1 be a bounded subdivision in D with Lipschitzian boundary and

$$BM(D_1) = \left\{ \text{matrices } K \in L^\infty(D)^{N^2}, \ \alpha I \leqslant K \leqslant \beta I, \ K\big|_{D \setminus D_1} = K_0, \right.$$

$$\left. \|K\|_{BV\left(D_1;\mathbb{R}^{N^2}\right)} \leqslant M \right\} \quad . \tag{12}$$

Then there exist an optimal matrice $K^* \in B(M)$ such that

$$J(K^*) \leq J(K) \quad , \quad \forall K \in BM \quad .$$

Proof. The proof derives from the compact embedding of $BV(D_1)$ in $L^1(D)$, see R. Temam [1], and the Proposition 7. More precisely let K_n be a sequence lying in BM, then there exists a subsequence, still denoted K_n, such that $K_n \longrightarrow K$ in $L^1(D_1)$ ∎

REMARK 9. (see [2])

The fact that K is the identity in (12) could be change for $K\big|_{D\setminus D_1} = K_0$ where K_0 is any given element in $D\setminus D_1$ with $K_0 \in L^2(D\setminus D_1)$, $\alpha I \leq K_0 \leq \beta I$. This kind of hypothesis means that we are not identifying K in the large but only on a compact region D_1 (D_1 being itself arbitrary).

On practical wiewpoint the cost functional J can be regularized as follows :

$$\sigma > 0, \quad J_\sigma(K) = J(K) + \sigma \|K\|_{BV\left(D_1 \,;\, \mathbb{R}^{N^2}\right)} \quad . \tag{13}$$

PROPOSITION 10.

Let $\sigma > 0$ and J be continuously defined on $L^2(Q)$ and let :

$$K_{ad} = \left\{\text{matrices } K \in L^\infty\left(D;\mathbb{R}^{N^2}\right),\ \alpha I \leq K \leq \beta I,\ K = I_d \text{ in } D\setminus D_1\right\} \quad . \tag{14}$$

Then there exists an optimal element $K^* \in K_{ad}$,

$$J_\sigma(K^*) \leq J_\sigma(K), \quad \forall K \in K_{ad} \qquad ∎ \tag{15}$$

REMARK 11.

If K was equal to given K_0 in the large we should modify J_σ as follows :

$$J_\sigma(K) = J(K) + \sigma\|K-K_0\|_{BV\left(D_1 \,;\, \mathbb{R}^{N^2}\right)} \quad . \tag{16}$$

REMARK 12.

In fact the norm in $BV\left(D_0,\ \mathbb{R}^{N^2}\right)$ could be replaced in the definition of K_{ad} and J_σ by any norm such that the boundness of this norm implies the convergence of a subsequence in $L^2(D_0)$. In particular one could chose the $H^\epsilon(D_0)$ norm, $0 < \epsilon < \frac{1}{2}$ which permit the discontinuities to K as well as the BV norm.

In the particular situation of interfaces the matrice K(x) will be a characteristic function

$$K(x) = \alpha\, \chi_\Omega(x) + \beta(1 - \chi_\Omega(x))$$

where Ω is a measurable subset of D_0.
The BV norm on K is then reduced to the perimeter of Ω relatively D_0 ;

$$P_{D_0}(\Omega) = \sup \left\{ \int_\Omega \operatorname{div}(g)\, dx \;\middle|\; g \in C^1_0\left(D_0 ; \mathbb{R}^N\right),\ \sup \|g(x)\| \leqslant 1 \right\} .$$

If the optimum domain is smooth enough we obtain the necessary condition as free boundary conditions, see J.P. Zolesio [2], [3].

REFERENCES

[1] R. Temam, Problèmes Mathématiques en Plasticité, Dunod, Paris, 1983.

[2] J.P. Zolesio, Existence and Uniqueness Results for Domain Variational Bernouilli-like Free Boundary Problems. In proc. of IFIP Conference "Modelling and Inverse Problems of Control for Distributed Parameter Systems", ed. Kurzhanski, Springer-Verlag. (to appear).

[3] J.P. Zolesio, Existence free boundary Problem, in "Distributed Parameter for Systems", F. Kappel, K. Kunish, W. Shapacker, eds., LNCIS n° 102, pp. 333-343, Spinger Verlag, 1986.

SHAPE HESSIAN BY THE VELOCITY METHOD: A LAGRANGIAN APPROACH*

M.C. DELFOUR
Centre de recherches mathématiques et
Département de mathématiques et de statistique,
Université de Montréal, C.P. 6128 Succ. A,
Montréal, Québec, Canada, H3C 3J7

J.P. ZOLÉSIO
Laboratoire de physique théorique,
Université des Sciences et Techniques du Languedoc,
Place Eugène Bataillon,
34060 - Montpellier, Cédex 02, France

ABSTRACT. In this paper we announce new results on the Shape Hessian of a shape functional by the Velocity (Speed) Method. We review and extend the Velocity Method and clarify its connections with methods using first or second order Perturbations of the Identity. We show that all these methods yield the same Shape gradient but a different and unequal Shape Hessian since each one depends on the choice of "connection". For autonomous velocity fields the Velocity Method yields a canonical bilinear Hessian. Expressions obtained by other methods can be recovered by adding to this canonical term the Shape gradient acting on the acceleration of the velocity field associated with the choice of perturbation of the identity. In the second part of the paper we give an application of the Lagrangian Method with Function Space Embedding to compute the Shape gradient and Hessian of a simple cost function associated with the non-homogeneous Dirichlet problem.

1. INTRODUCTION.

In this paper we announce new results on the *Shape Hessian* by the *Velocity (Speed) Method* (cf. J. CÉA [1,2,3] and J.P. ZOLÉSIO [1,2]) and apply the *Lagrangian Method* with *Function Space Embedding* to compute the Shape gradient and Hessian of a simple cost function associated with the non-homogeneous Dirichlet problem.

We describe a general method which applies to *differentiable semiconvex* cost functionals with applications to more general problems than the simple illustrative example we have chosen to consider. We also emphasize the use of the Function Space Embedding method (cf. DELFOUR-ZOLÉSIO [1,2,3,4,7]) combined with the implicit use of Lagrange multipliers. Therefore this paper complements our previous work where we have used a variational formulation for the Neumann problem (cf. ZOLÉSIO-DELFOUR [8]). In Shape Sensitivity Analysis the size of the computations can be quite large. Therefore it is extremely important to understand the fundamental structure of the Shape gradient and Hessian in order to simplify the computations and obtain mathematically meaningful expressions.

In the process we make a revision of the Velocity (Speed) Method and show how to associate with Methods of *Perturbation of the Identity* (first and second order) an appropriate non-autonomous family of velocity fields. For the Shape gradient, the different

* The research of the first author has been supported in part by a Killam fellowship from Canada Council and by National Sciences and Engineering Research Council of Canada operating grant A-8730 and a FCAR grant from the Ministère de l'Education du Québec.

methods yield expressions which may look different but are all equal. However this is no longer true for the Shape Hessian. In fact we shall show in section 2.4 that different perturbations of the identity can yield final expressions which are not equal. It turns out that we can introduce an infinity of definitions based on perturbations of the identity. However we shall show that they always contain a ***canonical bilinear term*** plus the Shape gradient of the functional acting in the direction of an ***acceleration*** field which is characteristic of the chosen perturbation. The canonical bilinear term exactly coincides with the second order Shape derivative obtained by the Velocity (Speed) Method for autonomous velocity fields. Moreover each expression obtained by a method of perturbation of the identity can be strictly recovered by adding to the canonical term the Shape gradient acting in the direction of an appropriate acceleration field. Therefore we propose to refer to this canonical term as the ***Shape Hessian***.

The above considerations clarify the fundamental concepts and reduce their complexity, but they do not eliminate all the associated computations. We need methods which provide both quick formal computations and appropriate mathematical justifications. We use Lagrangian methods combined with the use of theorems on the derivative of a MinMax with respect to a parameter. Such methods are well known and extensively used in Mechanical Sciences, Mathematical Programming and Optimal Control Theory. Their application to Shape Sensitivity Analysis is not completely straightforward since it leads to the "time-dependence" of the underlying function spaces appearing in the MinMax formulation. This phenomenon seems to be specific to that class of problems. Two techniques are available to get around this difficulty: the ***Function Space Parametrization*** and the ***Function Space Embedding*** methods. The first one has been used in DELFOUR-ZOLÉSIO [8,9], the second one will be used here.

It is fair to say that the use of Shape Hessians for discretized Finite Element Models and finitely parametrized shapes have been used in many places in the Engineering and Mechanics literatures. Some numerical expertise is available (cf. for instance, A. BERN [1], BERN-CHENOT-DEMAY-ZOLÉSIO [1]) and it is suspected that the really performing algorithms are not available in the open literature since they are marketable industrial products.

Few papers have dealt with the second variation of a Shape cost function for linear partial differential equations models. To our knowledge the first one by N. FUJII [1] used a second order perturbation of the identity along the normal to the boundary for second order linear elliptic problems. An extremely interesting paper by ARUMUGAN-PIRONNEAU [1,2] used the Shape second variation to solve the "ribblet problem". Finally J. SIMON [1] presented a computation of the second variation using a first order perturbation of the identity. The first general approach to the computation of Shape Hessians can be found in DELFOUR-ZOLÉSIO [8,9]. It uses the Velocity (Speed) Method and includes a simple illustrative example for the Neumann problem.

In conclusion, we would like to reiterate that the Velocity method and methods using first and second order perturbations of the identity lead to three different second order Shape derivatives which are not equal. The Velocity method with autonomous velocity fields provides the canonical bilinear Shape Hessian and all the other derivatives can be recovered by special choices of non-autonomous velocity fields.

2. SHAPE DERIVATIVES: DEFINITIONS AND PROPERTIES.

In this section we recall and extend the definitions of a Shape gradient and a Shape Hessian based on the Velocity Method (cf. J.P. ZOLÉSIO [1,2], DELFOUR-ZOLÉSIO [1,2]) and discuss their relationship to various methods based on perturbations of the identity operator.

2.1. Velocity (Speed) method and Perturbations of the Identity Operator. Let $V : [0,\tau] \times \mathbf{R}^N \to \mathbf{R}^N$ be a given velocity field for some fixed $\tau > 0$. The map V can be viewed as a family $\{V(t)\}$ of non-autonomous velocity fields on $\mathbf{R}^N$ defined by

$$x \mapsto V(t)(x) \overset{\text{def}}{=} V(t,x) : \mathbf{R}^N \mapsto \mathbf{R}^N. \tag{1}$$

Assume that

$$(V) \quad \begin{cases} \forall x \in \mathbf{R}^N, \quad V(\cdot,x) \in C^0([0,\tau];\mathbf{R}^N) \\ \exists c > 0, \forall x,y \in \mathbf{R}^N, \quad \|V(\cdot,y) - V(\cdot,x)\|_{C^0([0,\tau];\mathbf{R}^N)} \le c|y-x|, \end{cases}$$

where $V(.,x)$ is the function $t \mapsto V(t,.x)$. Associate with V the solution $x(t;X)$ of the ordinary differential equation

$$\frac{dx}{dt}(t) = V(t,x(t), \quad t \in [0,\tau], \quad x(0) = X \in \mathbf{R}^N. \tag{2}$$

and introduce the homeomorphisms

$$X \mapsto T_t(V)(X) \overset{\text{def}}{=} x(t;X) : \mathbf{R}^N \to \mathbf{R}^N. \tag{3}$$

and the maps

$$(t,X) \mapsto \tau(t,X) \overset{\text{def}}{=} T_t(V)(X) : [0,\tau] \times \mathbf{R}^N \to \mathbf{R}^N, \tag{4}$$

$$(t,x) \mapsto T_V^{-1}(t,x) \overset{\text{def}}{=} T_t^{-1}(V)(x) : [0,\tau] \times \mathbf{R}^N \to \mathbf{R}^N. \tag{5}$$

NOTATION 2.1. In the sequel we shall drop the V in $\tau(t,X)$ and $T_t(V)$ whenever no confusion is possible.

THEOREM 2.1. *(i) Under hypothesis (V) the maps T and T^{-1} have the following properties*

$$(T1) \quad \begin{cases} \forall X \in \mathbf{R}^N, \quad T(\cdot,X) \in C^1([0,\tau];\mathbf{R}^N) \\ \exists c > 0, \forall X,Y \in \mathbf{R}^N, \quad \|T(\cdot,Y) - T(\cdot,X)\|_{C^1([0,\tau];\mathbf{R}^N)} \le c|Y-X|, \end{cases}$$

$$(T2) \quad \forall t \in [0,\tau], \quad X \mapsto T_t(X) = T(t,X) : \mathbf{R}^N \to \mathbf{R}^N \quad \text{is bijective},$$

$$(T3) \quad \begin{cases} \forall x \in \mathbf{R}^N, \quad T^{-1}(\cdot,x) \in C^0([0,\tau];\mathbf{R}^N) \\ \exists c > 0, \forall x,y \in \mathbf{R}^N, \quad \|T^{-1}(\cdot,y) - T^{-1}(\cdot,x)\|_{C^0([0,\tau];\mathbf{R}^N)} \le c|y-x|. \end{cases}$$

(ii) If there exists a real $\tau > 0$ and a map $T : [0,\tau] \times \mathbf{R}^N \to \mathbf{R}^N$ verifying hypotheses (T1), (T2) and (T3), then the map

$$(t,x) \mapsto V(t,x) = \frac{\partial T}{\partial t}(t, T_t^{-1}(x)) : [0,\tau] \times \mathbf{R}^N \to \mathbf{R}^N, \tag{6}$$

verifies hypothesis (V), where T_t^{-1} is the inverse of $X \mapsto T_t(X)$. □

This first theorem is an equivalence result which says that we can either start from a family of velocity fields $\{V(t)\}$ on $\mathbf{R}^N$ or a family of transformations $\{T_t\}$ of $\mathbf{R}^N$ provided that the map $V, V(t,x) = V(t)(x)$, verifies (V) or the map $T, T(t,X) = T_t(X)$, verifies (T1), (T2) and (T3).

When we start from V, we obtain the velocity method. Given an initial domain Ω, the family of homeomorphisms $T_t(V)$defines a family of transformed domains

$$\Omega_t = T_t(V)(\Omega) = \{T_t(V)(X) : X \in \Omega\}. \tag{7}$$

In examples where we start from T, it is usually possible to verify hypotheses (T1),(T2) and (T3) and construct the corresponding velocity field V defined in (6). For instance perturbations of the identity to the first or second order fall in that category:

$$T_t(X) = X + tU(X) + \frac{t^2}{2}A(X) \;\; (A = 0 \text{ for the first order }),\; t \geq 0, X \in \mathbf{R}^N, \tag{8}$$

where U and A are sufficiently smooth transformations of $\mathbf{R}^N$. It turns out that for Lipschitz transformations U and A, hypotheses (T1), (T2) and (T3) are verified.

THEOREM 2.2. *Let U and A be two uniform Lipschitz transformations of $\mathbf{R}^N$:*

$$\exists c > 0,\; \forall X,\; Y \in \mathbf{R}^N,\; |U(Y) - U(X)| \leq c|Y - X|,\; |A(Y) - A(X)| \leq c|Y - X|.$$

(i) Let $\tau = \min\{1, 1/4c\}$ and T be given by (8). Then the velocity

$$(t,x) \mapsto V(t,x) = U(T_t^{-1}(x)) + tA(T_t^{-1}(x)) : [0,\tau] \times \mathbf{R}^N \to \mathbf{R}^N, \tag{9}$$

verifies hypotheses (V). □

REMARK 2.1. Observe that from (8) and (9)

$$V(0) = U,\; \dot{V}(0)(x) = \frac{\partial V}{\partial t}(t,x)|_{t=0} = A - [DU]U. \tag{10}$$

where DU is the Jacobian matrix of U. The term $\dot{V}(0)$ is an "acceleration" at $t = 0$ which will always be present even when $A = 0$. □

2.2. Shape gradient. In general a ***shape functional*** will be a map

$$\Omega \mapsto J(\Omega) : \mathcal{A} \subset \mathcal{P}(\mathbf{R}^N) \to \mathbf{R}. \tag{11}$$

defined on a subset $\mathcal{A}$ of the set $\mathcal{P}(\mathbf{R}^N)$ of all subsets of $\mathbf{R}^N$. Under the action of a velocity V verifying (V), the domain Ω is transformed into a new domain $\Omega_t(V) = T_t(V)(\Omega)$.

DEFINITION 2.1. *Given a velocity field V verifying (V), J is said to have an Eulerian semiderivative at Ω in the direction V if the following limit exists and is finite*

$$\lim_{t \searrow 0}[J(\Omega_t(V)) - J(\Omega)]/t. \tag{12}$$

When it exists, it is denoted $dJ(\Omega; V)$. □

This definition is quite general and may include situations where $dJ(\Omega; V)$ is not only a funtion of $V(0)$ but also of $V(t)$ in a neighbourhood of $t = 0$. This will not occur under some appropriate continuity hypothesis on the map $V \mapsto dJ(\Omega, V)$. To be more precise we introduce some notation. For any integers $k \geq 0$ and $m \geq 0$, and any compact subset K of $\mathbf{R}^N$

$$\mathcal{V}_K^{m,k} = C^m([0,\tau]; \mathcal{D}^k(K, \mathbf{R}^N)) \cap \mathcal{L}, \tag{13}$$

where $\mathcal{D}^k(K, \mathbf{R}^N)$ is the space of all k-time continuously differentiable maps from $\mathbf{R}^N$ to $\mathbf{R}^N$ with compact support in K and

$$\mathcal{L} = \left\{V : [0,\tau] \times \mathbf{R}^N \to \mathbf{R}^N : V \text{verifies (V)}\right\}. \tag{14}$$

With the above definitions we introduce the space

$$\vec{\mathcal{V}}^{m,k} \stackrel{\text{def}}{=} \lim_{\substack{\to \\ K}} \left\{\mathcal{V}_K^{m,k} : \forall K \text{ compact in } \mathbf{R}^N\right\} \tag{15}$$

where $\underrightarrow{\lim}$ denotes the inductive limit endowed with its natural inductive limit topology. For autonomous fields, the above constructions reduce to

$$\mathcal{V}^k = \left\{\begin{array}{ll} \mathcal{D}^0(\mathbf{R}^N, \mathbf{R}^N) \cap \operatorname{Lip}(\mathbf{R}^N, \mathbf{R}^N), & \text{if } k = 0 \\ \mathcal{D}^k(\mathbf{R}^N, \mathbf{R}^N) & \text{if } k \geq 1 \end{array}\right\} \tag{16}$$

where Lip $(\mathbf{R}^N, \mathbf{R}^N)$ denotes the space of transformations of $\mathbf{R}^N$ which are uniformly Lipschitzian. In all cases (V) will be verified.

THEOREM 2.3. *Let Ω be a domain in $\mathbf{R}^N$ and $m \geq 0$ and $k \geq 0$ be integers. Assume that for all V in $\vec{\mathcal{V}}^{m,k}$, $dJ(\Omega; V)$ exists and that the map*

$$V \mapsto dJ(\Omega; V) : \vec{\mathcal{V}}^{m,k} \to \mathbf{R} \tag{17}$$

is continuous. Then

$$\forall V \in \vec{\mathcal{V}}^{m,k}, \quad dJ(\Omega; V) = dJ(\Omega; V(0)). \quad \square$$

In the above analysis we have chosen to follow the classical framework of the Theory of distributions (cf. L. SCHWARTZ [1]) and perturb the domain Ω by velocity fields V with compact support.

DEFINITION 2.2. *Let Ω be a domain in $\mathbf{R}^N$.*

*(i) The shape funtional J is said to be **shape differentiable** at Ω if the Eulerian semiderivative $dJ(\Omega; V)$ exists for all V in $\mathcal{D}(\mathbf{R}^N, \mathbf{R}^N)$ and the map*

$$V \mapsto dJ(\Omega; V): \ \mathcal{D}(\mathbf{R}^N, \mathbf{R}^N) \to \mathbf{R} \tag{19}$$

is linear and continuous.

*(ii) The map (19) defines a vector distribution $G(\Omega)$ which will be called the **shape gradient** of J at Ω.*

(iii) When $G(\Omega)$ is continuous on $\mathcal{D}^k(\mathbf{R}^N, \mathbf{R}^N)$ for some finite $k \geq 0$, we say that $G(\Omega)$ is of order k. □

The next theorem gives additional properties of shape differentiable functionals.

THEOREM 2.4. *(Generalized Hadamard's structure theorem). Let Ω be a domain in $\mathbf{R}^N$ with boundary Γ and assume that J is shape differentiable.*
(i) The support of $G(\Omega)$ is contained in Γ. Moreover when $G(\Omega)$ is of finite order its support is compact.
(ii) If $G(\Omega)$ is of finite order k and Ω is an open domain in $\mathbf{R}^N$ with boundary Γ in C^{k+1}, then there exists a scalar distribution $g(\Omega)$ in $\mathcal{D}^k(\Gamma)'$ such that

$$dJ(\Omega; V) = < g(\Omega), \ V \bullet n >_{\mathcal{D}^k(\Gamma)} \tag{20}$$

where n is the unit outward normal to Ω on Γ and $V \bullet n$ denotes the scalar product of V and n in $\mathbf{R}^N$. □

REMARK 2.2. When Γ is compact $\mathcal{D}^k(\Gamma)$ coincides with $C^k(\Gamma)$. □

2.3. Shape Hessian. We first study the second order Eulerian semiderivative $d^2J(\Omega; V; W)$ of a functional $J(\Omega)$ for two non-autonomous vector fields V and W. A first theorem shows that under some natural continuity hypotheses, $d^2J(\Omega; V; W)$ is the sum of two terms: the "canonical term $d^2J(\Omega; V(0); W(0))$" plus the first order Eulerian semiderivative $dJ(\Omega; \dot{V}(0))$ at Ω in the direction $\dot{V}(0)$ of the time-partial derivative $\partial_t V(t, x)$ at $t = 0$.

As for first order Eulerian semiderivatives, this first theorem reduces the study of second order Eulerian semiderivatives to the autonomous case. So we shall specialize to fields V and W in $\mathcal{D}^k(\mathbf{R}^N, \mathbf{R}^N)$ and give the equivalent of Hadamard's structure theorem for the "canonical term".

2.3.1. Non-autonomous case. The basic framework introduced in sections 2.1 and 2.2 has reduced the computation of the Eulerian semiderivative of $J(\Omega)$ to the computation of the derivative

$$j'(0) = dJ(\Omega; V(0)) \tag{21}$$

of the function

$$j(t) = J(\Omega_t(V)). \tag{22}$$

For $t \geq 0$, we naturally obtain

$$j'(t) = dJ(\Omega_t(V); V(t)). \tag{23}$$

This suggests the following definition.

DEFINITION 2.3. *Let V and W belong to $\mathcal{L}$ and assume that for all $t \in [0, \tau]$, $dJ(\Omega_t(W); V(t))$ exists for $\Omega_t(W) = T_t(W)(\Omega)$. The functional J is said to have a **second order Eulerian semiderivative** at Ω in the directions (V, W) if the following limit exists*

$$\lim_{t \searrow 0} [dJ(\Omega_t(W); V(t)) - dJ(\Omega; V(0))]/t. \tag{24}$$

When it exists, it is denoted $d^2 J(\Omega; V; W)$. □

REMARK 2.3. This last definition is compatible with the second order expansion of $j(t)$ with respect to t around $t = 0$:

$$j(t) \cong j(0) + tj'(0) + \frac{t^2}{2} j''(0), \tag{25}$$

where

$$j''(0) = d^2 J(\Omega; V; V). \quad \square \tag{26}$$

REMARK 2.4. It is easy to construct simple examples with time-invariant fields V and W showing that $d^2 J(\Omega; V; W) \neq d^2 J(\Omega; W; V)$ (cf. DELFOUR-ZOLÉSIO [8]). □

The next theorem is the analogue of Theorem 2.3 and provides the canonical structure of the second order Eulerian semiderivative.

THEOREM 2.5. *Let Ω be a domain in $\mathbf{R}^N$ and $m \geq 0$ and $l \geq 0$ be integers. Assume that*

(i) $\forall V \in \vec{\mathcal{V}}^{m+1,l}$, $\forall W \in \vec{\mathcal{V}}^{m,l}$, $d^2 J(\Omega; V; W)$ exists,

(ii) $\forall W \in \vec{\mathcal{V}}^{m,l}$, $\forall t \in [0, \tau]$, J has a shape gradient at $\Omega_t(W)$ of order l,

(iii) $\forall\, U \in \mathcal{V}^l$, the map

$$W \mapsto d^2 J(\Omega; U; W) : \vec{\mathcal{V}}^{m,l} \to \mathbf{R} \tag{27}$$

is continuous. Then for all V in $\vec{\mathcal{V}}^{m+1,l}$ and all W in $\vec{\mathcal{V}}^{m,l}$

$$d^2 J(\Omega; V; W) = d^2 J(\Omega; V(0); W(0)) + dJ(\Omega; \dot{V}(0)), \tag{28}$$

where

$$\dot{V}(0)(x) = \lim_{t \searrow 0} [V(t, x) - V(0, x)]/t. \square \tag{29}$$

2.3.2. Autonomous case.

DEFINITION 2.4. *Let Ω be a domain in $\mathbf{R}^N$.*
*(i) The funtional $J(\Omega)$ is said to be **shape differentiable** at Ω if*

$$\forall V,\ \forall W \text{ in } \mathcal{D}(\mathbf{R}^N,\mathbf{R}^N),\ d^2J\ (\Omega; V; W) \text{ exists} \tag{30}$$

and the map

$$(V,W) \mapsto d^2J(\Omega;V;W) : \mathcal{D}(\mathbf{R}^N,\mathbf{R}^N) \times \mathcal{D}(\mathbf{R}^N,\mathbf{R}^N) \to \mathbf{R} \tag{31}$$

is bilinear and continuous. We denote by h the bilinear and continuous map (31).
(ii) Denote by $H(\Omega)$ the continous linear map on the tensor product $\mathcal{D}(\mathbf{R}^N,\mathbf{R}^N) \otimes \mathcal{D}(\mathbf{R}^N,\mathbf{R}^N)$, associated with h:

$$d^2J(\Omega;V;W) = \langle H(\Omega), V \otimes W\rangle = h(V,W), \tag{32}$$

where $V \otimes W$ is the tensor product of V and W defined as

$$(V \otimes W)_{ij}(x,y) = V_i(x)W_j(y),\quad 1 \le i,\ j \le N, \tag{33}$$

and $V_i(x)$ (resp. $W_j(y)$) is the i-th (resp. j-th) component of the vector V(resp. W) (cf. L. SCHWARTZ *[2]'s kernel theorem and* GELFAND-VILENKIN *[1]. $H(\Omega)$ will be called the **Shape Hessian** of J at Ω.*
(iii) When there exists an integer $\ell \ge 0$ such that $H(\Omega)$ is continuous on $\mathcal{D}^\ell(\mathbf{R}^N,\mathbf{R}^N)\otimes \mathcal{D}^\ell(\mathbf{R}^N,\mathbf{R}^N)$ we say that $H(\Omega)$ is of order ℓ. □

THEOREM 2.6. *Let Ω be a domain in $\mathbf{R}^N$ with boundary Γ and assume that J is twice shape differentiable at Ω.*
(i) $H(\Omega)$ has support in $\Gamma\times\Gamma$. Moreover the support of $H(\Omega)$ is compact when its order is finite.
(ii) If $H(\Omega)$ is of finite order ℓ, $\ell \ge 0$, Ω is an open domain in $\mathbf{R}^N$ with boundary Γ in $C^{\ell+1}$, then there exists a continuous linear map on the tensor product $\mathcal{D}^\ell(\Gamma,\mathbf{R}^N)\otimes\mathcal{D}^\ell(\Gamma)$ such that

$$d^2J(\Omega); V; W) = \langle h(\Omega), (\gamma_\Gamma V) \otimes ((\gamma_\Gamma W)\bullet n)\rangle \tag{34}$$

where $(\gamma_\Gamma V) \otimes ((\gamma_\Gamma W)\bullet n)$ is defined as the tensor product

$$((\gamma_\Gamma V) \otimes (\gamma_\Gamma W)\bullet n))_i(x,y) = (\gamma_\Gamma V_i)(x)((\gamma_\Gamma W)\bullet n)(y),\ x,y \in \Gamma, \tag{35}$$

$V_i(x)$ is the i-th component of $V(x)$ and

$$(\gamma_\Gamma(W)\bullet n)(y) = (\gamma_\Gamma W)(y)\bullet n(y),\ y \in \Gamma.\ \square \tag{36}$$

REMARK 2.5. Finally under the hypotheses of Theorem 5 and 6

$$\begin{aligned} d^2J(\Omega;V;W) =& \langle h(\Omega), (\gamma_\Gamma V(0)) \otimes ((\gamma_\Gamma W(0))\bullet n)\rangle \\ &+ \langle g(\Omega), (\gamma_\Gamma \dot{V}(0))\bullet n\rangle \end{aligned} \tag{37}$$

for all V in $\vec{\mathcal{V}}_D^{m+1,l}$ and W in $\vec{\mathcal{V}}_D^{m,l}$. □

2.4. Comparison with Methods of Perturbation of the Identity. At this juncture it is instructive to compare first and second order Eulerian semiderivatives obtained by the Velocity (Speed) Method with those obtained by first and second order perturbations of the identity: that is, when the transformations T_t are specified a priori by

$$T_t(X) = X + tU(X) + \frac{t^2}{2}A(X), \quad X \in \mathbf{R}^N, \tag{38}$$

where U and A are transformations of $\mathbf{R}^N$ verifying the hypotheses of Theorem 2.2. The transformation T_t in (38) is a *second order* perturbation when $A \neq 0$ and a *first order* perturbation when $A = 0$.

According to Theorem 2.2, first and second order Eulerian semiderivatives associated with (38) can be equivalently obtained by applying the Velocity (Speed) Method to the time-varying velocity fields V_{UA} given by (9)

$$dJ(\Omega; V_{UA}) = dJ(\Omega; V_{UA}(0)) = dJ(\Omega; U) \tag{39}$$

where we have used Remark 2.1 which says that

$$V_{UA}(0) = U \text{ and } \dot{V}_{UA}(0) = A - [DU]U. \tag{40}$$

Similarly if V_{WB} is another velocity field corresponding to

$$T_t(X) = X + tW(X) + \frac{t^2}{2}B(X), \quad X \in \mathbf{R}^N, \tag{41}$$

where W and B verify the hypotheses of Theorem 2.2, then

$$d^2J(\Omega; V_{UA}; V_{WB}) = d^2J(\Omega; V_{UA}(0); V_{WB}(0)) + dJ(\Omega; \dot{V}_{UA}(0)) \tag{42}$$

and

$$d^2J(\Omega; V_{UA}; V_{WB}) = d^2J(\Omega; U; W) + dJ(\Omega; A - [DU]U). \tag{43}$$

Expressions (39) and (43) are to be compared with the following expressions obtained by the Velocity (Speed) Method for two time-invariant vector fields V and W

$$dJ(\Omega; V) \text{ and } d^2J(\Omega; V; W). \tag{44}$$

For the Shape gradient the two expressions coincide; for the Shape Hessian we recognize the bilinear term in (43) and (44) but the two expressions differ by the term

$$dJ(\Omega; A - [DU]U). \tag{45}$$

Even for a first order perturbation ($A = 0$), we have a quadratic term in U.

This situation is analogous to the classical problem of defining second order derivatives on a manifold. The term (45) would correspond to the connexion while the bilinear term $d^2J(\Omega; V; W)$ would be the candidate for the ***canonical*** second order shape derivative. In this context we shall refer to the corresponding distribution $H(\Omega)$ as the *canonical Shape*

Hessian. All other second order shape derivatives will be obtained from $H(\Omega)$ by adding the gradient term $g(\Omega)$ acting as the appropriate acceleration field (connexion). □

REMARK 2.6. The method of perturbation of the identity can be made "more canonical" by using the following family of transformations

$$T_t(X) = X + tU(X) + \frac{t^2}{2}(A + [DU]U) \tag{46}$$

which yields

$$dJ(\Omega; U) \text{ for the gradient} \tag{47}$$

and

$$d^2J(\Omega; U; W) + dJ(\Omega; A) \text{ for the Hessian,} \tag{48}$$

where for a first order perturbation ($A = 0$) the second term disappears. □

REMARK 2.7. When Ω^* is an appropriately smooth domain which minimizes a twice Shape differentiable functional $J(\Omega)$ without constraints on Ω, the classical necessary conditions would be (at least formally)

$$dJ(\Omega^*; V) = 0, \quad \forall V, \tag{49}$$

$$d^2J(\Omega^*; W; W) \geq 0, \quad \forall W, \tag{50}$$

or equivalently for "smooth velocity fields V and W"

$$dJ(\Omega^*; V(0)) = 0, \quad \forall V \tag{51}$$

$$d^2J(\Omega^*; W(0); W(0)) + dJ(\Omega^*; \dot{V}(0)) \geq 0, \quad \forall W. \tag{52}$$

But in view of (51), condition (52) reduces to the following condition on the "canonical Shape Hessian"

$$d^2J(\Omega^*; W(0); W(0)) \geq 0, \quad \forall W. \; \Box \tag{53}$$

3. A SADDLE POINT FORMULATION OF THE DIRICHLET PROBLEM.

Let Ω be a bounded open domain in $\mathbf{R}^N$ with a sufficiently smooth boundary Γ. Let f and g be two fixed functions in $H^{\frac{1}{2}+\epsilon}(\mathbf{R}^N)$ and $H^{2+\epsilon}(\mathbf{R}^N)$, respectively, for some arbitrary small $\epsilon > 0$. Consider the solution y in $H^2(\Omega)$ to the non-homogeneous Dirichlet boundary value problem.

$$-\Delta y = f \text{ in } \Omega, y = g \text{ on } \Gamma. \tag{1}$$

We can also say that y is the solution of the weak equation

$$\int_\Omega (\Delta y + f)\psi \, dx + \int_\Gamma (y - g)\mu \, d\Gamma = 0 \tag{2}$$

for all ψ in $H^2(\Omega)$ and μ in $H^{\frac{1}{2}}(\Gamma)$, since the corresponding functional

$$L(\phi,\ \psi,\ \mu) = \int_\Omega (\Delta\phi + f)\psi\ dx + \int_\Gamma (\phi - g)\mu\ d\Gamma. \tag{3}$$

It has a unique saddle point $(\hat{\phi}, \hat{\psi}, \hat{\mu})$ which is completely characterized by the equations

$$\Delta\hat{\phi} + f = 0 \text{ in } \Omega, \tag{4}$$

$$\hat{\phi} - g = 0 \text{ in } \Gamma, \tag{5}$$

$$\int_\Omega \Delta\phi\ \hat{\psi}\ dx + \int_\Gamma \phi\hat{\mu}\ d\Gamma = 0,\ \forall\phi \in H^2(\Omega), \tag{6}$$

where the last equation yields

$$\Delta\hat{\psi} = 0 \text{ in } \Omega,\ \hat{\psi} = 0 \text{ on } \Gamma \textit{ and } \hat{\mu} = \frac{\partial\hat{\psi}}{\partial n} \text{ on } \Gamma. \tag{7}$$

Of course, this implies that the saddle point is unique and given by

$$(\hat{\phi}, \hat{\psi}, \hat{\mu}) = (y, 0, 0). \tag{8}$$

The purpose of the above computation was to find out the form of the multiplier $\hat{\mu}$

$$\hat{\mu} = \frac{\partial\hat{\psi}}{\partial n} \text{ on } \Gamma, \tag{9}$$

in order to rewrite the previous functional as a function of two variables instead of three:

$$L(\phi, \psi) = \int_\Omega (\Delta\phi + f)\psi\ dx + \int_\Gamma (\phi - g)\frac{\partial\psi}{\partial n}\ d\Gamma, \tag{10}$$

for (ϕ, ψ) in $H^2(\Omega) \times H^2(\Omega)$. It is also advantageous for shape problems to get rid of boundary integrals whenever it is possible. So noting that

$$\int_\Gamma (\phi - g)\frac{\partial\psi}{\partial n}\ d\Gamma = \int_\Omega \operatorname{div}[(\phi - g)\nabla\psi]\ dx, \tag{11}$$

we finally use the functional

$$L(\phi, \psi) = \int_\Omega \{(\Delta\phi + f)\psi + (\phi - g)\Delta\psi + \nabla(\phi - g) \bullet \nabla\psi\}\ dx \tag{12}$$

on $H^2(\Omega) \times H^2(\Omega)$. It is readily seen that it has a unique saddle point $(\hat{\phi}, \hat{\psi})$ in $H^2(\Omega) \times H^2(\Omega)$ which is completely characterized by the saddle point equations:

$$\Delta\hat{\phi} + f = 0 \textit{ in } \Omega, \quad \hat{\psi} = g \textit{ on } \Gamma, \quad \Delta\hat{\psi} = 0 \text{ in } \Omega, \quad \hat{\psi} = 0 \text{ on } \Gamma. \tag{13}$$

4. SHAPE GRADIENT FOR THE DIRICHLET PROBLEM .

4.1. Formulation and formal computations. Consider the cost function

$$J(\Omega) = \frac{1}{2}\int_{\Omega} |y(\Omega) - y_d|^2 \, dx \tag{1}$$

associated with the solution $y = y(\Omega)$ of the Dirichlet problem (3.1) and the fixed function y_d in $H^{\frac{1}{2}+\epsilon}(\mathbf{R}^N)$ for some arbitrary fixed $\epsilon > 0$.

As in section 3, we reformulate this problem as the saddle point of a functional by introducting the Lagrangian

$$\begin{aligned} G(\Omega, \phi, \psi) = & \frac{1}{2}\int_{\Omega} |\phi - y_d|^2 dx \\ & + \int_{\Omega} \{(\Delta\phi + f)\psi + (\phi - g)\Delta\psi + \nabla(\phi - g) \bullet \nabla\psi\} \, dx \end{aligned} \tag{2}$$

on $H^2(\Omega) \times H^2(\Omega)$. It is readily seen that $G(\Omega, \cdot, \cdot)$ has a unique saddle point$(\hat{\phi}, \hat{\psi})$ which is completely characterized by the following saddle point equations:

$$\Delta\hat{\phi} + f = 0 \text{ in } \Omega, \quad \hat{\phi} = g \text{ on } \Gamma \tag{3}$$

$$\int_{\Omega} \{(\hat{\phi} - y_d)\phi + \Delta\phi\hat{\psi} + \phi\Delta\hat{\psi} + \nabla\phi \bullet \nabla\hat{\psi}\} \, dx = 0, \quad \forall\, \phi \in H^2(\Omega). \tag{4}$$

But the last equation is equivalent to

$$\int_{\Omega} [(\hat{\phi} - y_d) + \Delta\hat{\psi}]\phi \, dx + \int_{\Gamma} \frac{\partial\phi}{\partial n}\hat{\psi} \, d\Gamma = 0, \quad \forall\phi \in H^2(\Omega) \tag{5}$$

or

$$\Delta\hat{\psi} + (\hat{\phi} - y_d) = 0 \text{ in } \Omega, \quad \hat{\psi} = 0 \text{ on } \Gamma, \tag{6}$$

by using the theorem on the surjectivity of the trace. In the sequel, we shall use the notation (y, p) for the saddle point $(\hat{\phi}, \hat{\psi})$. As a result, we have

$$J(\Omega) = \underset{\phi \in H^2(\Omega)}{\text{Min}} \quad \underset{\psi \in H^2(\Omega)}{\text{Max}} \quad G(\Omega, \phi, \psi). \tag{7}$$

We shall now use the above Lagrangian formulation combined with the Velocity method (cf. J.CÉA [1,2,3], J.P.ZOLÉSIO[1,2], DELFOUR-ZOLÉSIO [1,2,3,4,7]) to compute the Shape gradient of $J(\Omega)$. Recall that the domain Ω is perturbed by a velocity vector field V which defines a homeomorphism (cf. section 2.1)

$$T_t : \mathbf{R}^N \to \mathbf{R}^N, \; T_t(X) = x(t), \tag{8}$$

and a new domain

$$\Omega_t = T_t(\Omega). \tag{9}$$

The Shape semiderivative is defined as(cf. section 2.2)

$$dJ(\Omega; V) = \lim_{t \searrow 0} [J(\Omega_t) - J(\Omega)]/t \tag{10}$$

whenever the limit exists. It is easy to check that

$$J(\Omega_t) = \underset{\phi \in H^2(\Omega_t)}{\mathrm{Min}} \quad \underset{\psi \in H^2(\Omega_t)}{\mathrm{Max}} \quad G(\Omega_t, \phi, \psi). \tag{11}$$

There are two ways to get rid of the time dependence in the underlying function spaces (cf. DELFOUR-ZOLÉSIO **[1,2]**):
- the *Function Space Parametrization Method*
- the *Function Space Embedding Method.*

In the first case, we parametrize the functions in $H^2(\Omega_t)$ by elements of $H^2(\Omega)$ through the transformation

$$\phi \mapsto \phi \circ T_t^{-1} = H^2(\Omega) \to H^2(\Omega_t), \tag{12}$$

where "o" denotes the composition of the two maps and we introduce the *Parametrized Lagrangian*,

$$\tilde{G}(t, \phi, \psi) = G(T_t(\Omega), \phi \circ T_t^{-1}, \psi \circ T_t^{-1}) \tag{13}$$

on $H^2(\Omega) \times H^2(\Omega)$. In the Function Space Embedding Method, we introduce a large enough domain, D which contains all the transformations $\{\Omega_t : 0 \leq t \leq \bar{t}\ \}$ of Ω for some small $\bar{t} > 0$.

In this paper, we shall use the Function Space Embedding Method with $D = \mathbb{R}^N$

$$J(\Omega_t) = \underset{\Phi \in H^2(\mathbb{R}^N)}{\mathrm{Min}} \quad \underset{\Psi \in H^2(\mathbb{R}^N)}{\mathrm{Max}} \quad G(\Omega_t, \Phi, \Psi). \tag{14}$$

As can be expected the price to pay for the use of this method, is the fact that the set of saddle points

$$S(t) = X(t) \times Y(t) \ \subset H^2(\mathbb{R}^N) \times H^2(\mathbb{R}^N) \tag{15}$$

is not a singleton anymore since

$$X(t) = \{\Phi \in H^2(\mathbb{R}^N) : \Phi|_{\Omega_t} = y_t\} \tag{16}$$

$$Y(t) = \{\Psi \in H^2(\mathbb{R}^N) : \Psi|_{\Omega_t} = p_t\} \tag{17}$$

where (y_t, p_t) is the unique solution in $H^2(\Omega_t) \times H^2(\Omega_t)$ to the previous saddle point equations on Ω_t

$$\Delta y_t + f = 0 \ \text{ in } \ \Omega_t, \quad y_t = g \ \text{ on } \ \Gamma_t, \tag{18}$$

$$\Delta p_t + (y_t - y_d) = 0 \ \text{ in } \ \Omega_t, \quad p_t = 0 \ \text{ on } \ \Gamma_t. \tag{19}$$

We are now ready to apply the theorem of CORREA-SEEGER **[1]** which says that under appropriate hypotheses (to be checked in the next section)

$$dJ(\Omega; V) = \underset{\Phi \in X(0)}{\mathrm{Min}} \quad \underset{\Psi \in Y(0)}{\mathrm{Max}} \quad \partial_t G(\Omega_t, \Phi, \Psi). \tag{20}$$

Since we have already characterized $X(0)$ and $Y(0)$, we only need to compute the partial derivative of

$$G(\Omega_t, \Phi, \Psi) = \int_{\Omega_t} \{\frac{1}{2}|\Phi - y_d|^2 + (\Delta\Phi + f)\Psi + (\Phi - g)\Delta\Psi + \nabla(\Phi - g) \bullet \nabla\Psi\} dx. \quad (21)$$

If we assume that Ω_t is sufficiently smooth, then

$$f, y_d \in H^{\frac{1}{2}+\epsilon}(\mathbf{R}^N) \text{ and } g \in H^{2+\epsilon}(\mathbf{R}^N) \Rightarrow y \text{ and } p \in H^{\frac{5}{2}+\epsilon}(\Omega) \quad (22)$$

and we can choose to consider our saddle points $S(t)$ in $H^{\frac{5}{2}+\epsilon}(\mathbf{R}^N) \times H^{\frac{5}{2}+\epsilon}(\mathbf{R}^N)$ rather than $H^2(\mathbf{R}^N) \times H^2(\mathbf{R}^N)$. If Φ and Ψ belong to $H^{\frac{5}{2}+\epsilon}(\mathbf{R}^N)$, then

$$\partial_t G(\Omega_t, \Phi, \Psi) = \int_{\Gamma_t} \{\frac{1}{2}(\Phi - y_d)^2 + (\Delta\Phi + f)\Psi + (\Phi - g)\Delta\Psi + \nabla(\Phi - g) \bullet \nabla\Psi\} V \bullet n_t \, d\Gamma_t. \quad (23)$$

This expression is an integral over the boundary Γ which will not depend on Φ and Ψ outside of $\bar{\Omega}$. As a result the Min and the Max can be dropped in expression (20) which reduces to

$$dJ(\Omega; V) = \int_{\Gamma} \{\frac{1}{2}(y - y_d)^2 + (\Delta y + f)p + (y - g)\Delta p + \nabla(y - g) \bullet \nabla p\} V \bullet n \, d\Gamma. \quad (24)$$

But

$$p = 0 \text{ and } y - g = 0 \quad \Rightarrow \quad \nabla p = \frac{\partial p}{\partial n} n, \quad \nabla(y - g) = \frac{\partial}{\partial n}(y - g)\, n \quad \text{on } \Gamma \quad (25)$$

and finally

$$dJ(\Omega; V) = \int_{\Gamma} \{\frac{1}{2}(g - y_d)^2 + \frac{\partial}{\partial n}(y - g)\frac{\partial p}{\partial n}\} V \bullet n \, d\Gamma. \quad (26)$$

4.2. Verification of the Hypotheses. As we have seen the computations of the Shape gradient is both quick and easy. We now turn to the step by step verification of the hypotheses of the underlying theorem. Many of the constructions given below are "canonical" and can be repeated for different problems in different contexts.

THEOREM 4.1. (CORREA AND SEEGER [1]). *Let $\tau > 0$, the sets X and Y and the functional $L : [0, \tau] \times X \times Y \to \mathbf{R}$ be given. Denote by*

$$S(t) = X(t) \times Y(t) \subset X \times Y \quad (27)$$

the set of saddle points of the functional $L(t, \cdot\,, \cdot)$ on $X \times Y$. Assume that
(H1) $\forall t \in [0, \tau]$, $S(t) \neq \emptyset$.
and that
(H2) $\forall (x, y) \in [X(0) \times \bigcup_{0 \le t \le \tau} Y(t)] \cup [\bigcup_{0 \le t \le \tau} X(t) \times Y(0)]$, $\partial_t L(t, x, y)$ exists on $[0, \tau]$.

Moreover, assume that there exist topologies $\mathcal{T}_X$ on X and $\mathcal{T}_Y$ on Y such that

(H3) *for all sequences $t_n \to 0$ as $n \to \infty$, $0 \le t_n \le \tau$, there exists $(x_0, y_0) \in S(0)$ and a subsequence of $\{t_n\}$, still denoted $\{t_n\}$, such that $\forall n$, $\exists (x_n, y_n) \in S(t_n)$ and $(x_n, y_n) \to (x_0, y_0)$ in $\mathcal{T}_X \times \mathcal{T}_Y$,*

(H4) *for all y in $Y(0)$ (resp. x in $X(0)$)*
$\liminf_{\substack{t \searrow 0 \\ n \to \infty}} \partial_t L(t, x_n, y) \ge \partial_t L(0, x_0, y)$ (resp. $\limsup_{\substack{t \searrow 0 \\ n \to \infty}} \partial_t L(t, x, y_n) \le \partial_t L(0, x, y_0)$).

Then the function

$$g(t) = \min_{x \in X} \max_{y \in Y} L(t, x, y)$$

on $[0, \tau]$ has a derivative at $t = 0$ given by

$$dg(0) = \lim_{t \searrow 0} [g(t) - g(0)]/t = \operatorname*{Inf}_{x \in X(0)} \operatorname*{Sup}_{y \in B(0)} \partial_t L(0, x, y)$$
$$= \operatorname*{Sup}_{y \in B(0)} \operatorname*{Inf}_{x \in X(0)} \partial_t L(0, x, y). \quad \square$$

Let y_d and $f \in H^1(\mathbf{R}^N)$ and $g \in H^{\frac{5}{2}}(\mathbf{R}^N)$ so that

$$X = Y = H^3(\mathbf{R}^N). \tag{28}$$

The saddle points $S(t) = X(t) \times Y(t)$ are given by

$$X(t) = \{\Phi \in X : \Phi|_{\Omega_t} = y_t\} \tag{29}$$

$$Y(t) = \{\Psi \in Y : \Psi|_{\Omega_t} = p_t\} \tag{30}$$

The sets $X(t)$ and $Y(t)$ are not empty since it is always possible to construct a continuous linear extension

$$\Pi^m : H^m(\Omega) \to H^m(\mathbf{R}^N) \tag{31}$$

for each $m \ge 1$. For instance with $m = 1$ and a boundary Γ which is $W^{1,\infty}$, see AGMON-DOUGLIS-NIRENBERG [1,2] and V.M.BABIĆ [1] for $m > 1$ (see also J.NEČAS [1]). Using this Π^m, then we define the following extension

$$\Pi_t^m : H^m(\Omega_t) \to H^m(\mathbf{R}^N) \tag{32}$$

$$\Pi_t^m(\phi) = [\Pi^m(\phi \circ T_t)] \circ T_t^{-1}. \tag{33}$$

In the sequel m is fixed and equal to 3, so we shall drop the superscript m and define the extensions

$$Y_t = \Pi_t y_t \quad P_t = \Pi_t p_t \tag{34}$$

of y_t and p_t, respectively. Hence,

$$Y_t \in X(t) \text{ and } P_t \in Y(t) \Rightarrow S(t) \neq \emptyset. \tag{35}$$

So hypothesis $(H1)$ is verified. Hypothesis $(H2)$ follows by hypotheses on f, y_d and g. To check hypothesis $(H3)$ and $(H4)$, we need two general theorems which can be used in various contexts and problems.

THEOREM 4.2. *For $V \in \mathcal{D}^1(\mathbf{R}^N, \mathbf{R}^N)$ and $\Phi \in L^2(\mathbf{R}^N)$*

$$\lim_{t \searrow 0} \Phi \circ T_t = \Phi \quad \text{and} \quad \lim_{t \searrow 0} \Phi \circ T_t^{-1} = \Phi \text{ in } L^2(\mathbf{R}^N). \tag{36}$$

PROOF. (i) The space $\mathcal{D}(\mathbf{R}^N)$ of continuous functions with compact support in $\mathbf{R}^N$ is dense in $L^2(\mathbf{R}^N)$. So given $\epsilon > 0$, there exists Φ_ϵ in $\mathcal{D}(\mathbf{R}^N)$ such that

$$||\Phi - \Phi_\epsilon||_{L^2}^2 < \epsilon^2 / \max\{J_t^{-1} : 0 \le t \le \tau\}.$$

Hence,

$$||\Phi \circ T_t - \Phi|| \le ||\Phi_\epsilon \circ T_t - \Phi_\epsilon|| + ||\Phi \circ T_t - \Phi_\epsilon \circ T_t|| + ||\Phi - \Phi_\epsilon||. \tag{37}$$

But, $\forall t \in [0, \tau]$

$$\int_{\mathbf{R}^N} |\Phi \circ T_t - \Phi_\epsilon \circ T_t|^2 dx = \int_{\mathbf{R}^N} |\Phi - \Phi_\epsilon|^2 J_t^{-1} dx \le \epsilon^2$$

So the last two terms in (37) are less than 2ϵ. It remains to evaluate the first term for a fixed function Φ_ϵ. Φ_ϵ has a compact support K in $\mathbf{R}^N$. Now, choose a bounded open domain D which contains $T_t(K)$ for all t in $[0, \tau]$. Since Φ_ϵ is uniformly continuous on $\mathbf{R}^N$

$$\exists \delta > 0, \ |x - y| < \delta \Rightarrow |\Phi_\epsilon(y) - \Phi_\epsilon(x)| < \epsilon / m(D)^{\frac{1}{2}}.$$

But, T_t is also uniformly continuous and

$$\exists \eta > 0, \ \forall 0 \le t < \eta, \ \forall x \in D, \ |T_t x - x| < \delta.$$

By construction

$$\text{supp } (\Phi_\epsilon \circ T_t) = T_t \text{ (supp } \Phi_\epsilon) \subset D$$

and

$$\Phi_\epsilon = 0 \ \text{ and } \ \Phi_\epsilon \circ T_t = 0 \text{ outside of } D.$$

Finally,

$$\int_{\mathbf{R}^N} |\Phi_\epsilon(T_t x) - \Phi_\epsilon(x)|^2 \, dx = \int_D |\Phi_\epsilon(T_t x) - \Phi_\epsilon(x)|^2 \, dx \le \epsilon^2$$

and this implies that

$$\forall \epsilon > 0, \ \exists \eta > 0, \ \forall 0 \le t \le \eta, \ ||\Phi \circ T_t - \Phi||_{L^2(\mathbf{R}^N)} \le 3\epsilon.$$

(ii) For the second part of (36) we make a change of variable and use the result of part (i)

$$\int_{\mathbf{R}^N} |\Phi \circ T_t^{-1} - \Phi|^2 dx = \int_{\mathbf{R}^N} |\Phi - \Phi \circ T_t|^2 J_t dx \le \epsilon^2$$

This completes the proof. □

COROLLARY. *Under the hypotheses of Theorem 4.2 for $m \geq 1$, V in $\mathcal{D}^m(\mathbf{R}^N, \mathbf{R}^N)$ and $\Phi \in H^m(\mathbf{R}^N)$,*

$$\lim_{t>0\to 0} \Phi \circ T_t = \Phi \quad \textit{and} \quad \lim_{t>0\to 0} \Phi \circ T_t^{-1} = \Phi \textit{ in } H^m(\mathbf{R}^N). \quad \square \tag{38}$$

REMARK 4.1. In fact for $m \geq 1$ and $V \in \mathcal{D}^m(\mathbf{R}^N, \mathbf{R}^N)$ the transformation

$$S(t)\Phi = \Phi \circ T_t, \quad \forall \Phi \in H^m(\mathbf{R}^N),\ \forall t,\ 0 \leq t \leq \tau, \tag{39}$$

defines a strongly continuous semigroup of class C_0 on $H^m(\mathbf{R}^N)$ with infinitesimal generator

$$\mathcal{A}\Phi = \nabla\Phi \bullet V,\ D(\mathcal{A}) = \{\Phi \in H^m(\mathbf{R}^N) : \nabla\Phi \bullet V \in H^m(\mathbf{R}^N)\}. \quad \square$$

THEOREM 4.3. *Under the hypotheses of Theorem 4.2,*

$$y^t \to y^0 \ \text{ in } \ H^m(\Omega) - \text{strong (resp. weak)} \tag{40}$$

implies that

$$Y_t \to Y_0 \ \text{ in } \ H^m(\mathbf{R}^N) - \text{strong (resp. weak)}.$$

PROOF. The strong case is obvious. We prove the weak case for $m = 0$. By definition,

$$Y_t = (\Pi y^t) \circ T_t^{-1}$$

and for all Φ in $L^2(\mathbf{R}^N)$, we consider

$$\int_{\mathbf{R}^N} Y_t \Phi \, dx = \int_{\mathbf{R}^N} (\Pi y^t) \circ T_t^{-1} \Phi \, dx = \int_{\mathbf{R}^N} \Pi y^t \Phi \circ T_t \, J_t \, dx$$

We have shown in Theorem 4.2 that

$$\Phi \circ T_t \to \Phi \ \text{ in } \ L^2(\mathbf{R}^N) \text{ strong}.$$

In addition, $J_t \to 1$ and by linearity and continuity of Π

$$\Pi y^t \to \Pi y \ \text{ in } \ L^2(\mathbf{R}^N) - \text{weak}.$$

Hence,

$$\forall \Phi \in L^2(\mathbf{R}^N), \ \int_{\mathbf{R}^N} Y_t \Phi \, dx \to \int_{\mathbf{R}^N} \Pi y \Phi \, dx = \int_{\mathbf{R}^N} Y_0 \Phi \, dx.$$

This proves the weak convergence. $\square$

To verify hypothesis $H3$, we transform (y_t, p_t) on Ω_t to $(y^t, p^t) = (y_t \circ T_t, p_t \circ T_t)$ on Ω. The pair (y^t, p^t) is the transported pair of solutions from Ω_t to Ω. It is the unique solution in $H^1(\Omega) \times H^1(\Omega)$ of the system

$$-\text{div}[A(t)\nabla y^t] = J_t \, f \circ T_t \ \text{ in } \ \Omega, \ \ y^t = g \circ T_t \ \text{ on } \ \Gamma, \tag{41}$$

$$-\operatorname{div}[A(t)\nabla p^t] = J_t(y^t - y_d \circ T_t) \text{ in } \Omega, \quad p^t = 0 \text{ on } \Gamma, \tag{42}$$

where

$$A(t) = J_t(DT_t)^{-1*}(DT_t)^{-1}, \quad J_t = |\det DT_t|, \tag{43}$$

and DT_t is the Jacobian matrix of T_t.

For sufficiently smooth domains Ω and vector fields V, the pair $\{y^t, p^t\}$ is bounded in $H^1(\Omega) \times H^1(\Omega)$ as t goes to zero . Since $H^1(\Omega)$ is a Hilbert space, we can extract weakly convergent subsequences to some $(\bar{y}, \bar{p})$ in $H^1(\Omega) \times H^1(\Omega)$. However, by linearity of the equation with respect to (y^t, p^t) and continuity of the coefficients with respect to t, the limit point $(\bar{y}, \bar{p})$ will coincide with (y^0, p^0), since the system has a unique solution at $t = 0$. Then we go back to the equation for y^t and y and show that the convergence is strong in $H^1(\Omega)$. Finally by using the regularity of the data and the classical regularity theorems we show that $(y^t, p^t) \to (y, p)$ in $H^3(\Omega) \times H^3(\Omega)$.

For the verification of hypothesis H4, we go back to expression (3.25) which can be rewritten as a volume integral

$$\partial_t G(\Omega_t, \Phi, \Psi) = \int_{\Omega_t} \operatorname{div}\{[\frac{1}{2}(\Phi - y_d)^2 + (\Delta\Phi + f)\Psi + (\Phi - g)\Delta\Psi + \nabla(\Phi - g) \bullet \nabla\Psi]V\} \, dx \tag{44}$$

for $(\Phi, \Psi) \in H^3(\mathbf{R}^N) \times H^3(\mathbf{R}^N)$. Now introduce the map

$$\begin{cases} (\Phi, \Psi) \mapsto F(\Phi, \Psi) = [\frac{1}{2}(\Phi - y_d)^2 + (\Delta\Phi + f)\Psi + (\Phi - g)\Delta\Psi + \nabla(\Phi - g) \bullet \nabla\Psi]V \\ : H^3(\mathbf{R}^N) \times H^3(\mathbf{R}^N) \to (H^1(\mathbf{R}^N))^N. \end{cases}$$

It is bilinear and continuous. Finally the map

$$(t, F) \mapsto \int_{\Gamma_t} F \circ n_t \, d\Gamma = \int_{\Omega_t} F \, dx = \int_{\Omega} (divF) \circ T_t \, J_t^{-1} \, dx \quad : [0, \tau] \times H^1(\mathbf{R}^N) \to \mathbf{R} \tag{45}$$

is continuous. Then

$$(t, \Phi, \Psi) \mapsto \partial_t G(\Omega_t, \Phi, \Psi) = \int_{\Gamma_t} F(\Psi, \Psi) \bullet n_t \, d\Gamma_t \tag{46}$$

is continuous and hypothesis (H4) is verified. This completes the verification of the hypotheses.

5. SHAPE HESSIAN FOR THE DIRICHLET PROBLEM.

5.1. Formulation and formal computations. We proceed as in sections 3 and 4 and provide the mathematical justification in section 5.2. For the second derivative, we need two time invariant vector fields V and W on $\mathbf{R}^N$ and the expression of the first derivative $dJ(\Omega_t(W); V)$ where $\Omega_t(W)$ is the perturbation of the domain Ω by the vector field W:

$$dJ(\Omega_t(W); V) = \int_{\Omega_t(W)} \operatorname{div}\{[\frac{1}{2}(g - y_d)^2 + \nabla(y_t - g) \bullet \nabla p_t]V\} \, dx, \tag{1}$$

where (y_t, p_t) are the unique solutions in $H^3(\Omega_t(W)) \times H^3(\Omega_t(W))$ to the equations

$$\Delta y_t + f = 0 \text{ in } \Omega_t(W), \quad y_t = g \text{ on } \Gamma_t(W), \tag{2}$$

$$\Delta p_t + (y_t - y_d) = 0 \text{ in } \Omega_t(W), \quad p_t = 0 \text{ on } \Gamma_t(W). \tag{3}$$

Then, we express (1) as a MinMax over a new Lagrangian:

$$dJ(\Omega_t(W); V) = \underset{\Phi,\Psi \in H^3(\mathbf{R}^N)}{\text{Min}} \quad \underset{P,\Sigma \in H^2(\mathbf{R}^N)}{\text{Max}} G(\Omega_t, \Phi, \Psi, P, \Sigma), \tag{4}$$

where $G = G(\Omega_t, \Phi, \Psi, P, \Sigma)$ is given by

$$\begin{aligned} G = \int_{\Omega_t} &\{\text{div}[\frac{1}{2}(g - y_d)^2 + \nabla(\Phi - g) \bullet \nabla\Psi]V\} \\ &+ [\Delta\Phi + f]P + (\Phi - g)\Delta P + \nabla(\Phi - g) \bullet \nabla P \\ &+ [\Delta\Psi + \Phi - y_d]\Sigma + \Psi\Delta\Sigma + \nabla\Psi \bullet \nabla\Sigma\} \, dx. \end{aligned} \tag{5}$$

This new Lagrangian is affine in (P, Σ), but is not necessarly convex in (Φ, Ψ). However, it is semiconvex in (Φ, Ψ) and we shall see that CORREA-SEEGER [1] will still apply to our special Lagrangian where the sets $X(t) \times Y(t)$,

$$X(t) \subset H^3(\mathbf{R}^N) \times H^3(\mathbf{R}^N) \tag{6}$$

$$Y(t) \subset H^2(\mathbf{R}^N) \times H^2(\mathbf{R}^N) \tag{7}$$

will be given by the usual "saddle point equations":

$$\int_{\Omega_t} [\Delta\hat{\Phi} + f]P + (\hat{\Phi} - g)\Delta P + \nabla(\hat{\Phi} - g) \bullet \nabla P \, dx = 0, \ \forall P, \tag{8}$$

$$\int_{\Omega_t} [\Delta\hat{\Psi} + \hat{\Phi} - y_d]\Sigma + \hat{\Psi}\Delta\Sigma + \nabla\hat{\Psi} \bullet \nabla\Sigma \, dx = 0, \ \forall\Sigma, \tag{9}$$

$$\int_{\Omega_t} \text{div}\{[\nabla(\hat{\Phi} - g) \bullet \nabla\Psi]V\} + \Delta\Psi\hat{\Sigma} + \Psi\Delta\hat{\Sigma} + \nabla\Psi \bullet \nabla\hat{\Sigma} \, dx = 0, \ \forall\Psi, \tag{10}$$

$$\int_{\Omega_t} \text{div}\{[\nabla\Phi \bullet \nabla\hat{\Psi}]V\} + \Delta\Phi\hat{P} + \Phi\Delta\hat{P} + \nabla\Phi \bullet \nabla\hat{P} + \Phi\hat{\Sigma} \, dx = 0, \forall\Phi. \tag{11}$$

It is obvious that (8) and (9) yield

$$\hat{\Phi}|_{\Omega_t} = y_t \text{ and } \hat{\Psi}|_{\Omega_t} = p_t. \tag{12}$$

Similarly, equations (10) and (11) have solutions $(\hat{\Sigma}, \hat{P})$ in $H^2(\mathbf{R}^N) \times H^2(\mathbf{R}^N)$ such that

$$Y_t' = \hat{\Sigma}|_{\Omega_t}, \quad P_t' = \hat{P}|_{\Omega_t} \tag{13}$$

are unique in $H^2(\Omega_t) \times H^2(\Omega_t)$ and solution of

$$\Delta Y_t' = 0 \text{ in } \Omega_t(W), \quad Y_t' = -\frac{\partial}{\partial n_t}(y_t - g)V \bullet n_t \quad \text{on } \Gamma_t(W), \tag{14}$$

$$\Delta P_t' = 0 \text{ in } \Omega_t(W), \quad P_t' = -\frac{\partial p_t}{\partial n_t} V \bullet n_t \quad \text{on } \Gamma_t(W). \tag{15}$$

It can be shown that Y_t' and P_t' coincide with the "partial derivative" with respect to t of appropriate extensions of y_t and p_t from $\Omega_t(W)$ to $\mathbf{R}^N$.

Finally, the partial derivative of the Lagrangian G with respect to t is given by

$$\begin{aligned} \partial_t G = \int_{\Gamma_t} \{ & \text{div}\{[\frac{1}{2}(g - y_d)^2 + \nabla(\Phi - g) \bullet \nabla\Psi]V\} \\ & + [\Delta\Phi + f]P + (\Phi - g)\Delta P + \nabla(\Phi - g) \bullet \nabla P \\ & + [\Delta\Psi + \Phi - y_d]\Sigma + \Psi\Delta\Sigma + \nabla\Psi \bullet \nabla\Sigma\} W \bullet n_t \, d\Gamma_t \end{aligned} \tag{16}$$

for Φ, Ψ, P, Σ in $H^3(\mathbf{R}^N)$, y_d and f in $H^2(\mathbf{R}^N)$ and g in $H^{\frac{7}{2}}(\mathbf{R}^N)$. The immediate consequence of this computation is that y_t, p_t, Y_t', P_t' all belong to $H^3(\Omega_t)$. But, Y_t', P_t' in $H^3(\Omega_t)$ require that y_t and p_t belong to $H^4(\Omega_t)$. This is precisely why we chose the appppropriate smoothness of y_d, f and g.

Therefore, we must choose our saddle points $X(t) \times Y(t)$ in $(H^4(\mathbf{R}^N) \times H^4(\mathbf{R}^N)) \times (H^3(\mathbf{R}^N) \times H^3(\mathbf{R}^N))$,

$$X(t) = \{(\Phi, \Psi) \in H^4(\mathbf{R}^N) \times H^4(\mathbf{R}^N) : \Phi|_{\Omega_t} = y_t, \Psi|_{\Omega_t} = p_t\} \tag{17}$$

$$Y(t) = \{(P, \Sigma) \in H^3(\mathbf{R}^N) \times H^3(\mathbf{R}^N) : P|_{\Omega_t} = P_t', \Sigma|_{\Omega_t} = Y_t'\}. \tag{18}$$

Finally, since $\partial_t G$ is a functional on Ω_t, it will only use the restriction to Ω_t of the various functions in $X(t) \times Y(t)$. Therefore, the Min and the Max can be removed and

$$\begin{aligned} d^2 J(\Omega; V; W) = \int_{\Gamma} \{ & \text{div}[\frac{1}{2}(y - y_d)^2 + \frac{\partial}{\partial n}(y - g)\frac{\partial p}{\partial n}]V\} \\ & + [\Delta y + f]P_V' + (y - g)\Delta P_V' + \nabla(y - g) \bullet \nabla P_V' \\ & + [\Delta p + y - y_d]Y_V' + p\Delta Y_V' + \nabla p \bullet \nabla Y_V'\} W \bullet n \, d\Gamma. \end{aligned} \tag{19}$$

But,

$$\Delta y + f = 0, \; y = g, \; \Delta p + y - y_d = 0 \; \text{ and } \; p = 0 \text{ on } \Gamma, \tag{20}$$

and

$$d^2 J(\Omega; V; W) = \int_{\Gamma} \{ \text{ div } \{[\frac{1}{2}(g - y_d)^2 + \frac{\partial}{\partial n}(y - g)\frac{\partial p}{\partial n}]V\} + \frac{\partial}{\partial n}(y - g)\frac{\partial P_V'}{\partial n} + \frac{\partial p}{\partial n}\frac{\partial Y_V'}{\partial n}\} W \bullet n \, d\Gamma, \tag{21}$$

where we have added the subscript V to P' and Y' to emphasize the fact that they both depend on V.

The last step consists in the elimination of P'_V which will introduce Y'_W. To do that, we set $\psi = \hat{\Psi}_V|_{\Omega_t} = P'_V$ in equation (10) with $V = W$ and $t = 0$

$$\int_\Omega \text{div } \{[\nabla(y-g) \bullet \nabla\psi]W\} + \Delta\psi Y'_W + \psi\Delta Y'_W + \nabla\psi \bullet \nabla Y'_W \, dx = 0$$

$$\Rightarrow \int_\Omega \nabla(y-g) \bullet \nabla P'_V W \bullet n d\Gamma + \int_\Omega \Delta P'_V Y'_W + P'_V \Delta Y'_W + \nabla P'_V \bullet \nabla Y'_W \, dx = 0 \tag{22}$$

and $\phi = \hat{\Phi}_W|_{\Omega_t} = Y'_W$ in equation (11) with $t = 0$

$$\int_\Omega \text{div } \{[\nabla\Phi \bullet \nabla p]V\} + \Delta\Phi P'_V + \Phi\Delta P'_V + \nabla\Phi \bullet \nabla P'_V + \Phi Y'_V dx = 0$$

$$\Rightarrow \int_\Gamma \nabla Y'_W \bullet \nabla p \; V \bullet n \; d\Gamma + \int_\Gamma \Delta Y'_W P'_V + Y'_W \Delta P'_V + \nabla Y'_W \bullet \nabla P'_V + Y'_W Y'_V \, dx = 0 \tag{23}$$

This yields the following identity

$$\int_\Gamma \nabla(y-g) \bullet \nabla P'_V \; W \bullet n \; d\Gamma = \int_\Gamma \nabla Y'_W \bullet \nabla p \; V \bullet n \; d\Gamma + \int_\Gamma Y'_W Y'_V \, dx \tag{24}$$

or

$$\int_\Gamma \frac{\partial}{\partial n}(y-g)\frac{\partial P'_V}{\partial n} W \bullet n \; d\Gamma = \int_\Gamma \frac{\partial Y'_W}{\partial n}\frac{\partial p}{\partial n} V \bullet n \; d\Gamma + \int_\Gamma Y'_W Y'_V \, dx. \tag{25}$$

As a result

$$\begin{aligned} d^2J(\Omega; V; W) = & \int_\Gamma \{\text{div } [\tfrac{1}{2}(g-y_d)^2 + \frac{\partial}{\partial n}(y-g)\frac{\partial p}{\partial n}]V\} W \bullet n \; d\Gamma \\ & + \int_\Gamma \frac{\partial p}{\partial n} [\frac{\partial Y'_W}{\partial n} V \bullet n + \frac{\partial Y'_V}{\partial n} W \bullet n] \; d\Gamma + \int_\Gamma Y'_W Y'_V \, dx \end{aligned} \tag{26}$$

where Y'_V is the unique solution of

$$\Delta Y'_V = 0 \text{ in } \Omega, \quad Y'_V = -\frac{\partial}{\partial n}(y-g)V \bullet n \text{ on } \Gamma. \tag{27}$$

5.2. Verification of the hypothese. In section 5.1, we have boldly applied the conclusion of the theorem of Correa and Seeger to a Lagrangian which contains a cost functional which is not necessarily convex. This means that the corresponding Lagrangian functional does not necessarily have saddle points. Yet, the conclusions of the theorem extend to semiconvex cost functionals (section 5.2.1). The verification of the hypotheses will essentially be the same as for the gradient in section 4.2 (section 5.5.2).

5.2.1. Semiconvex cost functionals. Consider a Lagrangian functional of the form

$$G(t,x,y) = F(t,x) + b(t,x,y) \tag{28}$$

for a family of continuous bilinear forms $b(t,x,y)$ on $X \times Y$ and continuous cost functionals $F(t,x)$ on X. Formally, the saddle points equations are given by

$$x^t \in X, \ b(t,x^t,y) = 0, \ \forall y \in Y \tag{29}$$

$$y^t \in Y, \ dF(t,x^t;x) + b(t,x,y^t) = 0, \ \forall x \in X. \tag{30}$$

When $G(t,x,y)$ is convex in x and concave in y, (29)-(30) characterize the saddle points $X(t)\times Y(t) \subset X \times Y$ of $G(t,\cdot,\cdot)$. So, when $F(t,x)$ is not convex in x, equations (29)-(30) need not characterize saddle points of $G(t,\cdot,\cdot)$.

We say that the functional $F(t,x)$ is semiconvex in x if there exists a family of continuous convex functionals $C(t,x)$ on X such that $F(t,x) + C(t,x)$ is convex in x. This means that $F(t,\cdot) + C(t,\cdot)$ and $C(t,\cdot)$ both have directional derivatives and hence $F(t,\cdot)$ also has a directional derivative: the following limit exists

$$dF(t,x;x') = \lim_{\theta \searrow 0} \frac{F(t,x+\theta x') - F(t,x)}{\theta} \tag{31}$$

Denote by $X(t)$ the set of all solutions

$$x^t \in X, \quad b(t,x^t,y) = 0, \quad \forall y \in Y \tag{32}$$

and assume that

$$\forall x^t \in X(t), \quad F(t,x^t) = J(t), \tag{33}$$

that is $F(t,x^t)$ is only a function of t. We use $J(t)$ as the definition of our cost function.

Now, assume that $F(t,\cdot)$ is semiconvex and that

$$\forall x^t \in X(t), \ C(t,x^t) = J_0(t), \tag{34}$$

that is $C(t,x^t)$ is only a function of t. We, again, use $J_0(t)$ as the definition of the cost function associated with C. Finally, let

$$J_C(t) = F(t,x^t) + C(t,x^t) \tag{35}$$

which is also only a function of t. Then, it is obvious that

$$J(t) = J_C(t) - J_0(t) \tag{36}$$

and that (if $dJ_C(0)$ and $dJ_0(0)$ exist)

$$dJ(0) = dJ_C(0) - dJ_0(0). \tag{37}$$

So, we are back to the use of CORREA-SEEGER [1] for both J_C and J_0. Construct the Lagrangians

$$G_C(t,x,y) = F(t,x) + C(t,x) + b(t,x,y) \tag{38}$$

$$G_0(t,x,y) = C(t,x) + b(t,x,y) \tag{39}$$

and assume that if all the hypotheses are verified both for G_0 and G_C

$$J_0(t) = \min_{x\in X} \max_{y\in Y} G_0(t,x,y) \quad dJ_0(0) = \min_{x\in X(0)} \max_{y\in Y_0(0)} \partial_t G_0(0,x,y) \tag{40}$$

$$J_C(t) = \min_{x\in X} \max_{y\in Y} G_C(t,x,y) \quad dJ_C(0) = \min_{x\in X(0)} \max_{y\in Y_C(0)} \partial_t G_C(0,x,y), \tag{41}$$

where the saddle point equations for J_0 are given by

$$x^t \in X,\ B(t,x^t,y) = 0,\ \forall y, \tag{42}$$

$$y_0^t \in Y,\ dC(t,x^t;x) + b(0,x,y_0^t) = 0,\ \forall x, \tag{43}$$

and the saddle point equations for $J_C(t)$

$$x^t \in X,\ B(t,x^t,y) = 0,\ \forall y, \tag{44}$$

$$y_C^t \in Y,\ dC(t,x^t;x) + dF(t,x^t;x) + b(0,x,y_C^t) = 0,\ \forall x. \tag{45}$$

Assume that

$$dJ_0(0) = \partial_t C(0,x^0) + \partial_t b(0,x^0,y_0^0),\ \forall x^0 \in X(0),\ \forall y_0^0 \in Y_0(0), \tag{46}$$

and that $dC(0,x^0;x)$ and $dF(0,x^0;x)$ are independent of the point x^0 chosen in $X(0)$. By substracting (43) from (45), construct the new variable $y^0 = y_C^0 - y_0^0$, which is a solution of

$$y^0 \in Y,\ dF(0,x^0;x) + b(0,x,y^0) = 0,\ \forall x, \tag{47}$$

and the set

$$Y(0) = \{y_C^0 - y_0^0 : y_C^0 \in Y_C(0),\ y_0^0 \in Y_0(0)\}.$$

Now,

$$\begin{aligned} dJ_C(0) &= \operatorname*{Min}_{x\in X(0)} \operatorname*{Max}_{y\in Y_C(0)} \partial_t F(0,x) + \partial_t C(0,x) + \partial_t b(0,x,y) \\ &= \operatorname*{Min}_{x\in X(0)} \operatorname*{Max}_{\substack{y_0^0\in B_0(0)\\ y_0\in Y(0)}} \{\partial_t F(0,x) + \partial_t b(0,x,y^0) + \partial_t C(0,x) + \partial_t b(0,x,y_0^0)\} \end{aligned}$$

But $\forall\ (x^0,y_0^0) \in X(0) \times Y_0(0)$

$$\partial_t C(0,x^0) + \partial_t b(0,x^0,y_0^0) = dJ_0(0),$$

and finally,

$$dJ(0) = dJ_C(0) - dJ_0(0) = \operatorname*{Min}_{x\in X(0)} \operatorname*{Max}_{y\in Y(0)} [\partial_t F(0,x) + \partial_t b(0,x,y)]. \tag{48}$$

where the saddle point $(x^0, y^0) \in X(0) \times Y(0)$ are the solution of the "formal saddle point equations" (29)-(30) for $t = 0$.

In section 4.2, the cost functional is semiconvex since there exists a constant $C > 0$ large enough such that

$$dJ(\Omega_t(W); V) + C[||y_t||^2_{H^4(\Omega_t)} + ||p_t||^2_{H^4(\Omega_t)}] \tag{49}$$

is convex and continuous on $H^4(\Omega_t) \times H^4(\Omega_t)$. The functional

$$C(t, \phi, \psi) = c||\phi||^2_{H^4(\Omega_t)} + ||\psi||^2_{H^4(\Omega_t)} \tag{50}$$

is clearly convex and continuous on $H^4(\Omega_t) \times H^4(\Omega_t)$. This provides a complete justification to the use of the conclusions of Correa and Seeger.

5.2.2. Verification of the hypotheses. We have chosen to work in $H^4(\mathbf{R}^N) \times H^4(\mathbf{R}^N) \times H^3(\mathbf{R}^N) \times H^3(\mathbf{R}^N)$ and introduced appropriate hypotheses on f, y_d and g in section 5.1. From this point on the technique is the same as the one in section 4.2 for the gradient. Therefore, we shall not repeat it here.

REFERENCES

AGMON S., DOUGLIS A., AND NIRENBERG L. [1], *Estimates near the boundary for solutions of elliptic partial differential equations satisfying general boundary conditions, I.*, Comm. Pure Appl. Math. **12** (1959), 623-727.

[2], *Estimates near the boundary for solutions of elliptic partial differential equations satisfying general boundary conditions, II.*, Comm. Pure Appl. Math. **17** (1964), 35-92.

ARUMUGAM G. AND PIRONNEAU O. [1], *On the problems of riblets as a drag reduction device*, Optimal Control Applications and Methods **10** (1989), 93-112.

[2], "Sur le problème des "riblets", Rapport de recherche R87027," Publications du Laboratoire d'analyse numérique, Université Pierre et Marie Curie, Paris, France, 1987.

BABIĆ V.M. [1], *Sur le prolongement des fonctions (in Russian)*, Uspechi Mat. Nauk **8** (1953), 111 - 113.

BERN A. [1], "Thèse de l'École Nationale Supérieure des Mines de Paris," CEMEF, Sophia Antipolis, France, October 1987.

BERN A., CHENOT J.L., DEMAY Y. AND ZOLÉSIO J.P. [1], *Numerical computation of the free boundary in non-Newtonian stationary flows*, Proc. Sixth Int. Symp. on Finite Element Methods in Flow Problems (June 1986), 383-390, Publications INRIA, Rocquencourt, France.

CANNARSA P. AND SONER H.M. [1], *On the singularities of the viscosity solutions to Hamilton - Jacobi - Bellman Equations*, Indiana Univ. Math. J. **36** (1987), 501-524.

CÉA J. [1], *Problems of Shape Optimal Design*, in "Optimization of Distributed Parameter Structures, vol II," E.J. Haug and J. Céa, eds., Sijhoff and Noordhoff, Alphen aan den Rijn, The Netherlands, 1981, pp. 1005-1048.

[2], *Numerical Methods of Shape Optimal Design*, in "Optimization of Distributed Parameter Structures, vol II," E.J. Haug and J. Céa, eds., Sijhoff and Noordhoff, Alphen aan den Rijn, The Netherlands, 1981, pp. 1049-1087.

[3], *Conception optimale ou identification de formes: calcul rapide de la dérivée directionnelle de la fonction coût*, Mathematical Modelling and Numerical Analysis(Modélisation mathématique et analyse numérique **20** (1986), 371-402.

CORREA R. AND SEEGER A. [1], *Directional derivatives of a minimax function*, Nonlinear Analysis, Theory, Methods and Applications **9** (1985), 13-22.

DELFOUR M.C., PAYRE G. AND ZOLÉSIO J.-P. [1], *Shape Optimal Design of a Radiating Fin*, in "System Modelling and Optimization," P. Thoft-Christensen, ed., Springer-Verlag, Berlin, Heidelberg, 1984, pp. 810-818.

[2], *An optimal triangulation for second order elliptic problems*, in "Computer Methods in Applied Mechanics and Engineering,vol 50," 1985, pp. 231-261.

DELFOUR M.C. AND ZOLÉSIO J.P. [1], *Dérivation d'un MinMax et application à la dérivation par rapport au contrôle d'une observation non-différentiable de l'état*, C.R. Acad. Sc. Paris, **t.302, Sér. I, no. 16** (1986), 571-574.

[2], *Shape Sensitivity Analysis via MinMax Differentiability*, SIAM J. on Control and Optimization **26** (1988), 834-862.

[3], *Differentiability of a MinMax and Application to Optimal Control and Design Problems, Part I*, in "Control Problems for Systems Described as Partial Differential Equations and Applications," I. Lasiecka and R. Triggiani, eds., Springer-Verlag, New York, 1987, pp. 204–219.

[4], *Differentiability of a MinMax and Application to Optimal Control and Design Problems, Part II*, in "Control Problems for Systems Described as Partial Differential Equations and Applications," I. Lasiecka and R. Triggiani, eds., Springer-Verlag, New York, 1987, pp. 220-229.

[5], *Further Developments in Shape Sensitivity Analysis via a Penalization Method*, in "Boundary Control and Boundary Variations," J. P. Zolésio, ed., Springer-Verlag, Berlin, Heidelberg, New York, Tokyo, pp. 153-191.

[6], *Shape Sensitivity Analysis via a Penalization Method*, Annali di Matematica Pura ed Applicata CLI (1988), 179-212.

[7], *Analyse des problèmes de forme par la dérivation des Min Max*, in "Analyse Non Linéaire," H. Attouch, J.P. Aubin, F.H. Clarke and I. Ekeland, eds, Série Analyse Non Linéaire, Annales de l'Institut Henri-Poincaré, Special volume in honor of J.-J. Moreau, Gauthier-Villars, Bordas, Paris, France, 1989, pp. 211-228.

[8], *Anatomy of the shape Hessian*, Annali di Matematica Pura et Applicata (to appear).

[9], *Computation of the shape Hessian by a Lagrangian method*, in "Fifth Symp. on Control of Distributed Parameter Systems," A. El Jai and M. Amouroux, eds., Pergamon Press, to appear, pp. 85–90.

FUJII N. [1], *Domain optimization problems with a boundary value problem as a constraint*, in "Control of Distributed Parameter Systems 1986," Pergamon Press, Oxford, New York, 1986, pp. 5-9.

[2], *Second variation and its application in a domain optimization problem*, in "Control of Distributed Parameter Systems 1986," Pergamon Press, Oxford, New York, 1986, pp. 431-436.

GUELFAND M. AND VILENKIN N.Y. [1], "Les distributions, Applications de l'analyse harmonique(trad. par G. Rideau)," Dunod, Paris, 1967.

HADAMARD J. [1], *Mémoire sur le problème d'analyse relatif à l'équilibre des plaques élastiques encastrées*, in "Oeuvres de J. Hadamard,vol II," (original reference: Mem. Sav. Etrang. 33 (1907), mémoire couronné par l'Académie des Sciences)., C.N.R.S., Paris, 1968, pp. 515-641.

NEČAS J. [1], "Les méthodes directes en théorie des équations elliptiques," Masson (Paris) et Academia (Pragues), 1967.

PIRONNEAU O. [1], "Optimal Design for Elliptic Systems," Springer-Verlag, New York, 1984.

SCHWARTZ, L. [1] "Théorie des distributions," Hermann, Paris, 1966.

[2], *Théorie des noyaux*, in "Proceedings of the International Congress of Mathematicians, Vol 1," 1950, pp. 220-230.

"Théorie des noyaux, Proceedings of the International Congress of Mathematicians," 1950, pp. 220-230.

SIMON, J. [1], *Second variations for domain optimization problems*, "Control of Distributed Parameter Systems (Proc. 4th Int. Conf. in Vorau)," Birkhauser Verlag, July 1988 (to appear).

ZOLÉSIO J. P. [1], "Identification de domaines par déformation, Thèse de doctorat d'état," Université de Nice, France, 1979.

[2], *The Material Derivative (or Speed) Method for Shape Optimization*, in "Optimization of Distributed Parameter Structures, vol II," E.J. Haug and J. Céa, eds., Sijhoff and Nordhoff, Alphen aan den Rijn, 1981, pp. 1089-1151.

SHAPE SENSITIVITY ANALYSIS OF

HYPERBOLIC PROBLEMS

Jan SOKOLOWSKY - Jean-Paul ZOLESIO

Systems Research Institute
Polish Academy of Sciences
ul. Newelsha 6
01-447 WARSAW - POLAND

CNRS, Laboratoire
de Physique Mathématique
Place Eugène Bataillon
34060 MONTPELLIER - FRANCE

ABSTRACT.

We provide the new results on the shape sensitivity analysis of the wave equation as well as of the Maxwell's equations in bounded domains. The form of shape derivative as well as of the domain derivative is derided for the hyperbolic equations.

1. INTRODUCTION

In the present paper we use the material derivative method for the shape sensitivity analysis of hyperbolic problems. First we consider the wave equation, using the standard energy estimates. We obtain the form of material as well as shape derivatives for the scalar hyperbolic equation. Finally we derive the results on the shape sensitivity analysis for the system of Maxwell's equation. We refer the reader to [3] for the related results on shape sensitivity analysis of PDE's.

Notation

We use the standard notation throughout the paper. Let us recall the transformation $T_s(V) : R^N \rightarrow R^N$ associated to a given vector field

$$V(.,.) \in C\left(0, \epsilon ; \mathcal{D}^k(R^N;R^N)\right) ,$$

is defined as follows.

For a given element $X \in R^N$, $T_s(V)(X) = X(s)$, where

$$\begin{cases} \dfrac{dx}{dt} = V(t, x(t)) , & t \in (0,\epsilon) \\ x(0) = x \end{cases}$$

2. SHAPE SENSITIVITY ANALYSIS OF WAVE EQUATION

Let $\Omega \subset R^N$ be a given domain of class C^k, k integer, $k \geq 1$, denote $I = (0,T)$, $T > 0$, $Q = I \times \Omega \subset R^{N+1}$. For a given vector field $V \in C\left(0,\epsilon ; \mathcal{D}^k(R^N;R^N)\right)$, $T_s(V)$ denotes the associated transformation. We denote by t the time variable, $t \in I$, while s is the transformation parameter, $s \in [0,\epsilon)$.

Denote $\Omega_s = T_s(V)(\Omega)$, $s \in [0,\epsilon)$ and let $Q_s = I \times \Omega_s \subset R^{N+1}$.

For any s, $0 \leq s < \epsilon$ we are concerned with the partial differential equation of the hyperbolic type, i.e. the wave equation defined in the cylinder Q_s.

Let $f \in L^1\left(I, L^2(R^N)\right)$ be given, denote by H the wave operator,

$$H\varphi \equiv -\Delta\varphi + \frac{\partial^2 \varphi}{\partial t^2} \tag{2.1}$$

let y be the solution of the following mixed hyperbolic problem :

$$Hy = f \quad \text{in} \quad Q_s \tag{2.2}$$

$$y(0) = \partial_t\, y(0) = 0 \quad \text{in} \quad \Omega_s \tag{2.3}$$

$$y = 0 \quad \text{on} \quad \Sigma_s \tag{2.4}$$

where $\Sigma_s = I \times \Gamma_s$ is the lateral boundary of Q_s, and $\Gamma_s = \partial\Omega_s$.

We recall the classical a priori estimate for the solution of (2.2)-(2.4).

<u>Proposition 2.1</u>

For any φ in $C^{\infty}\left(\bar{Q}_s\right)$ such that $\varphi\big|_{\Sigma_s} = 0$ we have

$$\|\varphi\|_{L^{\infty}\left(I,H^1_0(\Omega_s)\right)} \leqslant \|H.\varphi\|_{L^1\left(I,L^2(\Omega_s)\right)} \tag{2.5}$$

and

$$\left\|\frac{\partial\varphi}{\partial t}\right\|_{L^{\infty}\left(I,L^2(\Omega_s)\right)} \leqslant C\ \|H\varphi\|_{L^1\left(I,L^2(\Omega_s)\right)} \tag{2.6}$$

Proof.

Define the energy term

$$E(t) = \frac{1}{2}\int_{\Omega_s}\left(|\nabla\varphi|^2 + \left(\frac{\partial\varphi}{\partial t}\right)^2\right)dx$$

Since $\varphi = 0$ on Γ_s, using the Green formula we get

$$E'(t) = \int_{\Omega_s} H\varphi\ \frac{\partial\varphi}{\partial t}\ dx \leqslant \|H\varphi\|_{L^2(\Omega_s)} \left\|\frac{\partial\varphi}{\partial t}\right\|_{L^2(\Omega_t)} .$$

Since $E(0) = 0$, it follows that for any $t \in I$,

$$E(t) \leqslant \int_0^t \|H\varphi\|_{L^2(\Omega_s)} \left\|\frac{\partial\varphi}{\partial t}\right\|_{L^2(\Omega_s)} dt \leqslant \int_0^T \|H\varphi\|_{L^2(\Omega_s)} \left\|\frac{\partial\varphi}{\partial t}\right\|_{L^2(\Omega_s)} dt$$

$$\leqslant \|H\varphi\|_{L^1\left(I,L^2(\Omega_s)\right)} \|\partial_t\varphi\|_{L^{\infty}\left(I,L^2(\Omega_s)\right)}$$

hence for any $t \in I$

$$\frac{1}{2}\int_{\Omega_s}\left(\frac{\partial}{\partial t}y\right)^2 dx \leqslant E(t) \leqslant \|H\varphi\|_{L^1\left(I,L^2(\Omega_s)\right)} \|\partial_t\varphi\|_{L^\infty\left(I,L^2(\Omega_s)\right)}$$

which leads to (2.6). We have also for any $t \in I$,

$$\frac{1}{2}\int_{\Omega_s} |\nabla y|^2 \, dx \leqslant \left\|\frac{\partial}{\partial t}\varphi\right\|_{L^\infty\left(I,L^2(\Omega_s)\right)} \|H\varphi\|_{L^1\left(I,L^2(\Omega_s)\right)}$$

hence (2.5) follows. ∎

Let us consider the problem (2.2)-(2.4) defined in the cylinder Q_s. We denote by

$$y_s \in L^\infty\left(I\ ;\ H_0^1(\Omega_s)\right)$$

the solution of such a problem.

<u>Proposition 2.2</u>

Let $f \in L^1\left(I,L^2_{loc}(R^N)\right)$ be a given element.

For any bounded domain of class C^k, $k \geqslant 1$, and for any $s \in [0,\epsilon)$ there exists a unique solution y of the problem (2.2)-(2.4) such that

$$y_s \in L^\infty\left(I,\ H_0^1(\Omega_s)\right) \tag{2.7}$$

$$\frac{\partial}{\partial t}y_s \in L^\infty\left(I,L^2(\Omega_s)\right) \tag{2.8}$$

Proof.

We use the Galerkine technique and a priori estimates (2.5), (2.6), hence (2.8) follows in the classical way. ∎

Proposition 2.3

Let y_s be the solution of problem (2.2)-(2.4), then the following estimate holds :

$$\|y_s\|_{L^\infty\left(I,H_0^1(\Omega_s)\right)} + \left\|\frac{\partial}{\partial t} y_s\right\|_{L^\infty\left(I,L^2(\Omega_s)\right)} \leq 2\|Hy_s\|_{L^1\left(I,L^2(\Omega_s)\right)} \quad \blacksquare \quad (2.9)$$

Denote by $y^s = y_s \circ T_s$ the element transported to the fixed domain Ω, defined by

$$y^s(t,x) = y_s(t, T^s(x))$$

which satisfies, for all s, $0 \leq s < \epsilon$,

$$y^s \in L^\infty\left(I, H_0^1(\Omega)\right)$$

$$\frac{\partial}{\partial t} y^s \in L^\infty(I,L^2(\Omega)) \quad . \qquad (2.10)$$

The transported wave operator H_s is defined on Q as follows :

$$(H.\varphi) \circ T_s = H_s .(\varphi \circ T_s) \qquad (2.11)$$

Obviously we have

$$\left(\frac{\partial^2}{\partial t^2} \varphi\right) \circ T_s = \frac{\partial^2}{\partial t^2} (\varphi \circ T_s) \quad .$$

Since the variables t and s are independent. Let us recall that

$$(\Delta\varphi) \circ T_s = \gamma(s)^{-1} \operatorname{div}(A(s) \, . \, \nabla(\varphi \circ T_s))$$

where

$$A(s) = \gamma(s)\ DT_s^{-1} \ .\ {}^*DT_s^{-1} \quad ,$$

thus

$$H_s \varphi = -\gamma(s)^{-1} \text{ div } (A(s).\nabla\varphi) + \frac{\partial^2}{\partial t^2}\varphi \quad . \tag{2.12}$$

By the change of variables $x = T_s(x)$ in the estimate (2.9) we obtain the following estimate for the transported solution y^s.

<u>Lemma 2.4</u>

There exists a constant $\alpha > 0$ such that for any s, $0 \leqslant s < \epsilon$,

$$\alpha \|y^s\|_{L^\infty\left(I,H_0^1(\Omega)\right)} + \alpha \left\|\frac{\partial}{\partial t} y^s\right\|_{L^\infty(I,L^2(\Omega))}$$

$$\leqslant 2 \int_0^T \left(\int_\Omega \left[-\frac{1}{\sqrt{\gamma(s)}} \text{ div } (A(s).\nabla y^s) + \sqrt{\gamma(s)}\ \frac{\partial^2}{\partial t^2} y^s\right]^2 dx\right)^{1/2} dt$$

$$= 2 \left\|\sqrt{\gamma(s)}\ H_s\ y^s\right\|_{L^1(I,L^2(\Omega))} \tag{2.13}$$

Proof.

(2.13) follows by the change of variables $x = T_s(x)$ in (2.9), α is a given constant such that

$$A(s,x) \geqslant \alpha\, I$$
$$\gamma(s,x) \geqslant \alpha$$

for all $s \in [0,\epsilon)$, $x \in \bar{\Omega}$. ∎

Denote $z^s = y^s - y$, in view of (2.13) we get :

$$\alpha \|z^s\|_{L^\infty\left(I,H^1_0(\Omega)\right)} + \alpha \left\|\frac{\partial}{\partial t} z^s\right\|_{L^\infty(I,L^2(\Omega))} \leqslant 2 \|\gamma(s)\|^{1/2}_{L^\infty(\Omega)} \left\|H_s . z^s\right\|_{L^1(I;L^2(\Omega))}$$

On the other hand

$$\begin{aligned} H_s . z^s &= H_s y^s - H_s y \\ &= (Hy_s) \circ T_s - H_s y \\ &= (f \circ T_s - f) + (H - H_s) y \quad . \end{aligned} \tag{2.14}$$

In order to estimate the norm of the element $\frac{1}{s} z^s$ (when s goes to zero) for $s > 0$, it is sufficient to consider the last two terms.

Proposition 2.5

Let $\varphi \in L^1(I,H^2(\Omega))$ such that $\frac{\partial}{\partial t} \varphi \in L^1(I,L^2(\Omega))$ be given. Then the mapping $s \to H_s . \varphi$ is differentiable in the norm of the space $L^1(I,L^2(\Omega))$, the derivative at $s = 0$ is given by

$$\left(\frac{d}{ds} H_s . \varphi\right)\Big|_{s=0} = \operatorname{div} V(0) \, \nabla\varphi - \operatorname{div} \left(A'(0) \, \nabla\varphi\right) \tag{2.15}$$

where

$$A'(0) = \operatorname{div} V(0) I - 2\epsilon(V(0))$$

$$\epsilon(V(0)) = \frac{1}{2} \left(DV(0) + {}^*DV(0)\right)$$

Proof.

Under our assumptions, it follows that the mapping $s \to \gamma(s)$ and $s \to A(s)$ are differentiable in $C^{k-1}(\bar{\Omega})$ and $C^{k-1}(\bar{\Omega};R^N)$, respectively, then the required result follows from (2.12). ∎

Proposition 2.6

Suppose that $f \in L^1\left(I, H^1_{loc}(R^N)\right)$, then the mapping $s \to f(t, T_s(x))$ is differentiable in the norm of the space $L^1(I, L^2(\Omega))$, the derivative at s=0 is given by

$$\frac{d}{ds}(f \circ T_s)_{s=0}(t,x) = \nabla_x f(t,x).V(0,x) \tag{2.16}$$

The proof of Proposition 2.6 is similar to that of Proposition 2.16 in chapter 2, section 2.14 in [3], and therefore is omitted here. ■

Lemma 2.7

Assume that $H'y \in L^1(0,T;L^2(\Omega))$.

The material derivative $\dot{y} \in L^\infty\left(0,T;H^1_0(\Omega)\right)$, $\partial\dot{y}/\partial t \in L^\infty(0,T;L^2(\Omega))$ of solution y of hyperbolic problem (2.2)-(2.4) satisfies the equation

$$H\dot{y} = \operatorname{div}(fV) - H'y, \quad \text{in } Q$$

$$\dot{y} = 0, \quad \text{on } \Sigma$$

$$\dot{y}(0) = 0, \quad \frac{\partial \dot{y}}{\partial t}(0) = 0, \quad \text{in } \Omega$$

Proof.

We shall show that

$$\frac{1}{s}(y^s - y) - \dot{y} \to 0, \text{ in } L^\infty\left(0,T;H^1_0(\Omega)\right) \text{ strongly}$$

$$\frac{1}{s}\left(\frac{\partial y^s}{\partial t} - \frac{\partial y}{\partial t}\right) - \frac{\partial \dot{y}}{\partial t} \to 0, \text{ in } L^\infty(0,T;L^2(\Omega)) \text{ strongly, with } s \to 0.$$

Denote

$$z^s \equiv \frac{1}{s}(y^s - y) - \dot{y}$$

then

$$H_s z^s = F_s, \quad \text{in } Q$$
$$z^s = 0, \quad \text{on } \Sigma$$
$$z^s(0) = 0, \quad \frac{\partial z^s}{\partial t}(0) = 0, \quad \text{in } \Omega$$

where

$$F_s \equiv \frac{1}{s}(\gamma(s)\ f \circ T_s - f) - \operatorname{div}(fV) + \frac{1}{s}(H - H_s)y - H'y$$

From Propositions (2.5)-(2.6) it follows

$$F_s \to 0 \quad \text{in } L^1(0,T;L^2(\Omega)) \text{ strongly, with } s \to 0.$$

hence using the a priori estimates we obtain

$$\|z^s\|_{L^\infty\left(0,T;H_0^1(\Omega)\right)} + \left\|\frac{\partial z^s}{\partial t}\right\|_{L^\infty(0,T;L^2(\Omega))} \leq C\left\|H_s z^s\right\|_{L^1(0,T;L^2(\Omega))} =$$

$$= C\ \|F_s\|_{L^1(0,T;L^2(\Omega))} \longrightarrow 0$$

that completes a proof of Lemma 2.7. ∎

Now we are in the position to derive the form of the shape derivative $y' = y'(\Omega;V)$ for the hyperbolic problem.

<u>Lemma 2.8</u>

Suppose that

$$H(\nabla y.V) \in L^1\ (0,T;L^2(\Omega))$$

then the shape derivative y' is given by the unique solution of the following hyperbolic problem

$$Hy' = 0, \quad \text{in} \quad Q$$

$$y' = -\frac{\partial y}{\partial n} \langle V, n \rangle_{R^N}, \quad \text{on} \quad \Sigma$$

$$y'(0) = 0, \ \frac{\partial y}{\partial t}(0) = 0, \quad \text{in} \quad \Omega$$

Proof.

Since $y' = \dot{y} - \nabla y.V$, it follows

$$Hy' = H\dot{y} - H(\nabla y.V) \in L^1(0,T;L^2(\Omega))$$

$$y'(0) = \frac{\partial y'}{\partial t}(0) = 0, \quad \text{in} \quad \Omega$$

$$y' = \dot{y} - y.V = -\frac{\partial y}{\partial n}\langle V,n\rangle_{R^N}, \quad \text{on} \quad \Sigma$$

thus

$$y' \in L^\infty(0,T;H^1(\Omega))$$

$$\frac{\partial y'}{\partial t} \in L^\infty(0,T;L^2(\Omega)) \quad .$$

Finally, in order to show that $Hy' = 0$, in Q, let us consider the integral identity

$$\int_0^T \int_{\Omega_s} \left\{ y_s \frac{\partial^2 \varphi}{\partial t^2} + \nabla y_s . \nabla \varphi \right\} dx\, dt = \int_0^T \int_{\Omega_s} f\varphi\, dx\, dt, \ \forall\, \varphi \in \mathcal{D}(R^{N+1}),\ \varphi\Big|_{Q_s} \in \mathcal{D}(Q_s\ )$$

for $s \in [0,\epsilon)$

differentiation with respect to s, at $s = 0$, leads to the integral identity

$$\int_0^T \int_\Omega \left\{ y' \frac{\partial^2 \varphi}{\partial t^2} + \nabla y' \cdot \nabla \varphi \right\} dx\, dt = 0, \qquad \forall \varphi \in \mathcal{D}(Q)$$ ∎

3. MATERIAL DERIVATIVE OF MAXWELL'S EQUATIONS

Let $\Omega \subset \mathbb{R}^3$ be a given domain of class C^k, k integer, $k \geq 1$. Denote $I = [0,T]$, $T > 0$, $Q = I\times\Omega \subset \mathbb{R}^4$. For a given field $V \in C\left(0,\epsilon,\mathcal{D}^k(\mathbb{R}^3;\mathbb{R}^3)\right)$, $T_s(V)$ denotes the associated transformation. As in the previous section t denotes the times variable, $t \in I$, while s is the transformation parameter, $s \in [0,\epsilon[$. Denote $\Omega_s = T_s(V)(\Omega)$, $s \in [0,\epsilon[$ and let $Q_s = I\times\Omega_s \subset \mathbb{R}^4$ be the cylindrical evolution domain with lateral boundary $\Sigma_s = I \times \Gamma_s$.

For any s let Ω_s be occuped by an electromagnetic medium of constant electric permitivity ϵ_0 and constant magnetic permeability μ_0. We further assume that the electrical charge density ρ and the current density in Ω_s are zero.

It is assumed that the time evolution of the electric and magnetic fields E and H is driven by an externally applied density of current $I(\Gamma_s)$ flowing tangentially in Γ_s. Denote $H(s,t,x)$ and $E(s,t,x)$ the magnetic and electrical fields defined on the time evolution domain Q_s. These functions satisfy the Maxwell's equations

$$\epsilon_0 \frac{\partial E}{\partial t} - \text{curl } H = 0, \ \mu_0 \frac{\partial H}{\partial t} + \text{curl } E = 0 \quad \text{in } \Omega_s, \ t > 0 \qquad (3.1)$$

$$\text{div } E = \text{div } H = 0 \quad \text{in } \Omega_s, \quad t > 0 \qquad (3.2)$$

with the boundary condition

$$n_s \times H = - I(\Gamma_s) \quad \text{on } \Gamma_s, \ t > 0 \qquad (3.3)$$

where n_s is the unit normal vector to Γ_s pointing into the exterior of Ω_s.

Let $E_0 = E_0(x)$, $H_0 = H_0(x)$ denote given function defined on $\mathbb{R}^3$. The initial conditions for problem (3.1)-(3.3) can be read as follows :

$$E(s,0,.) = E_o\Big|_{\Omega_s} \quad , \quad H(s,0,.) = H_o\Big|_{\Omega_s} \quad . \tag{3.4}$$

The tangential density of current $I(\Gamma_s)$ which is applied on the lateral surface Σ is defined as follows : let $G = G(t,x)$ be given smooth function defined on $I \times \mathbb{R}^3$, we set

$$I(\Gamma_s) = n_s \times \text{curl } G \quad \text{on} \quad \Gamma_s \quad . \tag{3.5}$$

Following J.E. Lagnese [2] we introduce the Hilbert spaces :

$$J(\Omega_s) = \text{closure in } L^2\left(\Omega_s;\mathbb{R}^3\right) \quad \text{of} \quad \left\{\chi \in C^\infty\left(\bar{\Omega}_s\right),\ \text{div } \chi = 0\right\} \tag{3.6}$$

for each integer k

$$J^k(\Omega_s) = J(\Omega_s) \cap H^k\left(\Omega_s;\mathbb{R}^3\right) \tag{3.7}$$

$$J^k_n(\Omega_s) = \left\{\chi \in J^k(\Omega_s),\ n_s \cdot \chi = 0 \quad \text{on} \quad \Gamma_s\right\} \tag{3.8}$$

$$J^k_t(\Omega_s) = \left\{\chi \in J^k(\Omega_s),\ n_s \times \chi = 0 \quad \text{on} \quad \Gamma_s\right\} \tag{3.9}$$

with the norms induced by $H^k\left(\Omega_s,\ \mathbb{R}^3\right)$. We further introduce

$$J^*_n(\Omega_s) = \left\{\chi \in J^2_n(\Omega_s),\ n_s \times \text{curl } \chi = 0 \quad \text{on} \quad \Gamma_s\right\} \tag{3.10}$$

The map $\varphi \mapsto \text{curl } \varphi$ is a continuous bijection of $J^k_t(\Omega_s)$ $\left(\text{respectively } J^k_n(\Omega_s)\right)$ onto $J^{k-1}_n(\Omega_s)$ $\left(\text{respectively } J^{k-1}_t(\Omega_s)\right)$. It follows that $J^1_t(\Omega_s)$ and $J^1_n(\Omega_s)$ can be renormed using

$$\|\varphi\|_{J^1_t(\Omega_s)} = \|\text{curl } \varphi\|_{L^2\left(\Omega_s;\mathbb{R}^3\right)} \tag{3.11}$$

$$\|\psi\|_{J^1_n(\Omega_s)} = \|\mathrm{curl}\ \psi\|_{L^2\left(\Omega_s;\mathbb{R}^3\right)} \tag{3.12}$$

Since the problem (3.1)-(3.3) is well-posed, we derive the form of the material derivative $\frac{d}{ds} E(s,t,\ T_s(V)(.))$ as well as $\frac{d}{ds} H(s,t,T_s(V)(.))$ (at $s = 0$).

For this purpose we use the results collected in Lagnese [2], we refer the reader to [1] for more details.

Since $\mathrm{div}\ H = 0$, there is a vector valued function $W = W(s,t,x)$ defined on Q_s, determined up to a gradient ∇f, such that

$$\mu_o H = \mathrm{curl}\ W \quad \text{in} \quad Q_s \tag{3.13}$$

from (3.1) and (3.13) we get

$$\mathrm{curl}\left(\frac{\partial}{\partial t} W\right) + \mathrm{curl}\ E = 0 \quad \text{in} \quad \Omega_s \quad . \tag{3.14}$$

Since $\mathrm{curl}\left(\frac{\partial}{\partial t} W + E\right) = 0$ it follows that $\frac{\partial}{\partial t} W + E = \nabla g$, for an element g. It can be show, [1], [2] that the function f can be chosen in such a way that $g = 0$ therefore W satisfies :

$$\epsilon_o\ \mu_o\ \frac{\partial^2}{\partial t^2} W + \mathrm{curl\ curl}\ W = 0 \quad \text{in} \quad Q_s \tag{3.15}$$

$$n_s \times \mathrm{curl}\ W = -\frac{1}{\mu_o} I\ (\Gamma_s) \quad \text{on} \quad \Gamma_s \tag{3.16}$$

$$W(0) = W_o \ ,\ \frac{\partial}{\partial t} W(0) + E_o = W_1 \tag{3.17}$$

where μ_0 curl $H_0 = - W_0$.
Conversely, if W is a solution of (3.15)-(3.17), then setting $\mu_o H = \mathrm{curl}\ W$,

$E = -\frac{\partial}{\partial t} W$ we see that (E,H) is satisfies (3.1)-(3.3) with the initial conditions

$$E_o = W_1 \quad , \quad H_o = \frac{1}{\mu_o} \text{curl } W_o \quad . \tag{3.18}$$

We shall make use of the following Green's formula :

For any φ and ψ in $J^2(\Omega_s)$ we have

$$a(s;\varphi,\psi) = \int_{\Omega_s} \text{curl}\varphi \text{ curl}\psi \, dx = \int_{\Omega_s} \text{curlcurl } \varphi \, \psi \, dx + \int_{\Gamma_s} \langle \text{curl}\varphi, n_n \times \psi \rangle d\Gamma_s \tag{3.19}$$

using the bilinear form a(s,.,.) which is coercive on $J^1_n(\Omega_s)$ the classical variational theory leads to the existence and uniqueness of the solution W of problem (3.15)-(3.17) such that

$$W \in C\left([0,\infty[\; ; \; J^1_n(\Omega_s)\right) \; , \; \frac{\partial}{\partial t} W \in C\left([0,\infty[, \; \hat{J}(\Omega_s)\right) \quad ,$$

$$W'' \in C\left([0,\infty[, \; \left(J^1_n(\Omega_s)'\right)\right) \quad ,$$

where

$\hat{J}(\Omega_s)$ = closure in $L^2\left(\Omega_s \; ; \; \mathbb{R}^3\right)$ of

$$\left\{x \in \mathcal{D}\left(\Omega_s \; ; \; \mathbb{R}^3\right) \quad , \quad \text{div } x = 0\right\}$$

We derive the form of material derivative. Using an a priori estimate we shall prove the existence of the material derivative in $L^\infty\left(0,T, \; J^1_n(\Omega_s)\right) \cap W^{1,\infty}\left(0,T, \; L^2(\Omega_s)\right)$. Let E(t) be the energy defined by

$$E(t) = \frac{1}{2} \int_{\Omega_s} \left(\epsilon_o \mu_o \; (W_t)^2 + (\text{curl } W)^2\right) dx \tag{3.20}$$

we get

$$E'(t) = \int_{\Omega_s} (\epsilon_o \mu_o W_{tt} + \text{curl curl } W) W_t \, dx$$

$$+ \int_{\Gamma_s} \langle \text{curl } W(t,x), W_t(t,x) \times n_s(x) \rangle \, d\Gamma_s(x) \tag{3.21}$$

in view of the following identity

$$\langle \text{curl } \gamma, \psi \times n \rangle = \langle \text{curl } \varphi \times n, \psi \rangle$$

and since curl $W \times n = 0$ on Γ_s, we get $E'(t) = 0$, i.e. the problem (3.15)-(3.16) is conservative.

Remark.

By the Green formula it follows that if Z belongs to set

$$J_{curl} = \left\{ Z \in L^2(Q_s) \text{ s.t. } \epsilon_o \mu_o Z_{tt} + \text{curl curl } Z \in L^2(Q_s) \right\}$$

then curl $Z \times n_s$ is well defined as an element of $H^{-1/2}(\Sigma_s)$.

Let us define $Z = Z(s,t,x)$ as follows

$$Z(s,t,x) = W(s,t,x) + \frac{1}{\mu_o} G(t,x) \tag{3.22}$$

then the problem (3.15)-(3.17) is equivalent to

$$\epsilon_o \mu_o \frac{\partial^2}{\partial t^2} Z + \text{curl curl } Z = F \quad \text{in } Q_s \tag{3.23}$$

$$n_s \times \text{curl } Z = 0 \quad \text{on } \Sigma_s \tag{3.24}$$

$$Z(0) = W_o + \frac{1}{\mu_o} G(0,.) \quad , \quad \frac{\partial}{\partial t} Z(0) = E_o + \frac{1}{\mu_o} \frac{\partial}{\partial t} G(0,.) \quad \text{in } \Omega_s \tag{3.25}$$

where the right hand side F is given by

$$F = \epsilon_0 \frac{\partial^2}{\partial t^2} G + \frac{1}{\mu_0} \text{curl curl } G \qquad (3.26)$$

We assume now that the element F given by (3.26) belongs to $L^1\left(0,T,L^2(\Omega_s)^3\right)$. Define

$$E_z(t) = \frac{1}{2}\int_{\Omega_s} \left(\epsilon_0\mu_0 \; (Z_t)^2 + (\text{curl } Z)^2\right) dx \quad . \qquad (3.27)$$

Using exactly the same argument as for (3.21) we get in view of (3.23) :

$$E_z(t)' = \int_{\Omega_s} F \, Z_t \; dx \qquad (3.28)$$

and as in the previous section we get

$$\|E_z\|_{L^\infty(0,T)} \leqslant E_z(0) + \|E_z\|^{1/2}_{L^\infty(0,T)} \|F\|_{L^1\left(0,T,L^2(\Omega_s)^3\right)} \quad . \qquad (3.29)$$

By (3.29) there exists M > 0 such that

$$\|E_z\|_{L^\infty(0,T)} \leqslant M \quad . \qquad (3.30)$$

Remark.

The estimates (3.29), (2.30) are analogous to the estimate (2.9) we obtained for the wave equation in the previous section. We proceed exactly in the same way from this point to obtain the result and we omit here the technical details and directly discribe the results.

It can be easily verified that

$$(\text{div } \psi) \circ T_s = \gamma(s)^{-1} \text{ div } \left(\gamma(s)\ DT_s^{-1}.\psi \circ T_s\right) \tag{3.31}$$

and with the well known identity

$$\text{curl curl } \varphi = - \Delta\varphi + \nabla\ (\text{div } \varphi) \tag{3.32}$$

and with (3.28),

$$(\text{curl curl } \varphi) \circ T_s = - \gamma(s)^{-1} \text{ div } (A(s).\nabla(\varphi \circ T_s))$$

$$+ {}^{*}DT_s^{-1}\ \nabla \left(\gamma(s)^{-1} \text{ div}\left(\gamma(s)\ DT_s^{-1}\ .\ \psi \circ T_s\right)\right) \tag{3.33}$$

So we introduce the material deviratíve

$$\dot{W}(t,x) = \frac{\partial}{\partial s}\ W(s,t,T_s(x))|_{s=0} \tag{3.34}$$

the solution of the following problem

$$\epsilon_o \mu_o\ \dot{W}_{tt} + \text{curl curl } \dot{W} = \dot{F} + \text{div}(A'(0).\nabla W)$$

$$-{}^{*}DV(0)^{-1}.\nabla(\text{div } W) + \nabla(\text{div } V(0) \text{ div } W)$$

$$- \nabla\left(\text{div } (\text{div } V(0)\ DV(0)^{-1}.W)\right) \quad \text{in} \quad I \times \Omega \tag{3.35}$$

with the boundary condition

$$n \times \text{curl } \dot{W} = n \times \left[({}^{*}DV(0).\nabla) \times W\right] - \dot{n} \times \text{curl } W \tag{3.36}$$

where $W = W(0,0,x)$, $x \in \Omega_0$.
We use here the change of variables formula :

$$(\text{curl } \Phi) \circ T_s = \left({}^{*}DT_s^{-1}.\nabla\right) \times (\varphi \circ T_s) \tag{3.37}$$

and appropriate initial values $\dot{W}(0)$ and $\frac{\partial}{\partial t} \dot{W}(0)$.

Propriété 3.1

Let $G = G(t,x)$ be given such that $F(s)$ and F belongs respectively to $L^1\left(0,T,\ L^2(\Omega_s)\right)$ and $L^1\left(0,T,\ L^2(\Omega)\right)$ then

$$\frac{1}{s}\left(W(s) \circ T_s - W\right) - \dot{W} \longrightarrow 0 \quad s \longrightarrow 0$$

in $L^\infty(0,T,J^1(\Omega))$ and

$$\frac{1}{s}\left(\left(\frac{\partial}{\partial t} W(s) \circ T_s - \frac{\partial}{\partial t} W\right) - \frac{\partial}{\partial t}\dot{W}\right) \longrightarrow 0\ ,\ s \longrightarrow 0$$

in $L^\infty(0,T,\ L^2(\Omega))$
where $W = W(0,t,x)$ and $\dot{W} = \frac{\partial}{\partial s} W(0,t,x)$ ∎

REFERENCES

[1] K.O. Friedrichs, Mathematical methods of electromagnetic theory. Courant Institute of Mathematical Sciences, New York University, New York, 1974.

[2] J.E. Lagnese, Exact boundary controlability of Maxwell's equations in a general region. SIAM J. Control and Optimization, Vol. 27, n°2 (1989) pp. 374-388.

[3] J. Sokolowski and J.P. Zolesio, Introduction to shape optimization. Shape sensitivity analysis. Book to appear.

Differential Stability of Perturbed Optimization with Applications to Parameter Estimation

Murali Rao
Department of Mathematics
University of Florida
201 Walker Hall
Gainesville, FL 32611
USA

Jan Sokolowski
Systems Research Institute
Polish Academy of Sciences
ul. Newelska 6
01-447 Warszawa
Poland

ABSTRACT

A method for the differential stability of solutions to a class of parametric optimization problems is proposed . The solution of the regularized parametric optimization problem is given in the form of fixed point of metric projection onto the set of admissible parameters . The new result on the differential stability of the metric projection in Sobolev space $H^2(\Omega)$ onto a set of admissible parameters is established . Stability results with respect to the perturbations of observation for the solutions to a parameter estimation problem for the parabolic equation are derived .

MOS subject classification : 49B22, 49A29, 49A22, 93B30

1. INTRODUCTION

We shall consider the differential stability of solutions to a class of parametric optimization problems [S-4] . Such problems arise in particular as coefficient estimation problems for PDE's e.g. in [C-1],[C-K-1],[R-S-2],[S-1],[S-2],[S-3],[S-4] . Since a parametric optimization problem is not in general convex , therefore an optimal solution , if it exists , does not depend continuously on data . We provide a method for the differential stability analysis of a class of regularized parametric optimization problems . The method is based on the observation , that an optimal solution for such problem is the fixed point of metric projection onto the set of admissible parameters . For the sake of simplicity we shall consider a model problem for the parabolic equation , however the method of stability analysis is general and can be used for elliptic , hyperbolic and some nonlinear partial differential equations . In [C-K] the Hölder continuity of the solution to a regularized version of a coefficient estimation problem for the second order elliptic equation with respect to the observation is shown using the second order sufficient optimality conditions in mathematical programming . In the present paper we obtain the results on the differential stability of solutions to a parametric optimization problem for the parabolic equation . In particular we show that the Hölder continuity of solution of optimization problem with respect to the parameter leads to a differential stability result , taking into account the structure of the optimization problem under consideration .We shall use the new results , presented in [R-S-2] , on the directional differentiability of metric projection onto a set of admissible parameters in Sobolev space . We shall show that the metric projection in $H^2(\Omega)$ onto the set defined by local constraints is directionally differentiable in the sense of Hadamard . To this end we use the concept of polyhedric convex set [M], [H-1], see Definition 1 below . We refer the reader to [R-S-1] for the results on differential stability of metric projection and the applications including the sensitivity analysis of the Kirchhoff plate with an obstacle . The related results on the sensitivity analysis of convex optimization problems are presented in [S-5],[S-6],[S-7] and on the shape sensitivity analysis of variational inequalities in [S-Z-1],[S-Z-2] . We use standard notation throughout the paper [A-1],[L-M] .

First we formulate the parametric optimization problem under considerations.

Let $\Omega \subset R^n$, $n = 1,2,3$ be a given domain, $F(.,.) \in L^2(0,T;H^{-1}(\Omega))$ and $y_0(.) \in L^2(\Omega)$ be given elements. Denote $Q = \Omega \times (0,T)$, $\Sigma = \partial\Omega \times (0,T)$. Let us consider the following model problem

<u>Problem (P):</u>

Find an element $a(.) \in L^\infty(\Omega)$,$c_2 \geq a(x) \geq c_1 > 0$ in Ω, which minimizes the following cost functional

$$J(a) = |y(a;.,.) - z(.,.)|^2_{L^2(Q)} \quad (1.1)$$

here $z(.,.) \in L^2(Q)$ is a given element, and $y(a;.,.) \in W(0,T)$ is given by a unique solution of the following state equation in the form of parabolic initial - boundary value problem :

$$y_t - \mathrm{div}\,(\, a\nabla y\,) = F \;,\; \text{in } Q \quad (1.2)$$

$$y = 0 \;,\; \text{on } \Sigma \quad (1.3)$$

$$y(x,0) = y_o(x) \;,\; \text{in } \Omega \quad (1.4)$$

Therefore we want to determine an unknown coefficient $a(.)$ on the basis of a given observation $z(.,.)$ of solution to the parabolic equation .

Problem (P) is in general ill - posed. A generalized solution can be obtained using G - convergence technique (see e.g. [S-4]).

2. LEAST SQUARE INVERSE PROBLEM

In order to ensure the existence of a solution to problem (P) a regularization technique can be used. We denote by K the set of admissible coefficients

$$K = \{ a \in H^2(\Omega) |\; \psi_2(x) \geq a(x) \geq \psi_1(x),\; x \in \Omega \} \quad (2.1)$$

where $\psi_2(x) \geq \psi_1(x) > c > 0$ are given elements such that set (2.1) is nonempty. The regularized version of problem (P) takes the following form .

<u>Problem (P_α):</u>

Find an element $a(.) \in K$ which minimizes the following cost functional

$$J_\alpha(a) = |y(a;.,.) - z(.,.)|^2_{L^2(Q)} + \alpha|a|^2_{H^2(\Omega)} \quad (2.2)$$

where $\alpha > 0$ is a given constant.

We shall consider the differential stability of solutions to problem (P_α) with respect to the perturbations of element z.

Let $\varepsilon \geq 0$ be parameter , $\varepsilon \in [0,\delta)$, denote

$$z_\varepsilon = z + \varepsilon v$$

where $v \in L^2(Q)$ is given .

Let us consider the parametric optimization problem with observation z_ε .

<u>Problem (P^ε_α):</u>

Find an element $a(.) \in K$ which minimizes the following cost functional

$$J_{\alpha,\varepsilon}(a) = |y(a;.,.) - z_\varepsilon(.,.)|^2_{L^2(Q)} + \alpha|a|^2_{H^2(\Omega)}$$

■

We denote by a^*_ε a solution of problem (P^ε_α) , such a solution exists provided $\alpha > 0$. From the necessary optimality conditions for problem (P^ε_α) it follows that the element a^*_ε is given by the fixed point . It can be shown that there exists α_0 such that for $\alpha > \alpha_0$ the solutions a^*_ε , $\varepsilon \in [0,\delta)$, are locally unique - it is sufficient to select α_0 in such a way that the nonlinear mapping associated to the fixed point is the contraction . We show the directional differentiability of the solutions to problem (P^ε_α) with respect to ε . Our main result reads as follows

THEOREM 1

Assume that there exists $\beta > 0$ such that

$$d^2J_\alpha(a^*_0;a,a) \geq \beta|a|^2_{H^2(\Omega)} \quad , \forall a \in S \qquad (2.3)$$

then for $\varepsilon > 0$, ε small enough

$$a^*_\varepsilon = a^*_0 + \varepsilon q + o(\varepsilon) \text{ , in } H^2(\Omega)$$

where $|o(\tau)|_{H^2(\Omega)}/\tau \downarrow 0$ with $\tau\downarrow 0$ and the element q is given by a unique solution of the following optimality system.

Find (u,w,q) such that the following system is satisfied

Linearized state equation :

$$u_t - \operatorname{div}(a^*_0\nabla u) = \operatorname{div}(q\nabla y^*_0) \text{ , in } Q$$

$$u = 0 \text{ , on } \Sigma$$

$$u(0) = 0 \text{ , in } \Omega$$

Linearized adjoint state equation :

$$w_t - \operatorname{div}(a^*_0\nabla w) = -\operatorname{div}(q\nabla p^*_0) + u - v \text{ , in } Q$$

$$w = 0 \text{ , on } \Sigma$$

$$w(0) = 0 \text{ , in } \Omega$$

Optimality conditions :

$q \in S = T_K(a^*_0)\cap[f^*_0 - a^*_0]^\perp$

$$\int_0^T \int_\Omega (a - q)[\nabla y_0^* . \nabla w + \nabla u . \nabla p_0^*] dxdt + \alpha(a - q, q)_{H^2(\Omega)} \geq 0$$

$$\forall a \in S \quad \blacksquare$$

Here $T_K(a)$ denotes the tangent cone to K at $a \in K$, $[f - a]^\perp$ is the hyperplane orthogonal in $H^2(\Omega)$ to the element $f - a$. The convex cone S takes form

$$S = \{ \varphi \in H^2(\Omega) \mid \varphi \geq 0 \text{ q.e. on } \Xi_1 , \varphi \leq 0 \text{ q.e. on } \Xi_2 , \int \varphi \, d\mu = 0 \}$$

q.e. means " quasi - everywhere " [H-2],

$$\Xi_i = \{ x \in \Omega \mid a_0^*(x) = \psi_i(x) \} , \quad i = 1,2$$

non - negative measure μ is defined by

$$(a_0^*, \varphi)_{H^2(\Omega)} + \frac{1}{\alpha} \int_0^T \int_\Omega \varphi \nabla y_0^* . \nabla p_0^* dxdt = \int \varphi \, d\mu , \quad \varphi \in H^2(\Omega)$$

and p_0^* is the adjoint state given by the unique solution of the following elliptic problem

$$p_{0,t}^* - \operatorname{div} (a_0^* \nabla p_0^*) = y_0^* - z, \text{ in } Q$$

$$p_0^* = 0 , \text{ on } \Sigma$$

$$p_0^*(T) = 0 , \quad \text{in } \Omega$$

y_0^* is the solution of state equation for the coefficient a_0^*.

REMARK

In order to prove THEOREM 1 we provide in section 3 the new results [R-S-2] on the directional differentiability of metric projection in $H^2(\Omega)$ onto the convex set K which are interesting on its own.

Before we provide the proof of THEOREM 1 , first we show that the local solution of problem (P_α^ε) is given by a fixed point of the metric projection onto K.

Let us recall that any local solution of problem (P_α^ε) satisfies the necessary optimality conditions in the form of the following optimality system

Find elements $(y_\varepsilon^*, p_\varepsilon^*, a_\varepsilon^*)$ such that the following non - linear system is satisfied

State equation:

$$y^*_{\varepsilon,t} - \operatorname{div}(a^*_\varepsilon \nabla y^*_\varepsilon) = F, \text{ in } Q \tag{2.4}$$

$$y^*_\varepsilon = 0 , \text{ on } \Sigma \tag{2.5}$$

$$y^*_\varepsilon(0) = 0 , \quad \text{in } \Omega \tag{2.6}$$

Adjoint state equation:

$$p^*_{\varepsilon,t} - \operatorname{div}(a^*_\varepsilon \nabla p^*_\varepsilon) = y^*_\varepsilon - z, \text{ in } Q \tag{2.7}$$

$$p^*_\varepsilon = 0 , \text{ on } \Sigma \tag{2.8}$$

$$p^*_\varepsilon(T) = 0 , \quad \text{in } \Omega \tag{2.9}$$

Optimality conditions:

$a^*_\varepsilon \in K$:

$$\int_0^T \int_\Omega (a - a^*_\varepsilon) \nabla y^*_\varepsilon . \nabla p^*_\varepsilon dxdt + \alpha(a - a^*_\varepsilon, a^*_\varepsilon)_{H^2(\Omega)} \geq 0 \tag{2.10}$$

$$\forall a \in K \qquad \blacksquare$$

The optimality system is derived using the following form of the gradient of cost functional $J_{\alpha,\varepsilon}(a)$ at a^*_ε

$$dJ_{\alpha,\varepsilon}(a^*_\varepsilon; a) = 2\int_0^T \int_\Omega a \nabla y^*_\varepsilon . \nabla p^*_\varepsilon dxdt + 2\alpha(a^*_\varepsilon, a)_{H^2(\Omega)} \tag{2.11}$$

It is also useful to derive the form of second derivative of cost functional $J_{\alpha,\varepsilon}(a)$ at a^*_ε . To this end we differentiate $dJ_{\alpha,\varepsilon}(a^*_\varepsilon; a)$ with respect to a^*_ε in a direction b and we obtain

$$d^2 J_{\alpha,\varepsilon}(a^*_\varepsilon; a, b) = 2\int_0^T \int_\Omega a[\nabla y^*_\varepsilon . \nabla w^*_\varepsilon + \nabla u^*_\varepsilon . \nabla p^*_\varepsilon] dxdt + 2\alpha(a, b)_{H^2(\Omega)} \tag{2.12}$$

where $w^*_\varepsilon, u^*_\varepsilon$ are given by unique solutions of the following linearized equations .

Linearized state equation :

$$u^*_{\varepsilon,t} - \operatorname{div}(a^*_\varepsilon \nabla u^*_\varepsilon) = \operatorname{div}(b \nabla y^*_\varepsilon) , \text{ in } Q \tag{2.13}$$

$$u^*_\varepsilon = 0 , \text{ on } \Sigma \tag{2.14}$$

$$u^*_\varepsilon(0) = 0 , \quad \text{in } \Omega \tag{2.15}$$

Linearized adjoint state equation:

$$w^*_{\varepsilon,t} - \operatorname{div}(a^*_\varepsilon \nabla w^*_\varepsilon) = - \operatorname{div}(b \nabla p^*_\varepsilon) + u^*_\varepsilon, \text{ in } Q \tag{2.16}$$

$$w^*_\varepsilon = 0 , \text{ on } \Sigma \tag{2.17}$$

$$w^*_\varepsilon(T) = 0 , \quad \text{in } \Omega \tag{2.18}$$

Let us consider the stability of a local optimal solution of problem (P_α^ε) . Observe that if the norm $|y(a_\varepsilon^*;.,.) - z_\varepsilon(.,.)|_{L^2(Q)}$ is sufficiently small, i.e. the observation $z_\varepsilon(.,.)$ is sufficiently close to the non - convex attainable set

$$\mathcal{Y} = \{ y \in L^2(0,T;H_0^1(\Omega)) \mid y_t - \operatorname{div} (a\nabla y) = F ,$$
$$\text{in } Q , \text{ for some } a \in K \}$$

then it can be verified that the norm of ∇p_ε^* in $L^2(Q;R^n)$ is small. We address the following question , how large should be α in order to have the Hessian of $J_{\alpha,\varepsilon}(.)$ positive definite . To this end we define the bilinear form

$$\mathcal{L}_\varepsilon(a,b) = \int_0^T \int_\Omega a\nabla y_\varepsilon^* . \nabla \eta_\varepsilon^*(b) dx dt \tag{2.19}$$

where

$$\eta_{\varepsilon,t}^* - \operatorname{div} (a_\varepsilon^* \nabla \eta_\varepsilon^*) = u_\varepsilon^* , \quad \text{in } Q$$
$$\eta_\varepsilon^* = 0 , \text{ on } \Sigma$$
$$\eta_\varepsilon^*(0) = 0 , \quad \text{in } \Omega$$

here u_ε^* solves (2.13)-(2.15) for any given b.

Clearly

$$\forall a,b \in K : |\mathcal{L}_\varepsilon(a,b)| \le M_1 |a|_{L^\infty(\Omega)} |b|_{L^\infty(\Omega)} \tag{2.20}$$

for $n \le 3$

$$\le M |a|_{H^2(\Omega)} |b|_{H^2(\Omega)} , \tag{2.21}$$

It follows by inspection that for $\alpha > M$ and the norm $|y(a_\varepsilon^*;.,.) - z_\varepsilon(.,.)|_{L^2(Q)}$ sufficiently small the first component of $d^2J_{\alpha,\varepsilon}(a_\varepsilon^*;a,a)$ is dominated by the second component and therefore there exists $\beta > 0$ such that

$$d^2J_{\alpha,\varepsilon}(a_\varepsilon^*;a,a) \ge \beta(a,a)_{H^2(\Omega)} \tag{2.22}$$

Condition (2.22) seems to be restrictive however it leads to the stability result for the optimal solution a_0^* with respect to $z(.)$ - see PROPOSITION 1 .

Finally we show that optimality system (2.4)-(2.10) is equivalent to the fixed point condition. Define an element $f_\varepsilon^* \in H^2(\Omega)$ as follows

$$\alpha(f^*_\varepsilon,\varphi)_{H^2(\Omega)} = -\int_0^T\int_\Omega \varphi\nabla y^*_\varepsilon.\nabla p^*_\varepsilon dxdt \tag{2.23}$$

the element f^*_ε is well defined since $\nabla y^*_\varepsilon.\nabla p^*_\varepsilon \in L^1(Q)$, $\int_0^T \nabla y^*_\varepsilon.\nabla p^*_\varepsilon dt \in L^1(\Omega)$, and for $n \le 3$ it follows that $\varphi \in C(\bar{\Omega})$ by the Sobolev imbedding theorem. Notice that the optimality condition for (P^ε_α) reads

$$a^*_\varepsilon = P_K(f^*_\varepsilon) \tag{2.24}$$

therefore a^*_ε is the fixed point since the element f^*_ε , in view of (2.4)-(2.10) , depends on a^*_ε . From (2.23) it follows by inspection that there exists α_0 such that the nonlinear mapping associated to the fixed point (2.24) is the contraction in Sobolev space $H^2(\Omega)$ for $\alpha > \alpha_0$ and for any $\varepsilon \in [0,\delta)$, henceforth for $\alpha > \alpha_0$ the fixed point (2.24) is locally unique and simple calculations show [S-8] that a^*_ε is Lipschitz continuous with respect to ε .
Since in general the projection P_K fails to be differentiable we need the appropriate results on the directional differentiability of projection P_K in order to study the differential stability of solutions to the optimality system for (P^ε_α) . On the other hand if the projection P_K is differentiable at f^ε_0 the Implicit function theorem can be used to establish the differentiability of the solution a^*_ε with respect to ε .

PROPOSITION 1 [S-8]

Suppose that there exists $\beta > 0$ such that the condition

$$d^2J_{\alpha,\varepsilon}(a^*_0;a,a) \ge \beta(a,a)_{H^2(\Omega)} \quad , \forall a \in S = T_K(a^*_0)\cap[f^*_0 - a^*_0]^\perp \tag{2.25}$$

is satisfied , then

$$| a^*_\varepsilon - a^*_0 |_{H^2(\Omega)} \le C\varepsilon \tag{2.26}$$

■

PROOF OF THEOREM 1

From (2.26) it follows that there exists an element $q \in H^2(\Omega)$, such that for $\varepsilon > 0$, ε small enough

$$a_\varepsilon^* = a_0^* + \varepsilon q + r(\varepsilon)\ , \text{ in } H^2(\Omega) \tag{2.27}$$

where $r(\varepsilon)/\varepsilon \to 0$ weakly , with $\varepsilon \downarrow 0$, therefore

$$f_\varepsilon^* = f_0^* + \varepsilon f_1 + o(\varepsilon)\ , \text{ in } H^2(\Omega) \tag{2.28}$$

where the element f_1 is defined as follows

$$\alpha(f_1,\varphi)_{H^2(\Omega)} = -\int_0^T \int_\Omega \varphi[\nabla y_0^*.\nabla w + \nabla u.\nabla p_0^*]dxdt \tag{2.29}$$

The elements u,w solve the linearized equations.

Linearized state equation :

$$u_t - \operatorname{div}(\ a_\varepsilon^* \nabla u\ ) = \operatorname{div}(\ q\nabla y_\varepsilon^*\ )\ , \quad \text{in } Q \tag{2.30}$$

$$u = 0\ , \quad \text{on } \Sigma \tag{2.31}$$

$$u(0) = 0\ , \quad \text{in } \Omega \tag{2.32}$$

Linearized adjoint state equation:

$$-w_t - \operatorname{div}(\ a_\varepsilon^* \nabla w\ ) = \operatorname{div}(\ q\nabla p_\varepsilon^*\ ) + u - v\ , \quad \text{in } Q \tag{2.33}$$

$$w = 0\ , \quad \text{on } \Sigma \tag{2.34}$$

$$w(T) = 0\ , \quad \text{in } \Omega \tag{2.35}$$

From (2.24), using COROLLARY 1 given below , it follows

$$a_\varepsilon^* = P_K(f_\varepsilon^*) = P_K(f_0^* + \varepsilon f_1 + o(\varepsilon))$$

by THEOREM 2

$$= P_K(f_0^*) + \varepsilon P_S(f_1) + o(\varepsilon)$$

in view of (2.27)

$$q = P_S(f_1)$$

which completes the proof of THEOREM 1 since (2.3) implies that q is unique .

3. DIFFERENTIAL STABILITY OF METRIC PROJECTION

In this section we provide the related results [R-S-2] on the differential stability of metric projection in $H^2(\Omega)$ onto the set K. We shall use the concept of polyhedric convex set [H-1], [M].
We assume for simplicity that the set of admissible parameters takes the form

$$K = \{\ a \in H^2(\Omega) \mid a(x) \ge \psi(x),\ x \in \Omega\}$$

where $\psi(.) = \psi_1(.) \in H^2(\Omega)$ is a given element.

Exactly the same argument applies in the case of convex set

$$K = \{ a \in H^2(\Omega) | \ \psi_2(x) \geq a(x) \geq \psi_1(x), \ x \in \Omega\}$$

where $\psi_2(x) \geq \psi_1(x) > c > 0$ are given in $H^2(\Omega)$.

Let $T_K(a)$ denotes the tangent cone to K at $a \in K$. It is clear that $T_K(a)$ is the closure in the space $H^2(\Omega)$ of the following convex cone

$$C_K(a) = \{\varphi \in H^2(\Omega) \mid \exists \ t > 0 \text{ such that } \ a + t\varphi \in K \}$$

For a given element $f \in H^2(\Omega)$, such that $a = P_K(f)$ let us define the following convex cone in the space $H^2(\Omega)$

$$S = T_K(a) \cap [f - P_K(f)]^{\perp} \qquad (3.1)$$
$$= T_K(a) \cap [f - a]^{\perp}$$

DEFINITION 1

The set K is polyhedric provided for any $f \in H^2(\Omega)$

$$T_K(a) \cap [f - a]^{\perp} = \text{cl}(C_K(a) \cap [f - a]^{\perp}) \qquad (3.2)$$

here cl stands for the closure, $a = P_K f$. ■

THEOREM 2 [R-S-2]

The set K is polyhedric.

COROLLARY 1

Let set K be polyhedric and $f(.) : [0,\delta) \to H^2(\Omega)$ be given mapping strongly right - differentiable at 0, then for $\varepsilon > 0$, ε small enough

$$P_K(f(\varepsilon)) = P_K(f(0)) + \varepsilon P_S(f'(0^+)) + o(\varepsilon) \qquad (3.3)$$

where we denote $a = P_K(f(0))$,

$$S = T_K(a) \cap [f(0) - a]^{\perp}$$
$$= \{ \varphi \in H^2(\Omega) \mid \varphi \geq 0 \ , \text{ q.e. on } \Xi \ , \int \varphi \, d\mu = 0 \}$$

$$\Xi = \{ x \in \Omega \mid a(x) = \psi(x) \}$$

and measure μ is defined by

$$(a - f,\varphi)_{H^2(\Omega)} = \int \varphi \, d\mu \ , \quad \varphi \in H^2(\Omega)$$

■

COROLLARY 1 follows from Theorem 2 [S-5] .

We derive the form of tangent cone $T_K(a)$ for any $a \in K$, and we show that the set K is polyhedric.

First we recall the notion of Capacity.

We define the Capacity (relative to $H^2(\Omega)$) of a compact set $F \subset \Omega$ as

$$C(F) = \inf \{ \int_\Omega (|\Delta\phi|^2 + |\phi|^2)dx \mid \phi \in C_0^\infty(\Omega),\ \phi \geq 1 \text{ on } F\}$$

For an arbitrary analytic set A the capacity of A is defined as the supremum of the capacities of compact subsets of A. A statement holds quasi-everywhere - q.e. in short - if it holds except for a set of points of capacity zero. Sets of capacity zero are necessarily of zero Lebesque measure.

It is known that [H-2] every element of $H^2(\Omega)$ has a quasi continuous version: given $f \in H^2(\Omega)$ and $\varepsilon > 0$ there is an open set of capacity $< \varepsilon$ such that the restriction of f to the complement of this open set is continuous. Two quasi continuous versions agree q.e. It is also known that if $f_n \to f$ in $H^2(\Omega)$ we can extract a subsequence of f_n such that $f_n \to f$ q.e. pointwise. Convergence in H^2 implies q.e. convergence. Hence if

$$\Xi = \{ x \mid u(x) = \psi(x)\} \tag{3.4}$$

then $\varphi \in T_K(u)$ implies $\varphi \geq 0$ q.e. on Ξ.

Notice that $\varphi \in H^2(R^n)$ implies $\varphi|_\Omega \in H^2(\Omega)$. It is well known that

$$H^2(R^n) = G[L^2(R^n)] \tag{3.5}$$

where G is the inverse of $-(\Delta - I)$ in $L^2(R^n)$:

$$-(\Delta - I)^{-1} = G \tag{3.6}$$

Now we characterize tangent cone.

THEOREM 3 [R-S-2]

For any $u \in K$ the tangent cone $T_K(u)$ takes the form

$$T_K(u) = \{ \varphi \in H^2(\Omega) \mid \varphi \geq 0 \text{ q.e. on } \Xi \} \tag{3.7}$$

For the convenience of the reader we provide here the proof of THEOREM 3 borrowed from [R-S-2].

PROOF OF THEOREM 3.

We denote $H^2 = H^2(\Omega)$. Clearly the left side of (3.2) is contained in the right side

$$T_K(u) \subset \{ \varphi \in H^2(\Omega) \mid \varphi \geq 0 \text{ q.e. on } \Xi \}$$

Let $V \in H^2$ and $V \geq 0$ q.e. on Ξ. We show that $V \in T_K(u)$.
For this let φ_0 be the projection of V onto the cone $T_K(u)$. Then from standard result on projection

$$(\varphi_0 - V, \varphi)_{H^2} \geq 0 \;, \quad \forall \varphi \in T_K(u) \tag{3.8}$$

As observed before for each $f \in L^2(R^n)$, $Gf|_\Omega \in H^2$. Hence there exists a unique $g \in L^2(R^n)$ such that for each $f \in L^2(R^n)$

$$(\varphi_0 - V, Gf|_\Omega)_{H^2} = \int gf \, dx \tag{3.9}$$

It is easily seen from definition that every non - negative function in H^2 belongs to $C_K(u)$. Hence

$$(\varphi_0 - V, \varphi)_{H^2} \geq 0 \;, \quad \text{if } \varphi \geq 0 \tag{3.10}$$

Thus if $f \in L^2(R^n)$ is such that $Gf \geq 0$ we have from (3.9)

$$\int gf \, dx \geq 0 \tag{3.11}$$

This implies g is " 1-excessive " i.e. $g \geq 0$ and $(\Delta - I)g \leq 0$.
From Standard Potential Theory [D] we conclude

$$g = G\mu + h \tag{3.12}$$

where μ is positive measure and h is 1-harmonic i.e. $(\Delta - I)h = 0$.
Since $0 \leq g \in L^2(R^n)$ and $h \leq g$, $h \in L^2(R^n)$. But then $0 = G(\Delta - I)h = -h$ i.e. $h = 0$. Thus $G\mu = g$. Using this in (3.9) we get because G is symmetric

$$(\varphi_0 - V, Gf|_\Omega)_{H^2} = \int G\mu . f \, dx = \int Gf \, d\mu \tag{3.13}$$

Now the left side depends only on the restriction of Gf to Ω. Hence μ must be concentrated on Ω. Further any $\varphi \in H^2$ is the restriction of a function Gf to Ω. We get thus from (3.13)

$$(\varphi_0 - V, \varphi)_{H^2} = \int \varphi \, d\mu \;, \quad \varphi \in H^2 \tag{3.14}$$

Now $u + \psi - u \geq \psi$ or that $\psi - u \in C_K(u)$. Using this in (3.14), using (3.8) and noting that $\psi - u < 0$ off Ξ we conclude that μ must be concentrated on Ξ. Again since $(\varphi_0 - V, \varphi_0)_{H^2} = 0$

$$0 \leq (\varphi_0 - V, \varphi_0 - V)_{H^2} = (\varphi_0 - V, -V)_{H^2} = -\int V \, d\mu \leq 0 \tag{3.15}$$

because μ is concentrated on Ξ and $V \geq 0$ on Ξ we conclude

$$(\varphi_0 - V, \varphi_0 - V)_{H^2} = 0 \qquad (3.16)$$

or that $\varphi_0 = V$. ∎

REFERENCES

[A-1] R.A. Adams *Sobolev Spaces* Academic Press, New York 1975

[C-1] G. Chavent, On the uniqueness of local minima for general abstract non - linear least square problems. Rapport de Recherche No.645, INRIA, Rocquencourt, France

[C-K-1] F. Colonius and K. Kunisch, Stability of Perturbed Optimization Problems with Applications to Parameter Estimation. (preprint)

[D] J.L. Doob, *Classical Potential Theory and its Probabilistic Counter Part* Springer Verlag , New York , 1985.

[D-Z] M. Delfour and J.P. Zolesio, Shape sensitivity analysis via Min Max differentiability, *SIAM J. Control and Optimization* (26)4 , 1988 .

[H-1] A. Haraux, How to differentiate the projection on a convex set in Hilbert space. Some applications to variational inequalities. *J. Math. Soc. Japan* 29(4) 1977 p.615-631

[H-2] L.I. Hedberg, Spectral Synthesis in Sobolev Spaces, and Uniqueness of Solutions of Dirichlet Problem. Acta Math. 147(1981) p.237-264

[L-M] J.L. Lions and E. Magenes, *Problemes aux limites non homogenes.* Dunod, Paris, 1968

[M] F. Mignot, Controle dans les inequations variationelles elliptiques. *J. Funct. Anal.* (22)1976 p.25-39

[R-S-1] M. Rao and J. Sokołowski, Sensitivity of unilateral problems in $H_0^2(\Omega)$ and applications (to appear)

[R-S-2] M. Rao and J. Sokołowski, Differential Stability of Solutions to Parametric Optimization Problems (to appear)

[S-1] J. Sokołowski, On Parametric Optimal Control for Weak Solutions of Abstract Linear Parabolic Equations. Control and Cybernetics, vol.4, no.3(1975) 59-84.

[S-2] J. Sokołowski, On Parametric Optimal Control for a Class of

Linear and Quasilinear Equations of Parabolic Type. Control and Cybernetics, vol.4, no. 1 (1975) 19-38.

[S-3] J. Sokołowski, Optimal control in coefficients of boundary value problems with unilateral constraints *Bulletin of the Polish Academy of Sciences, Technical Sciences*, vol.31, no.1-12, 1983 p.71-81

[S-4] J. Sokołowski, Parametric optimization problems for evolution initial-boundary value problems. In: *Analysis and Algorithms of Optimization Problems*, K. Malanowski and K. Mizukami (Eds.), Lecture Notes in Control and Information Sciences, Vol. 82, Springer Verlag, 1986 p.61-87

[S-5] J. Sokołowski, Differential stability of solutions to constrained optimization problems. *Appl. Math. Optim.* (13) 1985 p.97-115

[S-6] J. Sokołowski, Sensitivity analysis of control constrained optimal control problems for distributed parameter systems. *SIAM J. Control and Optimization* (25)6, 1987 p.1542-1556

[S-7] J. Sokołowski, Shape sensitivity analysis of boundary optimal control problems for parabolic systems. *SIAM Journal on Control and Optimization*. (26)4, 1988, p.763-787.

[S-8] J. Sokołowski, to appear.

[S-Z-1] J. Sokołowski and J.P. Zolesio, Shape sensitivity analysis of unilateral problems. *SIAM J. Math. Anal.* 5(18), 1987 p.1416-1437

[S-Z-2] J. Sokołowski and J.P. Zolesio, *Introduction to Shape Optimization . Shape sensitivity analysis* . (to appear)

A NUMERICAL METHOD FOR DRAG MINIMIZATION VIA THE SUCTION AND INJECTION OF MASS THROUGH THE BOUNDARY

Max D. Gunzburger
Department of Mathematics
Virginia Polytechnic Institute and State University
Blacksburg, VA 24061, USA

Lisheng Hou
Départment de Mathématiques et de Statistique
Université Laval
Quebec, G1K 7P4, Canada

Thomas P. Svobodny
Department of Mathematics and Statistics
Wright State University
Dayton, OH 45435, USA

We study the problem of minimizing the viscous drag on a body via the addition or removal of mass through the boundary. The control considered is the mass flux through all or part of the boundary; the functional to be minimized is the viscous dissipation. We use Lagrange multiplier techniques to derive a system of partial differential equations from which optimal, *i.e.*, minimum drag, solutions may be determined. Then, finite element approximations of solutions of the optimality system are defined and optimal error estimates are derived.

This work was supported by the Air Force Office of Scientific Research under grant numbers AFOSR-88-0197 for MDG and LH and AFOSR-85-0263 and AFOSR-86-0085 for TPS. The work of MDG was also partially performed under the auspices of the U.S. Department of Energy.

I – INTRODUCTION Let $(\mathbf{u}, p)$ denote the velocity and pressure fields and $\mathbf{g}$ the boundary velocity control. Let Ω denote the flow domain and Γ its boundary. Consider the functional

$$\mathcal{K}(\mathbf{u},\mathbf{g}) = \frac{\nu}{2}\int_\Omega |(\mathrm{grad}\,\mathbf{u}) + (\mathrm{grad}\,\mathbf{u})^T|^2\,d\Omega - \int_\Omega \mathbf{f}\cdot\mathbf{u}\,d\Omega + \frac{\nu}{2}\int_{\Gamma_c} |\mathrm{grad}_s\,\mathbf{g}|^2\,d\Gamma \tag{1.1}$$

where grad_s denotes the surface gradient operator and Γ_c denotes the portion of the boundary on which the control is allowed to act. Since the density is a constant, $\mathbf{g}$ is proportional to the mass flux. The functional (1.1) measures the drag due to viscosity; for a discussion of the relation between (1.1) and the viscous drag, see [Ser]. The appearance of the control $\mathbf{g}$ in (1.1) is necessary since we will not impose any *a priori* constraints on the size of that control. Problems such that the controls are constrained *a priori* to be bounded are treated in [GHS1].

The optimization problems we study are to seek states $(\mathbf{u}, p)$ and controls $\mathbf{g}$ such that $\mathcal{K}(\cdot,\cdot)$ is minimized, subject to the constraints

$$-\nu \operatorname{div}\left((\operatorname{grad}\mathbf{u}) + (\operatorname{grad}\mathbf{u})^T\right) + \mathbf{u}\cdot\operatorname{grad}\mathbf{u} + \operatorname{grad} p = \mathbf{f} \quad \text{in } \Omega\,, \tag{1.2}$$

$$\operatorname{div}\mathbf{u} = 0 \quad \text{in } \Omega \tag{1.3}$$

and

$$\mathbf{u} = \begin{cases} \mathbf{b} & \text{on } \Gamma_u \\ \mathbf{b}+\mathbf{g} & \text{on } \Gamma_c\,, \end{cases} \tag{1.4}$$

i.e., $(\mathbf{u}, p, \mathbf{g})$ satisfy the Navier-Stokes equations (1.2), the incompressibility condition (1.3), and the inhomogeneous boundary condition (1.4).

In (1.1)-(1.4), Ω denotes a bounded domain in $\mathbb{R}^d$, $d = 2$ or 3 with a boundary Γ; Γ_u and Γ_c are portions of Γ such that $\overline{\Gamma}_u \cup \overline{\Gamma}_c = \overline{\Gamma}$ and $\Gamma_u \cap \Gamma_c = 0$. When finite element approximations are considered, it is assumed that Ω is a convex polyhedral domain; otherwise, it is assumed that Γ is either convex or is of class $C^{1,1}$. In (1.3)-(1.6), ν denotes the (constant) kinematic viscosity, $\mathbf{f}$ a given body force and $\mathbf{b}$ a given velocity field defined on the boundary. Thus Γ_c and Γ_u denote the portions of Γ where velocity controls are and are not applied, respectively. In (1.3) we have absorbed the constant density into the pressure and the body force. If the variables in (1.1)-(1.4) are nondimensionalized, then ν is simply the inverse of the Reynolds number Re. Also note that since the density is a constant, the boundary condition (1.4) also specifies the mass flux at the boundary.

Some constaints are placed on candidate controls. Most notably, we will require that

$$\int_{\Gamma_c} \mathbf{g}\cdot\mathbf{n}\, d\Gamma = -\int_{\Gamma} \mathbf{b}\cdot\mathbf{n}\, d\Gamma = 0 \tag{1.5}$$

and, if Γ_c is not connected,

$$\mathbf{g} = 0 \quad \text{on } \partial\Gamma_c\,, \tag{1.6}$$

where $\partial\Gamma_c$ denotes the boundary of Γ_c, the latter viewed as a subset of Γ. The incompressibility constraint (1.3) necessitates the imposition of the compatibility contition given by the left equality in (1.5); we impose the right inequality only for the sake of simplifying the exposition. All our results hold equally well if the right equality in (1.5) is not assumed. The relation (1.6) is imposed in order to ensure that solutions of our optimization problems are "sufficiently" regular.

The only type of controls we allow are the velocity (or mass flux) on the boundary. Such a situation is common; *e.g.*, one often attempts, through the suction or injection of fluid through orifices on the boundary, to reduce the viscous drag on a body moving through a fluid. Control may be effected in other ways, *e.g.*, through the body force or the stress vector on the boundary. Such cases are treated in [GHS2]; see also [Lio].

In practical situations it is likely that the velocity is specified on only part of Γ_u. Thus, for example, one may also want to consider problems such that on part of Γ_u one specifies the stress force, or more generally, some components of the velocity and complementary components of the stress. In principle, there is no difficulty extending the results of this paper to such cases, provided the necessary existence, regularity and approximation results for analogous boundary value problems for the Navier-Stokes equations are available.

The plan of the paper is as follows. In the remainder of this section we introduce the notation that will be used throughout the paper. Then, in Section 2, we give a precise statement of our optimization problem and derive an optimality system. In Section 3 we consider finite element approximations and derive error estimates. Details concerning the results of this particular paper may be found in [GHS3].

1.1 – Notation Throughout, C will denote a positive constant whose meaning and value changes with context. Also, $H^s(\mathcal{D})$, $s \in \mathbb{R}$, denotes the standard Sobolev space of order s with respect to the set $\mathcal{D}$ where $\mathcal{D}$ is either the flow domain Ω, or its boundary Γ, or part of the that boundary. Of course, $H^0(\mathcal{D}) = L^2(\mathcal{D})$. Corresponding Sobolev spaces of vector valued functions will be denoted by $\mathbf{H}^s(\mathcal{D})$; *e.g.*, $\mathbf{H}^1(\Omega) = [H^1(\Omega)]^d$. Dual spaces will be denoted by $(\cdot)^*$.

Of particular interest will be the space

$$\mathbf{H}^1(\Omega) = \{v_j \in L^2(\Omega) \quad | \quad \frac{\partial v_j}{\partial x_k} \in L^2(\Omega) \quad \text{for } j,k = 1,\ldots,d\}$$

and the subspaces

$$\mathbf{H}_0^1(\Omega) = \{\mathbf{v} \in \mathbf{H}^1(\Omega) \quad | \quad \mathbf{v} = \mathbf{0} \quad \text{on } \Gamma\}$$

and

$$L_0^2(\Omega) = \{q \in L^2(\Omega) \quad | \quad \int_\Omega q \, d\Omega = 0\}.$$

For functions defined on Γ_c we will use the subspaces

$$\mathbf{H}_n^1(\Gamma_c) = \{\mathbf{g} \in \mathbf{H}^1(\Gamma_c) \quad | \quad \int_{\Gamma_c} \mathbf{g} \cdot \mathbf{n} \, d\Gamma = 0\},$$

$$\mathbf{H}_0^1(\Gamma_c) = \{\mathbf{g} \in \mathbf{H}^1(\Gamma_c) \quad | \quad \mathbf{g} = \mathbf{0} \quad \text{on } \partial\Gamma_c\}$$

whenever Γ_c is not connected, and

$$\mathbf{H}_c^1 = \begin{cases} \mathbf{H}_n^1(\Gamma_c) & \text{if } \Gamma_c \text{ is connected} \\ \mathbf{H}_n^1(\Gamma_c) \cap \mathbf{H}_0^1(\Gamma_c) & \text{otherwise} . \end{cases}$$

Norms of functions belonging to $H^s(\Omega)$, $H^s(\Gamma)$ and $H^s(\Gamma_c)$ are denoted by $\|\cdot\|_s$, $\|\cdot\|_{s,\Gamma}$ and $\|\cdot\|_{s,\Gamma_c}$, respectively. Norms for spaces of vector valued functions will be denoted by the same notation as that used for their scalar counterparts. Semi-norms will be denoted by $|\cdot|_s$.

We define, for $p, q \in L^2(\Omega)$ and $\mathbf{u}, \mathbf{v} \in \mathbf{L}^2(\Omega)$

$$(p,q) = \int_\Omega pq \, d\Omega \quad \text{and} \quad (\mathbf{u},\mathbf{v}) = \int_\Omega \mathbf{u}\cdot\mathbf{v} \, d\Omega \,, \tag{1.7}$$

respectively, for $p, q \in L^2(\Gamma)$ or $\mathbf{u}, \mathbf{v} \in \mathbf{L}^2(\Gamma)$,

$$(p,q)_\Gamma = \int_\Gamma pq \, d\Gamma \quad \text{and} \quad (\mathbf{u},\mathbf{v})_\Gamma = \int_\Gamma \mathbf{u}\cdot\mathbf{v}\, d\Gamma \,, \tag{1.8}$$

respectively, and for $p, q \in L^2(\Gamma_c)$ or $\mathbf{u}, \mathbf{v} \in \mathbf{L}^2(\Gamma_c)$,

$$(p,q)_{\Gamma_c} = \int_{\Gamma_c} pq \, d\Gamma \quad \text{and} \quad (\mathbf{u},\mathbf{v})_{\Gamma_c} = \int_{\Gamma_c} \mathbf{u}\cdot\mathbf{v}\, d\Gamma \,, \tag{1.9}$$

respectively. Thus, the inner products in $L^2(\Omega)$ and $\mathbf{L}^2(\Omega)$ are both be denoted by $(\cdot,\cdot)$, those in $L^2(\Gamma)$ and $\mathbf{L}^2(\Gamma)$ by $(\cdot,\cdot)_\Gamma$ and those in $L^2(\Gamma_c)$ and $\mathbf{L}^2(\Gamma_c)$ by $(\cdot,\cdot)_{\Gamma_c}$. The notation of (1.7)-(1.9) will also be employed to denote pairings between Sobolev spaces and their duals.

We will use the two bilinear forms

$$a(\mathbf{u},\mathbf{v}) = \frac{1}{2}\int_\Omega \left((\operatorname{grad}\mathbf{u}) + (\operatorname{grad}\mathbf{u})^T\right) : \left((\operatorname{grad}\mathbf{v}) + (\operatorname{grad}\mathbf{v})^T\right) d\Omega \quad \forall\, \mathbf{u},\mathbf{v} \in \mathbf{H}^1(\Omega)$$

and

$$b(\mathbf{v},q) = -\int_\Omega q \operatorname{div}\mathbf{v}\, d\Omega \quad \forall\, \mathbf{v} \in \mathbf{H}^1(\Omega) \text{ and } \forall\, p \in L^2(\Omega)$$

and the trilinear form

$$c(\mathbf{u},\mathbf{v},\mathbf{w}) = \int_\Omega \mathbf{u}\cdot\operatorname{grad}\mathbf{v}\cdot\mathbf{w}\, d\Omega \quad \forall\, \mathbf{u},\mathbf{v},\mathbf{w} \in \mathbf{H}^1(\Omega)\,.$$

For details concerning the notation employed and for relevant properties of the various forms introduced above, one may consult [Ada], [GiR], [Gun] and [Tem].

II – THE OPTIMALITY SYSTEM

2.1 – The optimization problem and the existence of Lagange multipliers We begin by giving a precise statement of the optimization problem we consider. Let $\mathbf{g} \in \mathbf{H}^1_c$ denote the boundary control and let $\mathbf{u} \in \mathbf{H}^1(\Omega)$ and $p \in L^2_0(\Omega)$ denote the state, *i.e.*, the velocity and pressure fields, respectively. The state and control variables are constrained to satisfy the system (1.2)-(1.4), which we recast into the following particular weak form (see, *e.g.*, [Bab], [GiR], [Gun] or [Tem]):

$$\nu a(\mathbf{u},\mathbf{v}) + c(\mathbf{u},\mathbf{u},\mathbf{v}) + b(\mathbf{v},p) + (\boldsymbol{\lambda},\mathbf{v})_\Gamma = (\mathbf{f},\mathbf{v}) \quad \forall\, \mathbf{v} \in \mathbf{H}^1(\Omega)\,, \tag{2.1}$$

$$b(\mathbf{u},q) = 0 \quad \forall\, q \in L^2_0(\Omega) \tag{2.2}$$

and

$$(\mathbf{u},\boldsymbol{\mu})_\Gamma - (\mathbf{g},\boldsymbol{\mu})_{\Gamma_c} = (\mathbf{b},\boldsymbol{\mu})_\Gamma \quad \forall\, \boldsymbol{\mu} \in \mathbf{H}^{-1/2}(\Gamma) \tag{2.3}$$

where $\mathbf{f} \in \mathbf{L}^2(\Omega)$ and $\mathbf{b} \in \mathbf{H}^1(\Gamma)$ are given functions.

The functional (1.1), using the notation introduced in Section 1.1, is given by

$$\mathcal{K}(\mathbf{u},\mathbf{g}) = \frac{\nu}{2}a(\mathbf{u},\mathbf{u}) - (\mathbf{f},\mathbf{u}) + \frac{\nu}{2}|\mathbf{g}|^2_{1,\Gamma_c} \tag{2.4}$$

The *admissibility set* $\mathcal{U}_{ad}$ is defined by

$$\mathcal{U}_{ad} = \{(\mathbf{u},\mathbf{g}) \in \mathbf{H}^1(\Omega) \times \mathbf{H}^1_c :$$
$$\mathcal{K}(\mathbf{u},\mathbf{g}) < \infty \text{ and there exists a } p \in L^2_0(\Omega) \text{ and } \lambda \in \mathbf{H}^{-1/2}(\Gamma) \text{ such that (2.1)-(2.3) are satisfied}\}. \quad (2.5)$$

Then, $(\hat{\mathbf{u}},\hat{\mathbf{g}}) \in \mathcal{U}_{ad}$ is called an *optimal solution* if there exists $\epsilon > 0$ such that

$$\mathcal{K}(\hat{\mathbf{u}},\hat{\mathbf{g}}) \le \mathcal{K}(\mathbf{u},\mathbf{g}) \quad \forall (\mathbf{u},\mathbf{g}) \in \mathcal{U}_{ad} \text{ satisfying } \|\mathbf{u}-\hat{\mathbf{u}}\|_1 + \|\mathbf{g}-\hat{\mathbf{g}}\|_{1,\Gamma_c} \le \epsilon. \quad (2.6)$$

In this setting, one can then prove the following result.

Theorem 2.1– *There exists an optimal solution* $(\hat{\mathbf{u}},\hat{\mathbf{g}}) \in \mathcal{U}_{ad}$. *Moreover, any optimal solution* $\hat{\mathbf{u}}$ *belongs to* $\mathbf{H}^{3/2}(\Omega)$ *and any* $\hat{p} \in L^2_0(\Omega)$ *and* $\hat{\lambda} \in \mathbf{H}^{-1/2}(\Gamma)$ *such that* $\{(\hat{\mathbf{u}},\hat{\mathbf{g}}) \in \mathcal{U}_{ad}, \hat{p}\}$ *is a solution of* (2.1)-(2.3) *satisfies* $\hat{p} \in H^{1/2}(\Omega) \cap L^2_0(\Omega)$ *and* $\hat{\lambda} \in \mathbf{L}^2(\Gamma)$. ∎

We wish to use the method of Lagrange multipliers to turn the constrained optimization problem (2.5) into an unconstrained one. We first must show that suitable Lagrange multipliers exist.

Let $B_1 = \mathbf{H}^1(\Omega) \times L^2_0(\Omega) \times \mathbf{H}^1_c \times \mathbf{H}^{-1/2}(\Gamma)$ and $B_2 = (\mathbf{H}^1(\Omega))^* \times L^2_0(\Omega) \times \mathbf{H}^{1/2}(\Gamma)$ and let the nonlinear mapping $M : B_1 \to B_2$ denote the constraint equations, *i.e.*, $M(\mathbf{u},p,\mathbf{g},\lambda) = (\mathbf{f},s,\mathbf{b})$ for $(\mathbf{u},p,\mathbf{g},\lambda) \in B_1$ and $(\mathbf{f},s,\mathbf{b}) \in B_2$ if and only if

$$\nu a(\mathbf{u},\mathbf{v}) + c(\mathbf{u},\mathbf{u},\mathbf{v}) + b(\mathbf{v},p) + (\lambda,\mathbf{v})_\Gamma = (\mathbf{f},\mathbf{v}) \quad \forall\, \mathbf{v} \in \mathbf{H}^1(\Omega), \quad (2.7)$$

$$b(\mathbf{u},q) = (s,q) \quad \forall\, q \in L^2_0(\Omega) \quad (2.8)$$

and

$$(\mathbf{u},\mu)_\Gamma - (\mathbf{g},\mu)_{\Gamma_c} = (\mathbf{b},\mu)_\Gamma \quad \forall\, \mu \in \mathbf{H}^{-1/2}(\Gamma). \quad (2.9)$$

Given $\mathbf{u} \in \mathbf{H}^1(\Omega)$, the operator $M'(\mathbf{u}) \in \mathcal{L}(B_1;B_2)$ may be defined as follows: $M'(\mathbf{u}) \cdot (\mathbf{w},r,\mathbf{k},\sigma) = (\bar{\mathbf{f}},\bar{s},\bar{\mathbf{b}})$ for $(\mathbf{w},r,\mathbf{k},\sigma) \in B_1$ and $(\bar{\mathbf{f}},\bar{s},\bar{\mathbf{b}}) \in B_2$ if and only if,

$$\nu a(\mathbf{w},\mathbf{v}) + c(\mathbf{w},\mathbf{u},\mathbf{v}) + c(\mathbf{u},\mathbf{w},\mathbf{v}) + b(\mathbf{v},r) + (\sigma,\mathbf{v})_\Gamma = (\bar{\mathbf{f}},\mathbf{v}) \quad \forall\, \mathbf{v} \in \mathbf{H}^1(\Omega), \quad (2.10)$$

$$b(\mathbf{w},q) = (\bar{s},q) \quad \forall\, q \in L^2(\Omega) \quad (2.11)$$

and

$$(\mathbf{w},\mu)_\Gamma - (\mathbf{k},\mu)_{\Gamma_c} = (\bar{\mathbf{b}},\mu)_\Gamma \quad \forall\, \mu \in \mathbf{H}^{-1/2}(\Gamma). \quad (2.12)$$

We are now prepared to show the existence of Lagrange multipliers.

Theorem 2.2– *Let* $(\hat{\mathbf{u}}, \hat{\mathbf{g}}) \in \mathbf{H}^1(\Omega) \times \mathbf{H}^1(\Gamma_c)$ *denote an optimal solution in the sense of* (2.6). *Then there exists a nonzero Lagrange multiplier* $(\boldsymbol{\xi}, \phi, \boldsymbol{\eta}) \in \mathbf{H}^1(\Omega) \times L_0^2(\Omega) \times \mathbf{H}^{-1/2}(\Gamma)$ *satisfying the Euler equations*

$$\begin{aligned} -\mathcal{K}'(\hat{\mathbf{u}},\hat{\mathbf{g}}) \cdot (\mathbf{w}, r, \mathbf{k}, \sigma) + &< M'(\hat{\mathbf{u}}) \cdot (\mathbf{w}, r, \mathbf{k}, \sigma), (\boldsymbol{\xi}, \phi, \boldsymbol{\eta}) > \\ &= 0 \quad \forall (\mathbf{w}, r, \mathbf{k}, \sigma) \in \mathbf{H}^1(\Omega) \times L_0^2(\Omega) \times \mathbf{H}_c^1 \times \mathbf{H}^{-1/2}(\Gamma) \end{aligned} \tag{2.13}$$

where $< \cdot, \cdot >$ *denotes the duality pairing between* $\mathbf{H}^1(\Omega) \times L_0^2(\Omega) \times \mathbf{H}^{-1/2}(\Gamma)$ *and* $(\mathbf{H}^1(\Omega))^* \times L_0^2(\Omega) \times \mathbf{H}^{1/2}(\Gamma)$. ∎

2.2 – The optimality system Using (2.10)-(2.12) and dropping the $(\hat{\cdot})$ notation for optimal solutions, we may rewrite (2.13) in the form

$$\nu a(\mathbf{w}, \boldsymbol{\xi}) + c(\mathbf{w}, \mathbf{u}, \boldsymbol{\xi}) + c(\mathbf{u}, \mathbf{w}, \boldsymbol{\xi}) + b(\mathbf{w}, \phi) + (\boldsymbol{\eta}, \mathbf{w})_\Gamma = \nu a(\mathbf{u}, \mathbf{w}) - (\mathbf{f}, \mathbf{w}) \quad \forall \mathbf{w} \in \mathbf{H}^1(\Omega), \tag{2.14}$$

$$b(\boldsymbol{\xi}, r) = 0 \quad \forall r \in L_0^2(\Omega), \tag{2.15}$$

$$(\sigma, \boldsymbol{\xi})_\Gamma = 0 \quad \forall \sigma \in \mathbf{H}^{-1/2}(\Gamma) \tag{2.16}$$

$$\nu(\operatorname{grad}_s \mathbf{g}, \operatorname{grad}_s \mathbf{k})_{\Gamma_c} = -(\mathbf{k}, \boldsymbol{\eta})_{\Gamma_c} \quad \forall \mathbf{k} \in \mathbf{H}_c^1. \tag{2.17}$$

Since for some $\boldsymbol{\lambda} \in \mathbf{H}^{-1/2}(\Gamma)$ optimal solutions satisfy the constraint (2.1)-(2.3), we see necessary conditions for an optimum are that (2.1)-(2.3) and (2.14)-(2.17) are satisfied. This system of equations will be called the *optimality system*.

Using equation (2.1),which is satisfied by optimal solutions, we may replace (2.14) by

$$\nu a(\mathbf{w}, \boldsymbol{\xi}) + c(\mathbf{w}, \mathbf{u}, \boldsymbol{\xi}) + c(\mathbf{u}, \mathbf{w}, \boldsymbol{\xi}) + b(\mathbf{w}, \phi) + (\boldsymbol{\eta}, \mathbf{w})_\Gamma = -b(\mathbf{w}, p) - c(\mathbf{u}, \mathbf{u}, \mathbf{w}) - (\mathbf{v}, \boldsymbol{\lambda})_\Gamma \quad \forall \mathbf{w} \in \mathbf{H}^1(\Omega);$$

then, we effect the replacement $\tilde{\phi} = \phi + \hat{p}$ to yield

$$\nu a(\mathbf{w}, \boldsymbol{\xi}) + c(\mathbf{w}, \mathbf{u}, \boldsymbol{\xi}) + c(\mathbf{u}, \mathbf{w}, \boldsymbol{\xi}) + b(\mathbf{w}, \tilde{\phi}) + (\boldsymbol{\eta}, \mathbf{w})_\Gamma = -c(\mathbf{u}, \mathbf{u}, \mathbf{w}) - (\mathbf{v}, \boldsymbol{\lambda})_\Gamma \quad \forall \mathbf{w} \in \mathbf{H}^1(\Omega). \tag{2.18}$$

The replacement of the right hand side of (2.14) by the right hand side of (2.18) facilitates the derivation of the regularity results of Theorem 2.3.

Then the optimality system in terms of the variables $\mathbf{u}, p, \boldsymbol{\lambda}, \mathbf{g}, \boldsymbol{\xi}, \phi$ and $\boldsymbol{\eta}$ is given by (2.1)-(2.3) and (2.15-(2.18). Integrations by parts may be used to show that this system constitutes a weak formulation of the boundary value problem (dropping the the $(\tilde{\cdot})$ notation for $\tilde{\phi}$)

$$-\nu \operatorname{div}\left((\operatorname{grad} \mathbf{u}) + (\operatorname{grad} \mathbf{u})^T\right) + \mathbf{u} \cdot \operatorname{grad} \mathbf{u} + \operatorname{grad} p = \mathbf{f} \quad \text{in } \Omega, \tag{2.19}$$

$$\operatorname{div} \mathbf{u} = 0 \quad \text{in } \Omega, \tag{2.20}$$

$$\mathbf{u} = \begin{cases} \mathbf{g} + \mathbf{b} & \text{on } \Gamma_c \\ \mathbf{b} & \text{on } \Gamma_u, \end{cases} \tag{2.21}$$

$$-\Delta_s \mathbf{g} + \beta \mathbf{n} = \eta = -(\phi - p)\mathbf{n}\nu([\mathrm{grad}(\xi - \mathbf{u})] + [\mathrm{grad}(\xi - \mathbf{u})]^T) \cdot \mathbf{n} \quad \text{on } \Gamma_c , \tag{2.22}$$

$$\int_{\Gamma_c} \mathbf{g} \cdot \mathbf{n}\, d\Gamma = 0 \quad \text{and if } \Gamma_c \text{ is not connected}, \quad \mathbf{g} = \mathbf{0} \text{ on } \partial\Gamma_c \tag{2.23}$$

$$-\nu \mathrm{div}\left((\mathrm{grad}\,\xi) + (\mathrm{grad}\,\xi)^T\right) + \xi \cdot (\mathrm{grad}\,\mathbf{u})^T - \mathbf{u} \cdot \mathrm{grad}\,\xi + \mathrm{grad}\,\phi = -\mathbf{u} \cdot \mathrm{grad}\,\mathbf{u} \quad \text{in } \Omega , \tag{2.24}$$

$$\mathrm{div}\,\xi = 0 \quad \text{in } \Omega \tag{2.25}$$

and

$$\xi = \mathbf{0} \quad \text{on } \Gamma . \tag{2.26}$$

Note that in (2.22) Δ_s denotes the surface Laplacian and in (2.24)

$$(\mathbf{u} \cdot \mathrm{grad}\,\xi)_i = \sum_{j=1}^{d} u_j \frac{\partial \xi_i}{\partial x_j} \quad \text{and} \quad (\xi \cdot (\mathrm{grad}\,\mathbf{u})^T)_i = \sum_{j=1}^{d} \xi_j \frac{\partial u_j}{\partial x_i} \quad \text{for } i = 1, \ldots, d.$$

Also, in (2.22), $\beta \in \mathbb{R}$ is an additional unknown constant that accounts for the integral constraint of (2.23).

The optimality system (2.19)-(2.26) consists of the Navier-Stokes system (2.19)-(2.21), the system (2.24)-(2.26) whose left hand side is the adjoint of Navier-Stokes operator linearized about $\mathbf{u}$, and the surface Lapalacian system (2.22)-(2.23).

Insofar as the regularity of solutions of the optimality system (2.19)-(2.26) is concerned, we have the following result. (Note that unless Γ_c is connected, we cannot concluded that $\mathbf{u}|_\Gamma$ is any smoother than an $\mathbf{H}^1(\Gamma)$ function, and in this case we cannot improve on the regularity result of Theorem 2.1. Thus, in the following theorem we assume that Γ_c is connected.)

Theorem 2.3– *Suppose that Γ_c is connected and that the given data satisfies $\mathbf{b} \in \mathbf{H}^{3/2}(\Gamma)$ and $\mathbf{f} \in \mathbf{L}^2(\Omega)$. Suppose that Ω is of class $C^{1,1}$. Then, if $(\mathbf{u}, p, \mathbf{g}, \xi, \phi) \in \mathbf{H}^1(\Omega) \times L^2_0(\Omega) \times \mathbf{H}^1_c \times \mathbf{H}^1(\Omega) \times L^2_0(\Omega)$ denotes a solution of the optimality system* (2.1)-(2.3) *and* (2.15)-(2.18), *or equivalently,* (2.19)-(2.26), *we have that $(\mathbf{u}, p, \mathbf{g}, \xi, \phi) \in \mathbf{H}^2(\Omega) \times H^1(\Omega) \times \mathbf{H}^{3/2}(\Gamma_c) \times \mathbf{H}^2(\Omega) \times H^1(\Omega)$. If the boundary is sufficiently smooth, we also may conclude that $\mathbf{g} \in \mathbf{H}^{5/2}(\Gamma_c)$.* ∎

The above result also holds for convex regions of $\mathbb{R}^2$, provided $\Gamma_c = \Gamma$. In general, if Γ_c is connected, we may show that if $\mathbf{f} \in \mathbf{H}^m(\Omega)$ and $\mathbf{b} \in \mathbf{H}^{m+\frac{3}{2}}(\Gamma)$ and Ω is sufficiently smooth, then $(\mathbf{u}, p, \mathbf{g}, \xi, \phi) \in \mathbf{H}^{m+2}(\Omega) \times H^{m+1}(\Omega) \times \mathbf{H}^{m+\frac{3}{2}}(\Gamma_c) \times \mathbf{H}^{m+2}(\Omega) \times H^{m+1}(\Omega)$. In particular, if $\mathbf{f}$ and $\mathbf{b}$ are of class $C^\infty(\overline{\Omega})$ and Ω is of class C^∞, then $\mathbf{u}, p, \mathbf{g}, \xi$ and ϕ are all $C^\infty(\overline{\Omega})$ functions as well.

III – FINITE ELEMENT APPROXIMATIONS A finite element discretization of the optimality system (2.1)-(2.3) and (2.15)-(2.18) is defined as follows. First one chooses families of finite dimensional subspaces $\mathbf{V}^{h_1} \subset \mathbf{H}^1(\Omega)$, $S^{h_1} \subset L^2(\Omega)$, $\mathbf{P}^{h_2} \subset \mathbf{H}^{-1/2}(\Gamma)$ and $\mathbf{Q}^{h_3} \subset \mathbf{H}^1(\Gamma_c)$. These families are parametrized by parameters h_1, h_2 and h_3 that tends to zero; commonly, these parameters are chosen to be some measure of appropriate interior and boundary grid sizes. We let $S_0^{h_1} = S^{h_1} \cap L_0^2(\Omega)$, $\mathbf{V}_0^{h_1} = \mathbf{V}^{h_1} \cap \mathbf{H}_0^1(\Omega)$, $\mathbf{Q}_c^{h_3} = \mathbf{Q}^{h_3}$ if Γ_c is connected and $\mathbf{Q}_c^{h_3} = \mathbf{Q}^{h_3} \cap \mathbf{H}_0^1(\Gamma_c)$ otherwise; also, we let $h = \max(h_1, h_2, h_3)$.

One may choose any pair of subspaces $\mathbf{V}^{h_1}$ and S^{h_1} that can be used for finding finite element approximations of solutions of the Navier-Stokes equations. Thus, concerning these subspaces, we make the following standard assumptions which are exactly those employed in well known finite element methods for the Navier-Stokes equations. First we have the approximation properties: there exist an integer k and a constant C, independent of h_1, $\mathbf{v}$ and q, such that

$$\inf_{\mathbf{v}^h \in \mathbf{V}^{h_1}} \|\mathbf{v} - \mathbf{v}^h\|_1 \le C(h_1)^m \|\mathbf{v}\|_{m+1} \quad \forall\, \mathbf{v} \in \mathbf{H}^{m+1}(\Omega)\,,\ 1 \le m \le k \tag{3.1}$$

and

$$\inf_{q^h \in S_0^{h_1}} \|q - q^h\|_0 \le C(h_1)^m \|q\|_m \quad \forall\, q \in H^m(\Omega) \cap L_0^2(\Omega)\,,\ 1 \le m \le k; \tag{3.2}$$

next, we assume the *inf-sup condition*, or *Ladyzhenskaya-Babuska-Brezzi condition*: there exists a constant C, independent of h_1, such that

$$\inf_{0 \ne q^h \in S_0^{h_1}} \ \sup_{0 \ne \mathbf{v}^h \in \mathbf{V}^{h_1}} \frac{b(\mathbf{v}^h, q^h)}{\|\mathbf{v}^h\|_1 \|q^h\|_0} \ge C. \tag{3.3}$$

This condition assures the stability of finite element discretizations of the Navier-Stokes equations. For thorough discussions of the approximation properties (3.1)-(3.2), see, *e.g.*, [Cia] and for like discussions of the stability condition (3.3), see, *e.g.*, [GiR] or [Gun]. These may also be consulted for a catalogue of finite element subspaces that meet the requirements of (3.1)-(3.3).

For the subspace $\mathbf{P}^{h_2}$, we have the approximation property: there exist an integer k and a constant C, independent of h_2 and $\boldsymbol{\mu}$, such that

$$\inf_{\boldsymbol{\mu}^h \in \mathbf{P}^{h_2}} \|\boldsymbol{\mu} - \boldsymbol{\mu}^h\|_{-1/2,\Gamma} \le C(h_2)^m \|\boldsymbol{\mu}\|_{m-\frac{1}{2}} \quad \forall\, \boldsymbol{\mu} \in \mathbf{H}^{m-\frac{1}{2}}(\Gamma)\,,\ 1 \le m \le k\,, \tag{3.4}$$

and the inverse assumption: there exists a constant C, independent of h_2 and $\boldsymbol{\mu}^h$ such that

$$\|\boldsymbol{\mu}^h\|_{s,\Gamma} \le C(h_2)^{s-q} \|\boldsymbol{\mu}\|_{q,\Gamma} \quad \forall\, \boldsymbol{\mu}^h \in \mathbf{P}^{h_2}\,,\ -1/2 \le q \le s \le 1/2\,. \tag{3.5}$$

Standard piecewise polynomial spaces defined with respect to the boundary Γ satisfy (3.4) and (3.5); see [Bab] and [Cia] for examples and details.

For the subspace $\mathbf{Q}_c^{h_3}$ we simply make the approximability assumption that there exist an integer k and a constant C, independent of h_3 and $\mathbf{k}$, such that

$$\inf_{q^h \in \mathbf{Q}_c^{h_3}} \|\mathbf{k} - \mathbf{k}^h\|_{s,\Gamma_c} \le C(h_3)^{m-s} \|\mathbf{k}\|_{m,\Gamma_c} \quad \forall\, \mathbf{k} \in \mathbf{H}^1(\Gamma_c)\,,\ 1 \le m \le k,\quad 0 \le s \le 1 \tag{3.6}$$

whenever Γ_c is connected; otherwise, we replace $\mathbf{H}^1(\Gamma_c)$ by $\mathbf{H}_0^1(\Gamma_c)$ in (3.6).

Once the approximating subspaces have been chosen, we seek $\mathbf{u}^h \in \mathbf{V}^{h_1}, p^h \in S_0^{h_1}, \lambda^h \in \mathbf{P}^{h_2}, \mathbf{g}^h \in \mathbf{Q}_c^{h_3}, \xi^h \in \mathbf{V}^{h_1}, \sigma^h \in S_0^{h_1}, \eta^h \in \mathbf{P}^{h_2}$ and $\beta \in \mathbb{R}$ such that

$$\nu a(\mathbf{u}^h, \mathbf{v}^h) + c(\mathbf{u}^h, \mathbf{u}^h, \mathbf{v}^h) + b(\mathbf{v}^h, p^h) + (\lambda^h, \mathbf{v}^h)_\Gamma = (\mathbf{f}, \mathbf{v}^h) + \quad \forall\, \mathbf{v}^h \in \mathbf{V}^{h_1}, \tag{3.7}$$

$$b(\mathbf{u}^h, q^h) = 0 \quad \forall\, q^h \in S^{h_1}, \tag{3.8}$$

$$(\mathbf{u}^h, \mu^h)_\Gamma - (\mathbf{g}^h, \mu^h)_{\Gamma_c} = (\mathbf{b}, \mu^h)_\Gamma \quad \forall\, \mu^h \in \mathbf{P}^{h_2}, \tag{3.9}$$

$$(\mathrm{grad}_s\, \mathbf{g}^h, \mathrm{grad}_s\, \mathbf{k}^h)_{\Gamma_c} + \beta \int_\Gamma \mathbf{k}^h \cdot \mathbf{n}\, d\Gamma = -(\eta^h, \mathbf{k}^h)_{\Gamma_c} \quad \forall\, \mathbf{k}^h \in \mathbf{Q}^{h_3}, \tag{3.10}$$

$$\int_{\Gamma_c} \mathbf{g}^h \cdot \mathbf{n}\, d\Gamma = 0, \tag{3.11}$$

$$\begin{aligned} \nu a(\mathbf{w}^h, \xi^h) + c(\mathbf{w}^h, \mathbf{u}^h, \xi^h) + c(\mathbf{u}^h, \mathbf{w}^h, \xi^h) + b(\mathbf{w}^h, \sigma^h) + (\mathbf{w}^h, \eta^h)_\Gamma \\ = -c(\mathbf{u}^h, \mathbf{u}^h, \mathbf{w}^h) - (\mathbf{w}^h, \lambda^h)_\Gamma \quad \forall\, \mathbf{w}^h \in \mathbf{V}^{h_1}, \end{aligned} \tag{3.12}$$

$$b(\xi^h, r^h) = 0 \quad \forall\, r^h \in S^{h_1} \tag{3.13}$$

and

$$(\xi^h, \sigma^h) = 0 \quad \forall\, \sigma^h \in \mathbf{P}^{h_2}. \tag{3.14}$$

Before stating the error estimate, we need to introduce the notion of a nonsingular solution; for details, see [BRR] or [GiR]. Let Λ denote a compact subset of $\mathbb{R}_+$. A solution $\big(\mathbf{u}(\nu), p(\nu), \xi(\nu), \phi(\nu)\big)$ of the problem (2.1)-(2.3) and (2.15)-(2.18) is nonsingular if the linear system

$$\nu a(\tilde{\mathbf{u}}, \mathbf{v}) + c(\tilde{\mathbf{u}}, \mathbf{u}, \mathbf{v}) + c(\mathbf{u}, \tilde{\mathbf{u}}, \mathbf{v}) + b(\mathbf{v}, \tilde{p}) + (\tilde{\lambda}, \mathbf{v})_\Gamma = (\tilde{\mathbf{f}}, \mathbf{v}) \quad \forall\, \mathbf{v} \in \mathbf{H}^1(\Omega),$$

$$b(\tilde{\mathbf{u}}, q) = (\tilde{s}, q) \quad \forall\, q \in L_0^2(\Omega),$$

$$(\tilde{\mathbf{u}}, \mu)_\Gamma - (\tilde{\mathbf{g}}, \mu)_{\Gamma_c} = (\tilde{\mathbf{b}}, \mu)_\Gamma \quad \forall\, \mu \in \mathbf{H}^{-1/2}(\Gamma),$$

$$\nu(\mathrm{grad}_s\, \tilde{\mathbf{g}}, \mathrm{grad}_s\, \mathbf{k})_{\Gamma_c} + (\mathbf{k}, \tilde{\eta})_{\Gamma_c} = (\tilde{\mathbf{0}}, \mathbf{k})_{\Gamma_c} \quad \forall\, \mathbf{k} \in \mathbf{H}_c^1,$$

$$\begin{aligned} \nu a(\mathbf{w}, \tilde{\xi}) + c(\mathbf{w}, \mathbf{u}, \tilde{\xi}) + c(\mathbf{u}, \mathbf{w}, \tilde{\xi}) + c(\mathbf{w}, \tilde{\mathbf{u}}, \xi) + c(\tilde{\mathbf{u}}, \mathbf{w}, \xi) + b(\mathbf{w}, \tilde{\phi}) + (\tilde{\eta}, \mathbf{w})_\Gamma \\ + c(\tilde{\mathbf{u}}, \mathbf{u}, \mathbf{w}) + c(\mathbf{u}, \tilde{\mathbf{u}}, \mathbf{w}) + (\mathbf{v}, \tilde{\lambda})_\Gamma = (\hat{\mathbf{f}}, \mathbf{w}) \quad \forall\, \mathbf{w} \in \mathbf{H}^1(\Omega), \end{aligned}$$

$$b(\tilde{\xi}, r) = (\hat{s}, r) \quad \forall\, r \in L_0^2(\Omega),$$

$$(\sigma, \tilde{\xi})_\Gamma = (\hat{\mathbf{b}}, \sigma) \quad \forall\, \sigma \in \mathbf{H}^{-1/2}(\Gamma)$$

has a unique solution $(\tilde{\mathbf{u}}, \tilde{p}, \tilde{\lambda}, \tilde{\mathbf{g}}, \tilde{\xi}, \tilde{\phi}, \tilde{\eta})$ for every $(\tilde{\mathbf{f}}, \tilde{s}, \tilde{\mathbf{b}}, \tilde{\mathbf{0}}, \hat{\mathbf{f}}, \hat{s}, \hat{\mathbf{b}})$ belonging to appropriate function spaces.

Using the methods of [Bab], [BRR] and [Cro] (see also [GiR]), one may derive the following result.

Theorem 3.1– *Assume that $h_2 \geq K h_1$ for a sufficiently small constant K, independent of h_1 and h_2. Assume that Λ is a compact interval of $\mathbb{R}_+$ and that there exists a branch $\{(\nu, (\mathbf{u}, p, \xi, \sigma)) \,:\, 1/\nu \in \Lambda\}$ of nonsingular solutions of the optimality system* (2.1)-(2.3) *and* (2.15)-(2.18). *Assume that the finite element spaces $\mathbf{V}^{h_1}$, $S_0^{h_1}$, $\mathbf{P}^{h_2}$*

and Q_c^{h} *satisfy the conditions* (3.1)-(3.6). *Then, there exists a unique branch* $(\mathbf{u}^h, p^h, \lambda^h, \mathbf{g}^h, \xi^h, \phi^h, \eta^h)$ *of solutions of the discrete optimality system* (3.7)-(3.14). *Moreover, if in addition the solution of the optimality system satisfies* $(\mathbf{u}(\nu), p(\nu), \xi(\nu), \phi(\nu)) \in \mathbf{H}^{m+1}(\Omega) \times \mathbf{H}^m(\Omega) \times \mathbf{H}^{m+1}(\Omega) \times \mathbf{H}^m(\Omega)$ *for* $1/\nu \in \Lambda$, *then there exists a constant* C, *independent of* h, *such that*

$$\|\mathbf{u} - \mathbf{u}^h\|_1 + \|p - p^h\|_0 + \|\lambda - \lambda^h\|_{-1/2,\Gamma} + \|\xi - \xi^h\|_1 + \|\phi - \phi^h\|_0 + \|\eta - \eta^h\|_{-1/2,\Gamma}$$
$$\leq Ch^{m-\frac{1}{2}}\left(\|\mathbf{u}\|_{m+1} + \|p\|_m + \|\xi\|_{m+1} + \|\phi\|_m\right)$$

If, in addition, $\mathbf{g} \in \mathbf{H}^{m+1}(\Gamma_c)$, *then*

$$\|\mathbf{u} - \mathbf{u}^h\|_1 + \|p - p^h\|_0 + \|\lambda - \lambda^h\|_{-1/2,\Gamma} + \|\xi - \xi^h\|_1 + \|\phi - \phi^h\|_0 + \|\eta - \eta^h\|_{-1/2,\Gamma}$$
$$\leq Ch^{m}\left(\|\mathbf{u}\|_{m+1} + \|p\|_m + \|\xi\|_{m+1} + \|\phi\|_m + \|\mathbf{g}\|_{m+1,\Gamma_c}\right)$$

and, for the approximation of the optimal control, we have that

$$\|\mathbf{g} - \mathbf{g}^h\|_{0,\Gamma_c} + h\|\mathbf{g} - \mathbf{g}^h\|_{1,\Gamma_c} \leq Ch^{m+1}\|\mathbf{g}\|_{m+1,\Gamma_c}.$$

REFERENCES

[Ada] Adams, R.; *Sobolev Spaces.* Academic, New York 1975.

[Bab] Babuska, I.; The finite element methods woth Lagrange multipliers. *Numer. Math.* **16** 1973, 179–192.

[BRR] Brezzi, F., Rappaz, J. and Raviart, P.-A.; Finite-dimensional approximation of nonlinear problems. Part I: branches of nonsingular solutions. *Numer. Math.* **36** 1980, 1–25.

[Cia] Ciarlet, P.; *The Finite Element Method for Elliptic Problems.* North-Holland, Amsterdam 1978.

[Cro] Crouzeix, M.; Approximation des problèmes faiblement non linéaires. To appear.

[GiR] Girault, V. and Raviart, P.-A.; *Finite Element Methods for Navier-Stokes Equations.* Springer, Berlin 1986.

[Gun] Gunzburger, M.; *Finite Element Methods for Incompressible Viscous Flows: A Guide to Theory, Practice and Algorithms.* Academic, Boston 1989.

[GHS1] Gunzburger, M., Hou, L. and Svobodny, T.; Boundary velocity control of incompressible flow with an application to viscous drag reduction. To appear.

[GHS2] Gunzburger, M., Hou, L. and Svobodny, T.; Analysis and finite element approximation of optimal control problems for the stationary Navier-Stokes equations with distribute and Neumann controls. To appear.

[GHS3] Gunzburger, M., Hou, L. and Svobodny, T.; Analysis and finite element approximation of optimal control problems for the stationary Navier-Stokes equations with Dirichlet controls. To appear.

[Lio] Lions, J.-L.; *Control of Distributed Singular Systems.* Bordas, Paris 1985.

[Ser] Serrin, J.; Mathematical principles of classical fluid mechanics, *Handbüch der Physik* VIII/1 (ed. by S. Flügge and C. Truesdell) Springer, Berlin 1959, 125–263.

[Tem] Temam, R.; *Navier-Stokes Equations.* North-Holland, Amsterdam 1979.

Using the physical properties of systems for control: an illustration

H.J.C. Huijberts *
Department of Applied Mathematics
University of Twente
P.O. Box 217
7500 AE Enschede
The Netherlands

August 13, 1990

Abstract

The importance of using the physical structure of a system for solving control problems is illustrated by means of the input-output decoupling problem with stability for Hamiltonian systems.

AMS Subject Classifications (1980): 93C10,93B50,93D15

Keywords: input-output decoupling, critical stability, clamped dynamics, constrained Hamiltonian systems.

1 Introduction

During the development of (finite dimensional) systems theory in the sixties and seventies there has been a tendency to neglect the natural structures imposed by the physical character of a system. At the end of the seventies and in the beginning of the eighties some authors have started to "put physics into control" (cf. [2,13,8], and also [12]). In [8] this is done by introducing the notion of a **Hamiltonian control system**. Roughly speaking, one can think of a Hamiltonian control system in the sense of [8] as a conservative mechanical system with co-located actuators and sensors. Prototypes of Hamiltonian control systems are rigid robotarm models and finite dimensional models of flexible robotarms.

In section 3 of this contribution we illustrate the importance of using the physical structure of a system for solving control problems by means of the input-output decoupling problem for Hamiltonian control systems, after we have briefly recollected the definition of a Hamiltonian control system in section 2.

2 Hamiltonian control systems

In this section we will give a definition of an affine Hamiltonian control system and introduce some tools for studying Hamiltonian systems (for details we refer to [8,1]).

*Attendance of this workshop was made possible by financial support of the Mechatronics Research Centre Twente

Definition 2.1 *Let S be a symplectic manifold with symplectic form ω. Then a system described by the equations*

$$\dot{x} = X_H(x) - \sum_{j=1}^{m} u_j X_{C_j}(x), \quad y_j = C_j(x) \ (j \in \underline{m}) \tag{1}$$

where $X_H, X_{C_1}, \cdots, X_{C_m}$ are Hamiltonian vector fields with Hamiltonian functions $H, C_1, \cdots, C_m$, defined by setting $\omega(X_H, -) = -dH$, $\omega(X_{C_j}, -) = -dC_j$ $(j \in \underline{m})$, is called an **affine Hamiltonian control system.**

□

Remark 2.2

1. *Throughout the set $\{1, \cdots, k\}$ $(k \in \mathbb{N})$ will be denoted by $\underline{k}$.*
2. *In the sequel the term Hamiltonian system will always mean an affine Hamiltonian control system.*

□

By Darboux's theorem there exist local coordinates $(q_1, \cdots, q_n, p_1, \cdots, p_n)$ for a symplectic manifold (S, ω), called **canonical coordinates**, such that $\omega = \sum_{i=1}^{n} dp_i \wedge dq_i$. In such coordinates $\dot{x} = X_H(x)$ reduces to the familiar expressions $\dot{q}_i = \frac{\partial H}{\partial p_i}(q,p)$, $\dot{p}_i = -\frac{\partial H}{\partial q_i}(q,p)$, and similarly for X_{C_j}. Note that this implies that a symplectic manifold is necessarily even dimensional. One particular subclass of Hamiltonian systems often encountered in practice, e.g. conservative mechanical systems, is defined below:

Definition 2.3 *Let S be a symplectic manifold of the form T^*Q with Q the configuration manifold with coordinates $(q_1, \cdots, q_n)$. Let furthermore $H(q,p) = K(q,p) + V(q)$, where $K(q,p) = \frac{1}{2} \sum_{i,j=1}^{n} g^{ij}(q) p_i p_j$ and the matrix $(g^{ij}(q))$ is positive definite for all q, and let $C_1, \cdots, C_m : Q \mapsto \mathbb{R}$. Then the Hamiltonian system (1) is called* **simple**. *The term $K(q,p)$ is called the* **kinetic energy** *and the term $V(q)$ is called the* **potential energy** *of the system.*

□

Given two real valued functions F, G on a symplectic manifold (S, ω), we define their **Poisson-bracket** by: $\{F, G\} := \mathcal{L}_{X_F} G = \omega(X_F, X_G)$ ($\mathcal{L}$ denoting the Lie-derivative). In canonical coordinates this reads: $\{F, G\} = \sum_{i=1}^{n} (\frac{\partial F}{\partial p_i}\frac{\partial G}{\partial q_i} - \frac{\partial F}{\partial q_i}\frac{\partial G}{\partial p_i})$. Furthermore we define inductively: $ad_F^0 G = G$, $ad_F^k G = \{F, ad_F^{k-1} G\}$ $(k = 1, 2, \cdots)$. The Poisson-bracket satisfies the **Jacobi-identity**: $\{F, \{G, H\}\} + \{H, \{F, G\}\} + \{G, \{H, F\}\} = 0$ for any $F, G, H : S \mapsto \mathbb{R}$.

For a Hamiltonian system (1) we define **characteristic numbers** $\rho_1(x), \cdots, \rho_m(x)$ in the following way: $\rho_i(x) = inf\{k \in \mathbb{N}$ *such that* $\{C_j, ad_H^k C_i\}(x) \neq 0$ *for some* $j\}$. Throughout this paper we will have as a standing assumption that the characteristic numbers are constant and finite, i.e. $\rho_i(x) = \rho_i < \infty$. If this assumption holds, we can define the **decoupling matrix** $A(x)$ for (1) as the matrix with elements $a_{ij}(x) = \{C_j, ad_H^{\rho_i} C_i\}(x)$ $(i, j \in \underline{m})$.

A submanifold $T \subset S$ is called a **symplectic submanifold** of S if ω restricted to T is non-degenerate. In the special case that a submanifold T is given by $T = \{x \mid F_1(x) = \cdots = F_{2k}(x) = 0\}$, where $F_1(x), \cdots, F_{2k}(x)$ are independent real valued functions on (S, ω), T is a symplectic submanifold of S if and only if:

$$(\forall x \in T)(\forall i \in \underline{2k})(\exists j \in \underline{2k})(\{F_i, F_j\}(x) \neq 0) \tag{2}$$

For the proof we refer to [10].

As a consequence we have:

Theorem 2.4 *Consider a Hamiltonian system (1) and assume that its decoupling matrix has full rank for all $x \in S$. Then:*

1. *The functions $C_1, \cdots, ad_H^{\rho_1} C_1, \cdots, C_m, \cdots, ad_H^{\rho_m} C_m$ are independent.*

2. $\sum_{i=1}^{m} (\rho_i + 1)$ *is even.*

3. *The submanifold $N^* = \{x \mid ad_H^k C_i(x) = 0;\ i \in \underline{m},\ k = 0, 1, \cdots, \rho_i\}$ is a symplectic submanifold of S.*

Proof For a proof of 1, see [6]. For a proof of 2, see [4]. 3 follows from (2),1,2 (see [9,10]).

□

3 The input-output decoupling problem with stability

We consider a Hamiltonian system (1). The **input-output decoupling problem** is defined as: find (if possible) a feedback $u = \alpha(x) + \beta(x)v$ with the (m, m)-matrix $\beta(x)$ invertible for all x, such that each of the new inputs v_i ($i \in \underline{m}$) influences one and only one output. It is well known (cf. [6]) that this problem is solvable for (1) if and only if the decoupling matrix $A(x)$ of (1), that has been defined in section 2, has full rank for all x. Hence we will assume in the sequel that $A(x)$ has full rank for all x. Furthermore we will assume that the distribution spanned by the input vector fields $X_{C_1}, \cdots, X_{C_m}$ is involutive. In fact the latter assumption is not a very restrictive assumption: it is already satisfied if all characteristic numbers are greater than 0 (cf. [5]).

By Theorem (2.4) we can choose new partial local coordinates $(\xi_1, \cdots, \xi_m)$, where $\xi_i = col(C_i, \cdots, ad_H^{\rho_i} C_i)$. Furthermore, since by Theorem (2.4) N^* is a symplectic submanifold, we can complete this set of partial local coordinates with partial local coordinates $(\bar{q}, \bar{p}) = (\bar{q}_1, \cdots, \bar{q}_d, \bar{p}_1, \cdots, \bar{p}_d)$ such that $(\xi_1, \cdots, \xi_m, \bar{q}, \bar{p})$ are coordinates for S and $(\bar{q}, \bar{p})$ are local canonical coordinates for N^* (where $2d = 2n - \sum_{i=1}^{m} (\rho_i + 1)$). Also, since the distribution spanned by the input vector fields is assumed to be involutive, we can choose $(\bar{q}, \bar{p})$ in such a way that $\mathcal{L}_{X_{C_j}}((\bar{q}, \bar{p})) = 0$ (cf. [5]).

Then in these new coordinates (1) becomes:

$$
\left\{
\begin{array}{rcll}
\dot{\xi}_{i0} & = & \xi_{i1} & \\
 & \vdots & & \\
 & \vdots & & (i = 1, \cdots, m) \\
\dot{\xi}_{i\rho_i - 1} & = & \xi_{i\rho_i} & \\
\dot{\xi}_{i\rho_i} & = & ad_H^{\rho_i + 1} C_i - \sum_{j=1}^{m} u_j \{C_j, ad_H^{\rho_i} C_i\} & \\
 & & & \\
\dot{\bar{q}}_i & = & \frac{\partial H}{\partial \bar{p}_i}(\bar{q}, \bar{p}, \xi) & \\
 & & & (i = 1, \cdots, d) \\
\dot{\bar{p}}_i & = & -\frac{\partial H}{\partial \bar{q}_i}(\bar{q}, \bar{p}, \xi) &
\end{array}
\right. \qquad (3)
$$

Remark 3.5 *From (3) we can also derive an interpretation of the characteristic numbers and the decoupling matrix. The fact that the i-th characteristic number equals ρ_i implies that the (ρ_i+1)-th time derivative of the i-th output is the first one that depends explicitly upon the inputs. The i-th row of the decoupling matrix gives the structure of this dependence.*

□

Now define $b(\bar{q},\bar{p},\xi) := col(ad_H^{\rho_1+1}C_1,\cdots,ad_H^{\rho_m+1}C_m)$ and apply the following control to (3):

$$u(\bar{q},\bar{p},\xi) = A^{-1}(\bar{q},\bar{p},\xi)[b(\bar{q},\bar{p},\xi) - v] \tag{4}$$

where $v = col(v_1,\cdots,v_m)$ denotes the new inputs.

Then it is easy to check that the resulting system is of the form

$$\begin{cases} \dot{\xi}_i = A_i\xi_i + B_iv_i & (i=1,\cdots,m) \\ \dot{\bar{q}}_i = \frac{\partial H}{\partial \bar{p}_i}(\bar{q},\bar{p},\xi) & \\ & (i=1,\cdots,d) \\ \dot{\bar{p}}_i = -\frac{\partial H}{\partial \bar{q}_i}(\bar{q},\bar{p},\xi) & \\ y_i = C_i\xi_i & (i=1,\cdots,m) \end{cases} \tag{5}$$

where

$$A_i = \begin{bmatrix} 0 & 1 & \cdots & \cdots & 0 \\ 0 & 0 & \ddots & & \vdots \\ \vdots & \vdots & \ddots & \ddots & \vdots \\ \vdots & \vdots & & \ddots & 1 \\ 0 & 0 & \cdots & \cdots & 0 \end{bmatrix}, \quad B_i = \begin{bmatrix} 0 \\ 0 \\ \vdots \\ 0 \\ 1 \end{bmatrix}$$

$$C_i = \begin{bmatrix} 1 & 0 & \cdots & \cdots & 0 \end{bmatrix}$$

It is obvious from (5) that applying the control (4) to (1) yields a decoupled system. The decoupled system consists of m independent linear systems and an unobservable nonlinear system without inputs on the symplectic manifold N^*. The dynamics of the nonlinear system are given by a Hamiltonian vector field that is parametrized by the coordinate functions ξ. Note that if we keep $\xi = 0$ the dynamics that is left on N^* is just the **clamped** dynamics (or zero dynamics) of the system, where clamped (or zero) dynamics is defined as the dynamics that is compatible with a zero output (cf. [7,11]). Furthermore, for all $i \in \underline{m}$ the pairs (A_i, B_i) are controllable and hence the linear subsystems can be made **asymptotically stable**, even by using feedback that does not violate the decoupling. A natural question is what can be said about the overall stability of the system (5). This question will be treated in the sequel.

Let us assume that the point $(\bar{q},\bar{p},\xi) = (0,0,0)$ is an equilibrium point of the system (5). In what follows we will be concerned with the **local** stability of this equilibrium point. Consider the clamped dynamics (the dynamics that is left when we put $\xi = 0$ in (5)):

$$\begin{cases} \dot{\bar{q}}_i = \frac{\partial \bar{H}}{\partial \bar{q}_i}(\bar{q},\bar{p}) & \\ & (i=1,\cdots,d) \\ \dot{\bar{p}}_i = -\frac{\partial \bar{H}}{\partial \bar{p}_i}(\bar{q},\bar{p}) & \end{cases} \tag{6}$$

where $\bar{H}(\bar{q},\bar{p}) = H(\bar{q},\bar{p},0)$.

Note that $(\bar{q},\bar{p})=(0,0)$ is an equilibrium point of (6). It is well known from **Centre Manifold Theory** (cf. [3]) that if the linear subsystems in (5) are made asymptotically stable the origin is a stable (asymptotically stable) (unstable) equilibrium point of (5) if and only if it is a stable (asymptotically stable) (unstable) equilibrium point of (6). Now the dynamics of (6) is governed by a Hamiltonian vector field. From **Liouville's theorem** (cf. [1]) we know that Hamiltonian vector fields are volume preserving. Hence we can only have asymptotic stability if $N^*=\{0\}$. In fact we have (for a proof, cf. [5]):

Theorem 3.6 *Consider the Hamiltonian system (1). Assume that the decoupling matrix has full rank and that the distribution spanned by the input vector fields is involutive. Let $\bar{H}$ be defined as the Hamiltonian H restricted to N^*. Assume that the origin is an equilibrium of (6). Finally, let coordinates $(\bar{q},\bar{p},\xi)$ for S be given as in this section. Then we have:*

1. *Application of a decoupling feedback (4) can only result in an asymptotically stable equilibrium point $(\bar{q},\bar{p},\xi)=(0,0,0)$ if $N^*=\{0\}$.*
2. *Apply any feedback to (5) that stabilizes the linear subsystems and preserves the decoupling. Then this feedback renders $(\bar{q},\bar{p},\xi)=(0,0,0)$ a stable equilibrium point if and only if the origin is a stable equilibrium point of (6).*
3. *If $\bar{H}$ has an isolated minimum in the origin, then the origin is a stable equilibrium of (6).*
4. *If moreover (1) is a simple Hamiltonian system (cf. Definition (2.3)), then for all $V(q)$ in an open and dense subset of $C^\infty(Q)$ we have: if the origin is a stable equilibrium point of (6), then $\bar{H}$ has an isolated minimum in the origin.*

□

Remark 3.7

1. *Theorem (3.2.4) gives a partial converse of Theorem (3.2.3). In general the converse of Theorem (3.2.3) does not hold, as can be easily seen if we take $\bar{H}(\bar{q},\bar{p})=\frac{1}{2}(\bar{q}_1^2+\bar{p}_1^2)-\frac{1}{2}(\bar{q}_2^2+\bar{p}_2^2)$, which describes two harmonic oscillators, one running backwards (cf. [1])*
2. *The theorem can be easily extended to a generalized version of system equations (1), where the input vector fields in (4.1) are not necessarily given by X_{C_j}, but are instead of the more general form $X_{P_j \circ (C_1,\cdots,C_m)}$, where the mapping $P=(P_1,\cdots,P_m):\mathbb{R}^m\mapsto\mathbb{R}^m$ is assumed to be an isomorphism (see [5]). For example this may happen in the case of robot manipulators. Generally the input torques correspond to the joint coordinates, but the outputs may be given in task space coordinates. In the case of an equal number of inputs and outputs the joint coordinates are usually related to the task coordinates by a transformation which is invertible except for some singular points.*

□

4 Conclusions

In this paper we have briefly illustrated the importance of using the physical structure of a system for solving control problems. This been done by means of the input-output decoupling problem with stability for Hamiltonian systems. It turned out that with the decoupling feedback we have chosen in section 2 at most critical stability can be achieved and that asymptotic stability can only be achieved if $N^*=\{0\}$. A question that arises very naturally is whether we can achieve asymptotic stability when $N^*\neq\{0\}$, by choosing another type of decoupling feedback. Up till now this question has not been answered and remains for further research.

References

[1] Abraham, R. & J.E. Marsden, *Foundations of mechanics* (2nd. edition), Benjamin/ Cummings, London, (1978).

[2] Brockett, R.W., *Control theory and analytical mechanics*, pp. 1-46 of *Geometric control theory* (C. Martin & R. Hermann, eds.), Vol. VII of **Lie Groups: History, Frontiers and Applications**, Math. Sci. Press, Brookline, (1977).

[3] Carr, J., *Applications of Centre Manifold Theory*, Springer, New York, (1981).

[4] Huijberts, H.J.C., *Nonlinear model matching with an application to Hamiltonian systems*, to appear in *Preprints of IFAC Symposium on Nonlinear Control Systems Design*, Capri.

[5] Huijberts, H.J.C. & A.J. van der Schaft, *Input-output decoupling with stability for Hamiltonian systems*, to appear in **Mathematics of Control, Signals and Systems**.

[6] Isidori, A., *Nonlinear control systems, an introduction*, Vol. 72 of **Lecture notes in control and information sciences**, Springer, Berlin, (1985).

[7] Isidori, A. & C.H. Moog, *On the nonlinear equivalent of the notion of transmission zeros*, pp. 146-158 of *Modelling and Adaptive Control* (C. Byrnes & A. Kurzhanski,eds.), Vol. 105 of **Lecture notes in control and information sciences**, Springer, Berlin, (1985).

[8] Van der Schaft, A.J., *System theoretic descriptions of physical systems*, **CWI Tract 3**, CWI, Amsterdam, (1984).

[9] Van der Schaft, A.J., *On feedback control of Hamiltonian systems*, pp. 273-290 in *Theory and applications of nonlinear control systems* (C. Byrnes & A. Lindquist, eds.), Elsevier, Amsterdam, (1986).

[10] Van der Schaft, A.J., *Equations of motion for Hamiltonian systems with constraints*, **J. Phys. A, Vol. 20**, pp. 3271-3277, (1987).

[11] Van der Schaft, A.J., *On clamped dynamics of nonlinear systems*, pp. 499-506 in *Analysis and Control of Nonlinear Systems* (C.I. Byrnes, C.F. Martin & R.E. Sacks, eds.), Elsevier, Amsterdam, (1988).

[12] Slotine, J.-J. E., *Putting physics in control*, Proceedings of **Colloque international automatique non lineaire, Nantes**, (1988).

[13] Willems, J.C., *System theoretic models for the analysis of physical systems*, **Richerche di Automatica, Vol. 10**, pp. 71-106, (1979).

Lecture Notes in Control and Information Sciences

Edited by M. Thoma and A. Wyner

Vol. 97: I. Lasiecka/R. Triggiani (Eds.)
Control Problems for Systems
Described by Partial Differential Equations
and Applications
Proceedings of the IFIP-WG 7.2
Working Conference
Gainesville, Florida, February 3-6, 1986
VIII, 400 pages, 1987.

Vol. 98: A. Aloneftis
Stochastic Adaptive Control
Results and Simulation
XII, 120 pages, 1987.

Vol. 99: S. P. Bhattacharyya
Robust Stabilization Against
Structured Perturbations
IX, 172 pages, 1987.

Vol. 100: J. P. Zolésio (Editor)
Boundary Control and Boundary Variations
Proceedings of the IFIP WG 7.2 Conference
Nice, France, June 10-13, 1987
IV, 398 pages, 1988.

Vol. 101: P. E. Crouch,
A. J. van der Schaft
Variational and Hamiltonian
Control Systems
IV, 121 pages, 1987.

Vol. 102: F. Kappel, K. Kunisch,
W. Schappacher (Eds.)
Distributed Parameter Systems
Proceedings of the 3rd International Conference
Vorau, Styria, July 6–12, 1986
VII, 343 pages, 1987.

Vol. 103: P. Varaiya, A. B. Kurzhanski (Eds.)
Discrete Event Systems:
Models and Applications
IIASA Conference
Sopron, Hungary, August 3-7, 1987
IX, 282 pages, 1988.

Vol. 104: J. S. Freudenberg/D. P. Looze
Frequency Domain Properties of Scalar
and Multivariable Feedback Systems
VIII, 281 pages, 1988.

Vol. 105: Ch. I. Byrnes/A. Kurzhanski (Eds.)
Modelling and Adaptive Control
Proceedings of the IIASA Conference
Sopron, Hungary, July 1986
V, 379 pages, 1988.

Vol. 106: R. R. Mohler (Editor)
Nonlinear Time Series and
Signal Processing
V, 143 pages, 1988.

Vol. 107: Y. T. Tsay, L.-S. Shieh, St. Barnett
Structural Analysis and Design
of Multivariable Systems
An Algebraic Approach
VIII, 208 pages, 1988.

Vol. 108: K. J. Reinschke
Multivariable Control
A Graph-theoretic Approach
274 pages, 1988.

Vol. 109: M. Vukobratović/R. Stojić
Modern Aircraft Flight Control
VI, 288 pages, 1988.

Vol. 110: B. J. Daiuto, T. T. Hartley,
S. P. Chicatelli
The Hyperbolic Map and Applications
to the Linear Quadratic Regulator
VI, 114 pages, 1989

Vol. 111: A. Bensoussan, J. L. Lions (Eds.)
Analysis and Optimization
of Systems
XIV, 1175 pages, 1988.

Vol. 112: Vojislav Kecman
State-Space Models of Lumped
and Distributed Systems
IX, 280 pages, 1988

Vol. 113: M. Iri, K. Yajima (Eds.)
System Modelling and Optimization
Proceedings of the 13th IFIP Conference
Tokyo, Japan, Aug. 31 – Sept. 4, 1987
IX, 787 pages, 1988.

Vol. 114: A. Bermúdez (Editor)
Control of Partial Differential Equations
Proceedings of the IFIP WG 7.2
Working Conference
Santiago de Compostela, Spain, July 6–9, 1987
IX, 318 pages, 1989

Vol. 115: H.J. Zwart
Geometric Theory for Infinite
Dimensional Systems
VIII, 156 pages, 1989.

Vol. 116: M.D. Mesarovic, Y. Takahara
Abstract Systems Theory
VIII, 439 pages, 1989

Lecture Notes in Control and Information Sciences

Edited by M. Thoma and A. Wyner

Vol. 117: K.J. Hunt
Stochastic Optimal Control Theory
with Application in Self-Tuning Control
X, 308 pages, 1989.

Vol. 118: L. Dai
Singular Control Systems
IX, 332 pages, 1989

Vol. 119: T. Başar, P. Bernhard
Differential Games and Applications
VII, 201 pages, 1989

Vol. 120: L. Trave, A. Titli, A. M. Tarras
Large Scale Systems:
Decentralization, Structure Constraints
and Fixed Modes
XIV, 384 pages, 1989

Vol. 121: A. Blaquière (Editor)
Modeling and Control of Systems
in Engineering, Quantum Mechanics,
Economics and Biosciences
Proceedings of the Bellman Continuum
Workshop 1988, June 13–14, Sophia Antipolis, France
XXVI, 519 pages, 1989

Vol. 122: J. Descusse, M. Fliess, A. Isidori,
D. Leborgne (Eds.)
New Trends in Nonlinear Control Theory
Proceedings of an International
Conference on Nonlinear Systems,
Nantes, France, June 13–17, 1988
VIII, 528 pages, 1989

Vol. 123: C. W. de Silva, A. G. J. MacFarlane
Knowledge-Based Control with
Application to Robots
X, 196 pages, 1989

Vol. 124: A. A. Bahnasawi, M. S. Mahmoud
Control of Partially-Known
Dynamical Systems
XI, 228 pages, 1989

Vol. 125: J. Simon (Ed.)
Control of Boundaries and Stabilization
Proceedings of the IFIP WG 7.2 Conference
Clermont Ferrand, France, June 20–23, 1988
IX, 266 pages, 1989

Vol. 126: N. Christopeit, K. Helmes
M. Kohlmann (Eds.)
Stochastic Differential Systems
Proceedings of the 4th Bad Honnef Conference
June 20–24, 1988
IX, 342 pages, 1989

Vol.127: C. Heij
Deterministic Identification
of Dynamical Systems
VI, 292 pages, 1989

Vol. 128: G. Einarsson, T. Ericson,
I. Ingemarsson, R. Johannesson,
K. Zigangirov, C.-E. Sundberg
Topics in Coding Theory
VII, 176 pages, 1989

Vol. 129: W. A. Porter, S. C. Kak (Eds.)
Advances in Communications and
Signal Processing
VI, 376 pages, 1989.

Vol. 130: W. A. Porter, S. C. Kak,
J. L. Aravena (Eds.)
Advances in Computing and Control
VI, 367 pages, 1989

Vol. 131: S. M. Joshi
Control of Large Flexible Space Structures
IX, 196 pages, 1989.

Vol. 132: W.-Y. Ng
Interactive Multi-Objective Programming
as a Framework for Computer-Aided Control
System Design
XV, 182 pages, 1989.

Vol. 133: R. P. Leland
Stochastic Models for Laser Propagation
in Atmospheric Turbulence
VII, 145 pages, 1989.

Vol. 134: X. J. Zhang
Auxiliary Signal Design in Fault
Detection and Diagnosis
XII, 213 pages, 1989

Vol. 135: H. Nijmeijer, J. M. Schumacher (Eds.)
Three Decades of Mathematical System Theory
A Collection of Surveys at the Occasion of the
50th Birthday of Jan C. Willems
VI, 562 pages, 1989

Vol. 136: J. Zabczyk (Ed.)
Stochastic Systems and Optimization
Proceedings of the 6th IFIP WG 7.1
Working Conference,
Warsaw, Poland, September 12–16, 1988
VI, 374 pages. 1989

Lecture Notes in Control and Information Sciences

Edited by M. Thoma and A. Wyner

Vol. 137: S. L. Shah, G. Dumont (Eds.)
Adaptive Control Strategies for Industrial Use
Proceedings of a Workshop
Kananaskis, Canada, 1988
VI, 360 pages. 1989

Vol. 138: D. C. McFarlane, K. Glover
Robust Controller Design Using Normalized Coprime Factor Plant Descriptions
X, 206 pages. 1990

Vol. 139: V. Hayward, O. Khatib (Eds.)
Experimental Robotics I
The First International Symposium
Montreal, June 19-21, 1989
XVII, 613 pages. 1990

Vol. 140: Z. Gajic, D. Petkovski, X. Shen
Singularly Perturbed and Weakly Coupled Linear Control Systems
A Recursive Approach
VII, 202 pages. 1990

Vol. 141: S. Gutman
Root Clustering in Parameter Space
VIII, 153 pages. 1990

Vol. 142: A. N. Gündeş, C. A. Desoer
Algebraic Theory of Linear Feedback Systems with Full and Decentralized Compensators
V, 176 pages. 1990

Vol. 143: H.-J. Sebastian, K. Tammer (Eds.)
System Modelling and Optimization
Proceedings of the 14th IFIP Conference
Leipzig, GDR, July 3-7, 1989
X, 960 pages. 1990

Vol. 144: A. Bensoussan, J. L. Lions (Eds.)
Analysis and Optimization of Systems
Proceedings of the 9th International Conference, Antibes, June 12-15, 1990
XII, 992 pages. 1990

Vol. 145: M. B. Subrahmanyam
Optimal Control with a Worst-Case Performance Criterion and Applications
X, 133 pages. 1990

Vol. 146: D. Mustafa, K. Glover
Minimum Entropy H_∞ Control
IX, 144 pages. 1990

Vol. 147: J. P. Zolésio (Ed.)
Stabilization of Flexible Structures
Third Working Conference,
Montpellier, France, January 1989
V, 327 pages, 1990